How Does Forest Management Affect Soil Dynamics?

How Does Forest Management Affect Soil Dynamics?

Guest Editors

Cristian Oneț
Vlad Stoian

MDPI

Basel • Beijing • Wuhan • Barcelona • Belgrade • Novi Sad • Cluj • Manchester

Guest Editors

Cristian Oneț
Department of
Environmental Engineering
University of Oradea
Oradea
Romania

Vlad Stoian
Department of Microbiology
University of Agricultural
Sciences and Veterinary
Medicine Cluj-Napoca
Cluj-Napoca
Romania

Editorial Office
MDPI AG
Grosspeteranlage 5
4052 Basel, Switzerland

This is a reprint of the Special Issue, published open access by the journal *Forests* (ISSN 1999-4907), freely accessible at: https://www.mdpi.com/journal/forests/special_issues/OK0O5AY9JS.

For citation purposes, cite each article independently as indicated on the article page online and as indicated below:

Lastname, A.A.; Lastname, B.B. Article Title. *Journal Name* **Year**, *Volume Number*, Page Range.

ISBN 978-3-7258-5971-9 (Hbk)
ISBN 978-3-7258-5972-6 (PDF)
https://doi.org/10.3390/books978-3-7258-5972-6

Cover image courtesy of Vlad Stoian

Contents

About the Editors

Cristian Oneț

Cristian Oneț is a Lecturer at the Department of Environmental Engineering, Faculty of Environmental Protection, University of Oradea, Romania. His research focuses on soil microbiology and environmental quality, forest management practices and their influence on soil dynamics, nutrient cycling, and microbial activity. He investigates the effects anthropogenic interventions on ecosystems, aiming to identify sustainable strategies for maintaining health and sustaining resilience. Another important component of his work addresses water quality and pollution control, including studies on wastewater characteristics and groundwater contamination in industrial areas.

Vlad Stoian

Vlad Stoian is an Associate Professor at the Department of Microbiology, Faculty of Agriculture, University of Agricultural Sciences and Veterinary Medicine (USAMV) Cluj-Napoca, Romania. His research focuses on soil microbiology, microbial ecology, and plant–microbe interactions, contributing to sustainable agricultural practices. He is particularly interested in exploring the role of arbuscular mycorrhizal fungi in improving plant growth, nutrient uptake, and resilience under different fertilization regimes. He has also contributed to the fields of microbial ecology and functional microbiology, studying how soil microbial communities respond to long-term management practices and environmental stressors.

forests

MDPI

Editorial

How Does Forest Management Affect Soil Dynamics?

Cristian Oneț [1,*] and Vlad Stoian [2,*]

1 Department of Environmental Engineering, Faculty of Environmental Protection, University of Oradea, 26 General Magheru Street, 410048 Oradea, Romania
2 Department of Microbiology, Faculty of Agriculture, University of Agricultural Sciences and Veterinary Medicine of Cluj-Napoca, 3-5 Calea Mănăștur, 400372 Cluj-Napoca, Romania
* Correspondence: cristyonet@yahoo.com (C.O.); vlad.stoian@usamvcluj.ro (V.S.)

Forest management practices can have both positive and negative effects on the dynamics of soil properties and can significantly influence the soil structure, nutrient cycling, organic matter content, and microbial activity. Sustainable management approaches aim to minimize soil disturbance, maintain soil fertility, and promote long-term ecosystem health and resilience.

This Special Issue presents a collection of nine papers—seven based on original data and two comprehensive reviews. The publications collected in this "How does Forest Management affect Soil Dynamics?" Special Issue address these challenges comprehensively and in an interdisciplinary manner. The studies were carried out in a wide range of locations such as Taiwan, China, Chile, and Romania. They present results derived from experimental studies on important topics such as the insights into how roots anchor plants within soil matrices, the influence of plants' introduction into an agroforestry system on the microbial resource limitations, the effects of thinning infected trees and cultivating resistant pines on soil microbial diversity and function, the impact of acid rain and vegetation clearing on soil biological function, the effects of natural regeneration following severe disturbance, changes in soil microbial communities in extreme desert areas due to various long-term management techniques, improving microbial carbon dynamics in pine forests, and how enzyme stoichiometry highlights microbial nutrient relief through nitrogen fertilization. Further, the Special Issue presents two review papers that contribute to our understanding of the forest soil carbon responses to timber harvesting and prescribed burns and of the microbial networks in forest soils (a 20-year research review).

Fan et al. [1] investigated the soil reinforcement capabilities of the Pachi bamboo root system through in situ shear and pullout tests. They determined that the root diameter is positively correlated with the tensile strength and elastic modulus. Pachi bamboo, which grows in dense clusters with deep clumping roots, significantly enhances the slope stability—especially at shallow depths—due to its extensive root network. The study confirmed that root reinforcement is closely linked to the number, size, and cross-sectional area of bamboo culms in a cluster. Simple formulas were developed to estimate the root resistance based on these culm characteristics. Over a 10-year span, the culm density within clusters increased, further improving the soil stabilization. These results highlight the practical potential of Pachi bamboo for slope protection and in ecological engineering.

In a complex study, Zhang et al. [2] highlighted how different interventions impact soil microbes and provided valuable insights for sustainable pine forest ecosystem management following nematode infestation. After pine forests were affected by pine wood nematode, different management strategies—removing infected trees (sanitation-thinned plots) or planting disease-resistant species (*Pinus thunbergii* and *Pinus taeda*)—led to notable differences in the soil microbial communities. The disease-resistant pine plots showed higher enzyme activities and distinct bacterial and fungal community structures compared to the sanitation-thinned plots. The soil moisture, pH, and potassium levels were key factors influencing the microbial community composition. Additionally, wood-decomposing and ectomycorrhizal fungi were more abundant in the resistant pine plots. The authors developed formulas to quantify the microbial community changes based on these environmental factors and management practices.

Another study examined how simulated acid rain and the removal of understory vegetation affected the soil biological activity in a *Cinnamomum camphora* plantation. He et al. [3] determined that acid rain significantly changed the soil organic carbon, CO_2 emissions, enzyme activity, and microbial community structure. Removing understory vegetation reduced the soil moisture and nutrient availability, disrupted the enzyme balance, and shifted microbial communities. The bacterial diversity increased but with decreased stability, while fungal communities were more resilient, due to their metabolic traits. The study revealed that bacterial instability was linked to carbon limitation, whereas fungal stability related to phosphorus limitation. Overall, the findings highlight the crucial role of understory vegetation in maintaining soil health and emphasize the need for integrated forest management to protect soil ecosystems in subtropical plantations.

Ortiz et al. [4] determined that passive post-disturbance management of native Nothofagus forests in south-central Chile led to similar nutrient levels, water cycling capacity, and a decline in soil carbon sequestration over 45 years. The dominance of the opportunistic grass *Chusquea* sp. limits understory diversity and ecosystem recovery despite providing soil protection. The soils showed resilience, but active scarification and agroforestry are recommended to enhance regeneration and productivity. The authors highlighted that further research is needed on the carbon dynamics and microbial communities to better understand soil carbon stabilization.

Soil microbiome transition areas were analyzed by Zhang et al. [5] in relation to different long-term management methods over a period of 13 years. The research was oriented to respond to vegetation loss due to extreme management through excessive cutting or plant burning. Both bacterial and fungal communities were influenced by soil organic carbon, while the microbial community structure exhibited shifts from the control to floodwater irrigation. The results reinforce the need for the application of carefully regulated cutting and burning practices, to optimize plant regeneration and soil enrichment with nutrients from organic matter decomposition by microorganisms, which will ensure long-term resilience and productivity in dessert ecosystems.

Rui et al. [6] studied soil extracellular enzyme activity and stoichiometry in a complex experiment on the cultivation of *Panax notoginseng* (Sanqi) within the *Pinus armandii* forest. Their research aimed to explore the impact of *P. armandii* monoculture and the Sanqi–*P. armandii* agroforestry system in terms of the soil quality and the interactive effects of the season, plant introduction, soil compartments, and nutrient limitation on soil microorganisms. Their findings showed that N, rather than P, restricts the microbial metabolism under both cultivation systems. Based on the results, the authors recommend the application of organic fertilizers to support the sustainable development of Sanqi–*P. armandii* agroforestry system and to alleviate microbial N limitations.

The microbial activity in the soil of a Moso bamboo forest under a gradient of N application was the focus of research conducted by Chu et al. [7]. The results showed that the application of N fertilizer alleviated the C and N limitation of microorganisms. Another important aspect was that N application altered the soil nutrient resources and modulated the microbial strategy for nutrient acquisition. Their findings provide a good theoretical base for the development of new fertilizer application strategies based on microbial nutrient requirements, to obtain sustainability in the management of Moso bamboo forests.

A global meta-analysis on the impact of clearcutting, thinning, and prescribed burning on soil carbon dynamics was conducted by Ono and Noormets [8]. The authors analyzed a database of 414 observations from 110 studies to quantify the impact of the management type on soil respiration and its associated biophysical and soil variables. Both clearcutting and prescribed burning produced a decline in soil respiration, although acting in a different manner—through the removal of crowns, which reduces the carbon supply, and through detritus combustions, which also kill roots and microorganisms. Thinning does not significantly affect the soil respiration components, due to its minor impact on belowground compartments and the compensatory growth of the remaining trees. Long-term field experiments are important for future research, to increase the understanding of carbon stabilization.

A review of the key research from 2003 to 2023 was conducted by Oneț et al. [9], aiming to respond to a series of questions related to the forest soil microbiome and its responses to ecological disturbances. Their findings show that forest microbiomes are shaped by both the soil and plant type. The high impact on microbial community assemblage is related to the nutrient levels, soil fertility, successional stages of the forest, and disturbance patterns. The forest microbiome presents variable dynamics in relation to seasonal conditions, applied management on forest species, and long-term environmental shifts. For a deeper understanding of the complex soil microbiome, the authors propose long-term interdisciplinary studies to forecast the shifts in microbial communities due to environmental and anthropogenic disturbances.

We would like to thank all the authors for their contributions to these important topics. Understanding microbial systems, community structures, and responses to management interventions is essential for gaining a deeper understanding of ecosystem functioning. It is our expectation that this Special Issue of *Forests* will offer a robust platform for future scholarly inquiry into this vital area of research.

Conflicts of Interest: The authors declare no conflicts of interest.

References

1. Fan, C.-C.; Chen, C.-H.; Chen, C. Exploring the Root–Soil Anchoring Dynamics of *Bambusa pachinensis* (Pachi Bamboo) Root System. *Forests* **2025**, *16*, 832. [CrossRef]
2. Zhang, X.; Liu, Z.; Cao, M.; Dai, T. Effects of Thinning of the Infected Trees and Cultivating of the Resistant Pines on Soil Microbial Diversity and Function. *Forests* **2025**, *16*, 813. [CrossRef]
3. He, Z.; Liu, Y.; Lin, Y.; Kong, X.; Lin, H.; He, X. The Effect of Acid Rain and Understory Vegetation Removal on the Biological Activity of the Soils of the *Cinnamomum camphora* (Linn) Presl Plantation. *Forests* **2025**, *16*, 525. [CrossRef]
4. Ortiz, J.; Panichini, M.; Neira, P.; Henríquez-Castillo, C.; Gallardo Jara, R.E.; Rodriguez, R.; Mutis, A.; Ramos, C.; Espejo, W.; Puc-Kauil, R.; et al. How Natural Regeneration After Severe Disturbance Affects Ecosystem Services Provision of Andean Forest Soils at Contrasting Timescales. *Forests* **2025**, *16*, 456. [CrossRef]
5. Zhang, Y.; Du, Y.; Mu, Z.; Islam, W.; Zeng, F.; Zhang, Z.; Gonzalez, N.C.T. Response of Soil Microbial Communities in Extreme Arid Deserts to Different Long-Term Management Methods. *Forests* **2025**, *16*, 306. [CrossRef]
6. Rui, R.; Hei, J.; Li, Y.; Wan, X.; Wang, S.; He, X. Alleviating Microbial Carbon Limitation in *Pinus armandii* Forests Through *Panax notoginseng* Cultivation. *Forests* **2025**, *16*, 158. [CrossRef]
7. Chu, H.; Su, W.; Zhou, Y.; Wang, Z.; Long, Y.; Sun, Y.; Fan, S. Enzyme Activity Stoichiometry Suggests That Fertilization, Especially Nitrogen Fertilization, Alleviates Nutrient Limitation of Soil Microorganisms in Moso Bamboo Forests. *Forests* **2024**, *15*, 1040. [CrossRef]

8. Ono, M.; Noormets, A. Impacts of Harvesting and Prescribed Burning on Forest Soil Carbon Dynamics: A Global Meta-Analysis. *Forests* **2025**, *16*, 1555. [CrossRef]
9. Onet, A.; Grenni, P.; Onet, C.; Stoian, V.; Crisan, V. Forest Soil Microbiomes: A Review of Key Research from 2003 to 2023. *Forests* **2025**, *16*, 148. [CrossRef]

forests

MDPI

Article

Exploring the Root–Soil Anchoring Dynamics of *Bambusa pachinensis* (Pachi Bamboo) Root System

Chia-Cheng Fan [1,*], Chung-Hao Chen [2] and Chunhsiung Chen [3]

1 Department of Construction Engineering, National Kaohsiung University of Science and Technology, Kaohsiung 824, Taiwan
2 CECI Engineering Consultants, Inc., Taipei 114, Taiwan; stone19910804@ceci.com.tw
3 Kuocheng Construction Co., Kaohsiung 800, Taiwan; chenver.hinet@msa.hinet.net
* Correspondence: ccfan@nkust.edu.tw

Abstract: Bamboo is widely distributed throughout the world, particularly in tropical and subtropical regions. This study aims to investigate the biomechanical properties of the root system of *Bambusa pachinensis* (Pachi bamboo). The root system of Pachi bamboo grows densely in clusters, with most roots growing vertically and potentially penetrating more than one meter into the soil after growing for several years. Owing to these characteristics, Pachi bamboo is considered a promising plant species for soil reinforcement. However, research on its root reinforcement capabilities remains limited. In situ shear and pullout tests were conducted to assess the root reinforcement of the fibrous root system. The root diameters of Pachi bamboo are typically less than 4 mm, and its tensile strength is notably lower than that of tree roots. This study establishes a method for estimating the root reinforcement of Pachi bamboo based on the number and cross-sectional area of the culms in a single bamboo cluster. The relationship between the maximum tensile force (F_{ult}) and root diameter (D) is $F_{ult} = (3.65)D^{2.59}$, where F_{ult} is in Newtons (N), and D is in millimeters (mm). The relationship between the pullout resistance (P_{ult}) and the shear resistance (S_{ult}) with the number of culms (SN) is $P_{ult} = 46.5(SN)$ and $S_{ult} = 0.53(SN) + 5$, where P_{ult} is in Newtons (N), and S_{ult} is in kilopascals (kPa). These results suggest a positive contribution of the number of culms to mechanical resistance.

Keywords: Pachi bamboo; forest management; root biomechanics; root reinforcement

1. Introduction

Plant species and root morphology are crucial factors in a plant's ability to reinforce the soil. The complex and diverse root system morphologies of different plant species have been a focal point of research in recent decades, reflecting the collaborative efforts of the scientific community. This collective work has led to the development of methods for estimating the shear resistance provided by roots in the soil, based on simplified root reinforcement models [1–3] and field experiments on rooted soils [4,5]. Another key factor influencing the reinforcement effect of a plant root system is the soil–root bonding strength, which directly affects the pullout resistance of the root system in soil [6,7]. Additionally, the contribution of plant root systems to slope stability has been extensively analyzed and discussed by researchers, further emphasizing the collaborative nature of this field [8–13].

Tap and oblique roots in a root system are considered advantageous for enhancing soil resistance [5,14]. Dense, thin roots offer good tensile resistance, while long tap roots provide bending resistance within the soil [15,16]. Stokes et al. (2009) [16] also suggested that dense, thin roots offer better resistance to soil than fewer, thicker roots, even when the root area ratios (RARs) are identical. Previous analytical studies have indicated that the critical root length required for effective root anchorage in the soil increases with root diameter [17]. Planting root systems with favorable characteristics on slopes is essential for providing better reinforcement and stability in soil bioengineering techniques [18].

Root forces generated in root-permeated soils subjected to deformation are crucial for mobilizing the resistance provided by root systems in the soil. The mechanical properties of plant roots are essential parameters for evaluating root reinforcement in the soil. Root systems vary significantly across plant species. Bamboo, for instance, is a species with many thin roots. Studies have shown that bamboo root systems can improve soil erosion control and conservation [19–22]. Furthermore, bamboo root systems act as anchors for the soil, preventing shallow landslips in forest slopelands [19,23]. The mechanical properties of bamboo fibers have been extensively studied over the past decades [24–28]. Additionally, research on the mechanical performance of bamboo fiber-reinforced composites has been conducted under various treatments and conditions [27,29–31]. However, there has been limited research on the quantitative contribution of bamboo roots to reinforcement in the soil.

Bamboo is widely distributed worldwide [32]. It is used in soil bioengineering techniques, and it has an effective ability to reduce soil detachment and movement. Bamboos generally grow faster than other forest plant species and thrive in various soil conditions, including sandy, loamy, lateritic, and alluvial soils [32]. Bamboo root systems are classified into two types: clumping and running. Clumping bamboos have culms that grow close together, whereas running bamboos feature a horizontal root system (rhizome) that spreads widely from the shoot. *Bambusa pachinensis* (Pachi bamboo), a clumping species first reported in Taiwan in 1916 [33], is characterized by a dense and thin root system. This bamboo species has also been identified in Southeastern China [34].

The root distribution of Pachi bamboo is a fibrous root system, characterized by a dense network of fine roots and the absence of a distinct taproot. Fibrous root systems generally exhibit relatively shallow rooting depths compared to those of woody species [35,36]. Despite this shallowness, they contribute to soil strength at a different scale, primarily through the combined effects of root depth and biomass [36]. Moreover, specific root traits—such as the presence of root hairs and the distribution of root diameters—can significantly influence shear strength by altering failure mechanisms within the soil–root matrix [36].

Roots that intersect a potential sliding surface enhance the soil's shear resistance through a mechanism known as basal root reinforcement [37,38]. This process is most effective when the rooting depth extends beyond the depth of the sliding surface [37]. The root system of Pachi bamboo is clumping in nature, consisting of numerous fine roots along with long, slender roots. With a rooting depth reaching the meter scale, Pachi bamboo has the potential to contribute meaningfully to basal root reinforcement in shallow slope failures, while also playing a vital role in controlling surface soil erosion. To assess the contribution of Pachi bamboo root systems to root reinforcement, it is essential to understand the mechanical properties of these roots.

Pullout resistance and shear resistance represent the capacities of root systems to withstand different types of applied forces [39]. The tensile strength of individual roots contributes to the mobilization of both pullout and shear resistance within the root system. It is important to recognize that pullout resistance, shear resistance, and tensile strength of individual roots each serve distinct but complementary functions in the overall root reinforcement mechanism. This research aims to investigate the mechanical properties of the Pachi bamboo root system. By performing in situ shear, pullout, and tensile tests on Pachi bamboo roots, we aim to establish a correlation between the root reinforcement of the bamboo root system and the characteristics of the bamboo. It is assumed that the root reinforcement of Pachi bamboo systems is correlated with the characteristics of bamboo culms. The findings of this study can be applied to assess the stability of sloping lands or riverbanks planted with Pachi bamboo, thereby enhancing the practicality and relevance of our research and offering valuable insights for engineering and environmental conservation practices.

2. Materials and Methods

2.1. Test Site

The test site, located on the west campus of Kaohsiung University of Science and Technology in Kaohsiung, Taiwan, is covered with sparse grass, *Axonopus affinis*. There were also small amounts of other grass species at the test site.

The soil at the site had a unit weight ranging from 14.9 to 15.8 kN/m^3 and a dry unit weight between 12.2 and 13.1 kN/m^3. According to the Unified Soil Classification System (USCS), the soil is classified as SP (poorly graded sand). These soil conditions provided an ideal environment for conducting the in situ tests on the Pachi bamboo root system.

2.2. Pachi Bamboo

Pachi bamboo culms are slender and grow in close clusters, with heights reaching between 3 and 8 m. Field investigations revealed that the root system of Pachi bamboo is dense and classified as clumping bamboo. Most of the roots grow vertically, penetrating deeply into the soil, while the lateral roots at shallower depths remain short. At other sites, Pachi bamboo roots extended to at least 1.6 m in length after four years of growth, as illustrated in Figure 1. Several roots, visible in the lower center of the image, were exposed within an excavated trench. A person holding a tape measure is shown in the photo to provide a scale reference for the exposed rooting depths.

Figure 1. Depth of the Pachi bamboo root system after growing for four years.

The Pachi bamboo at the test site had grown for two years. After the in situ shear test, we measured the root characteristics of Pachi bamboo. A small shovel was used to carefully excavate the soil around the entire root system until it was fully exposed. The investigation of the below-ground root system can be challenging and time-consuming, and roots may break during excavation [40]. The length, diameter, and orientation of each root in the Pachi bamboo root system were recorded. Root orientation (θ) was measured as the departure angle relative to the shear direction, ranging from 0° to 180°. Root diameters were measured using digital calipers with a resolution of 0.01 mm.

2.3. Mechanical Tests on Bamboo Root Systems and Testing Apparatus

The roots of Pachi bamboo exhibit a fibrous structure. In situ shear tests are well-suited for evaluating the reinforcement of these roots when subjected to shear forces at specific depths [39]. In situ shear tests were performed on soil samples permeated by the Pachi bamboo root system. The shear box used had a width of 0.6 m and a thickness of 0.5 m, with a shearing rate of 0.01 m/min. A test pit was excavated around the Pachi bamboo to match the dimensions of the shear box. The shear box and its components were designed for easy assembly in the field. A turbine wheel hoisting jack was employed to apply shear force to the back side of the shear box. Shear displacement and pushing force were measured using extensometers and load cells, with data recorded using a data logging system. The shear plane was located at a depth of 0.3 m. A total of 5 in situ shear tests were conducted on Pachi bamboo clusters in this study. Moreover, shear tests for five bare soil specimens were also carried out. A photo of the in situ shear test is shown in Figure 2a.

(a) (b)

Figure 2. The apparatus of in situ mechanical tests. (**a**) In situ shear test; (**b**) in situ pullout test.

Due to the fibrous nature of the Pachi bamboo root system, conducting pullout tests on single roots is largely impractical. Therefore, testing the entire root system under external forces near the ground surface is a more effective approach for assessing its anchorage performance. With many culms above ground and the roots deeply entwined in the soil, it is more practical to evaluate the pullout resistance of Pachi bamboo based on the bamboo culms themselves. In this study, we conducted pullout tests on five bamboo cluster samples, each with varying numbers of culms. The pullout test apparatus used a motor to apply the pulling force at a rate of 0.01 m/min. A steel wire was used to apply the pulling force to the root system beneath the bamboo shoots, and the pulling force could be applied vertically. A total of five pullout tests were performed on the Pachi bamboo clusters in this study. The soil moisture content for each specimen was measured after the experiments. A photo of the in situ pullout test is shown in Figure 2b.

Tensile tests on individual roots serve as the fundamental method for determining their tensile properties. In addition, the tensile properties of plant roots are crucial for providing shear resistance in the soil [1,41]. In this study, thirty-seven tensile tests were conducted on individual roots with uniform diameters ranging from 0.6 mm to 2.7 mm. These roots were collected from various bamboo root systems, and the root diameter near the rupture location was measured. The relationship between root diameter and the tensile strength of individual bamboo roots was established using the power function [42].

3. Results

3.1. Geometric Characteristics of Pachi Bamboo Root Systems

Figure 3 illustrates the root system of a Pachi bamboo cluster, characterized by numerous near-vertical roots and short lateral roots. The longest root in the 2.2-year-old bamboo cluster measured approximately 0.65 m, while the lateral roots were abundant and concentrated in the top 0.15 m of the soil. Most root diameters were less than 4 mm, with the majority ranging between 1.5 and 2 mm. Figure 4 presents the statistical distribution of number of roots and cross-sectional areas in various orientations within the root system based on 5 root system specimens after the pullout tests. Around 42% of the roots were near-vertical, positioned at angles between 70° and 110° relative to the horizontal plane. Notably, near-vertical roots accounted for approximately 45% of the total cross-sectional root area in the root system. The roots of Pachi bamboo are thin—typically less than 3–4 mm in diameter—and highly flexible. When subjected to shear or pullout forces, these roots may adjust their orientation within the soil matrix. Due to their slender and pliable nature, Pachi bamboo roots are well-suited to mobilize tensile forces in response to external loading, enhancing their contribution to soil reinforcement.

Figure 3. The Pachi bamboo root system at the test site (the numerals shown in the figures are in units of centimeters; the blue tag in the photo is used to facilitate the counting of roots with different diameters).

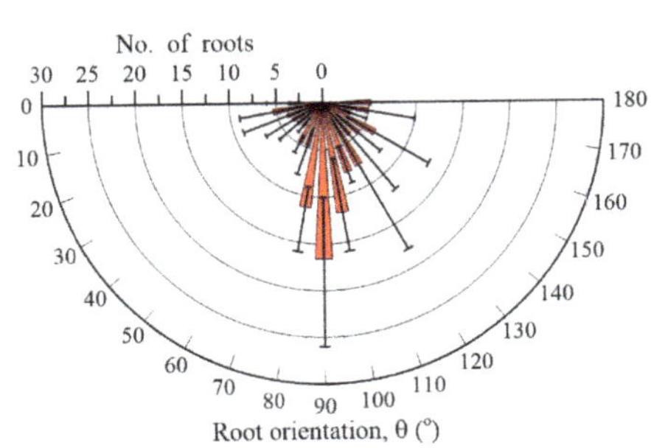

(a)

Figure 4. *Cont.*

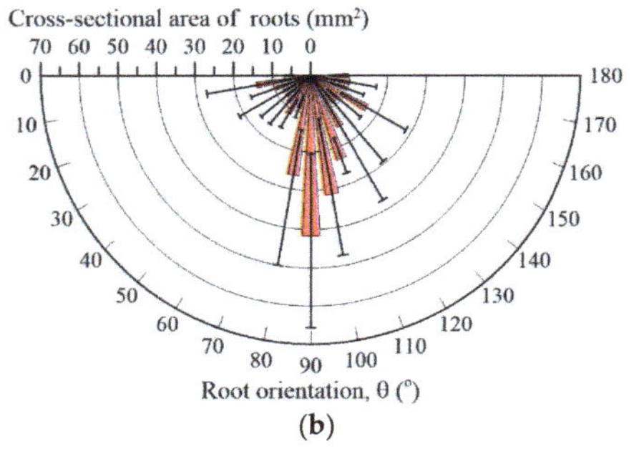

Figure 4. The geometrical distribution of the Pachi bamboo root system. (**a**) Number of roots vs. root orientation; (**b**) cross-sectional area vs. root orientation (the black line at the end of the red line represents the error bar).

3.2. Tensile Properties of Pachi Bamboo Roots

Potential errors may arise in the measurement of root diameter [43]. It is preferable to present the mechanical properties of roots in terms of force instead of stress [43]. Figure 5a illustrates the relationship between the maximum tensile force (F_{ult}) and the diameter (D) of Pachi bamboo roots. The elongation strain at the maximum tensile force varied between 10% and 17%. The relationship between the maximum tensile force (F_{ult}) and root diameter (D) is expressed mathematically as

$$F_{ult} = (3.65)D^{2.59} \quad R^2 = 0.48 \tag{1}$$

where F_{ult} is in N, and D is in mm.

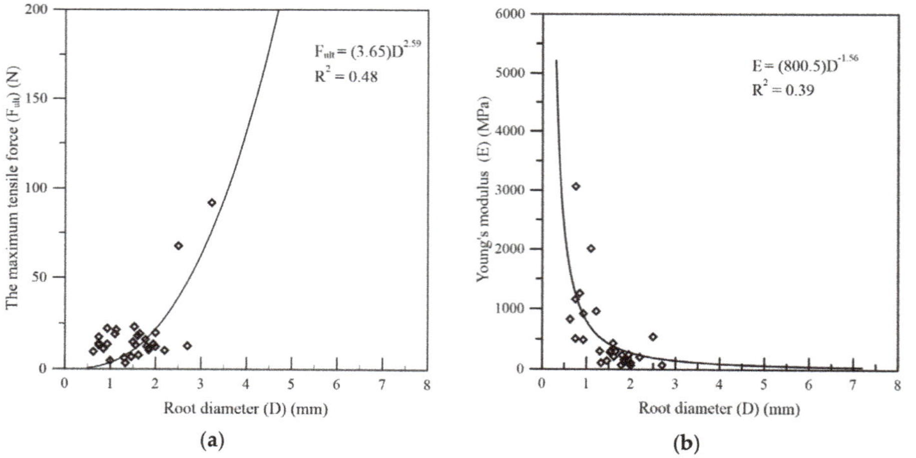

Figure 5. Tensile properties of Pachi bamboo roots. (**a**) Maximum tensile force; (**b**) elastic modulus.

In addition, Figure 5b presents the relationship between elastic modulus (E) of Pachi bamboo roots and their diameter (D). This relationship is described by the following equation:

$$E = (800.5)D^{-1.56} \quad R^2 = 0.39 \tag{2}$$

where E is in MPa.

3.3. Pullout Resistance of Pachi Bamboo Root Systems in Soil

Figure 6 illustrates the relationship between pullout force and pullout displacement for Pachi bamboo root systems for five specimens. Table 1 summarizes the pullout test results and the root system characteristics of the five bamboo cluster samples. The maximum pullout forces for bamboo roots with varying numbers of culms ranged from 285 N to 1061 N, and the shear displacements occurred at about 56 mm to 95 mm. Among the five Pachi bamboo cluster samples, 45%–59% of the roots were near-vertical.

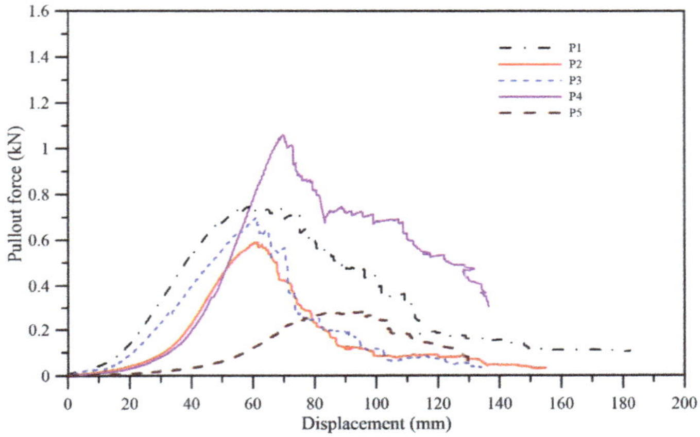

Figure 6. Pullout force vs. pullout displacement of the Pachi bamboo root system.

Table 1. Pullout test results on the root system of Pachi bamboo.

Test No.	No. of Bamboo Culms	The Maximum Pullout Force (N)	The Cross-Sectional Area of Culms (mm^2)	No. of Roots	No. of Near-Vertical Roots	The Cross-Sectional Area of Roots with Rupture (mm^2)	The Cross-Sectional Area of Roots (mm^2)
P1	12	748	715	185	105	109	419
P2	12	592	324	102	60	208	223
P3	18	703	962	99	45	194	200
P4	9	1061	509	110	54	322	322
P5	7	285	372	110	57	237	237

3.4. Shear Resistance of Pachi Bamboo Root Systems in Soils

Figure 7 depicts the relationship between shear stress and shear displacement for five shear tests on Pachi bamboo root systems. The soil moisture contents for specimen S1R, S2R, S3R, S4R, and S5R are 6%, 23.7%, 35.2%, 31.3%, and 32.5%, respectively. Table 2 summarizes the shear test results alongside the root system characteristics of the five bamboo clusters. The soil moisture content for S1R specimen is considerably low compared with other specimens, and its maximum shear stress reached 47.5 kPa. The S1R data will be excluded from the following study on the correlation between the shear resistance and the quantity of bamboo clusters and the root system.

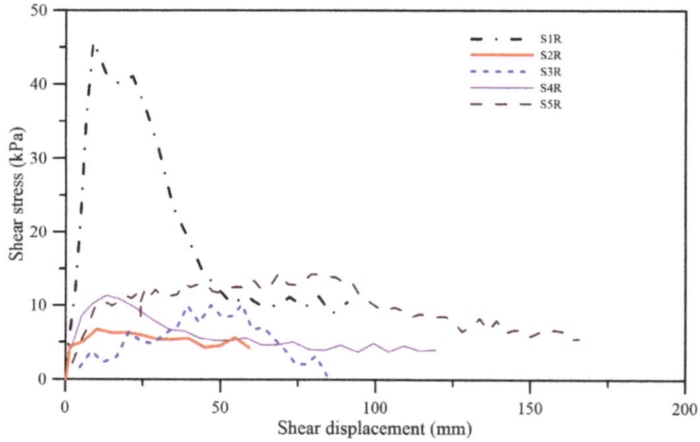

Figure 7. Shear stress vs. shear displacement of the soil permeated with Pachi bamboo root system.

Table 2. In situ shear test results of the Pachi bamboo root system.

Test No.	No. of Bamboo Culms	The Maximum Shear Stress (kPa)	The Cross-Sectional Area of Culms (mm^2)	No. of Roots	No. of Near-Vertical Roots	The Cross-Sectional Area of Roots (mm^2)	The Cross-Sectional Area of Near-Vertical Roots (mm^2)
S1R	4	47.5	353.4	48	21	115	21
S2R	5	6.3	373.1	19	9	102	46
S3R	5	10.7	605.1	17	8	52	32
S4R	18	11.7	1370	136	43	362	118
S5R	12	15	1111	69	39	185	90

In addition, we carried out four shear tests for bare soil specimens. The soil moisture contents for these specimens were 7.5% 9.1%, 33.4%, and 33.3%, and their shear strengths were 27.5 kPa, 17.8 kPa, 10.32 kPa, and 10.45 kPa, respectively. The soil moisture content considerably affects the shear strength of the soil. The soil moisture contents for the shear tests on bamboo root soil specimens were in the range of 23~35%. Thus, the shear strength of the bare soil with a soil moisture content of 33.3% can be used as a reference shear strength for rooted soil.

3.5. Relationships Between the Characteristics of Pachi Bamboo and Root Reinforcement

3.5.1. Pullout Resistance of Pachi Bamboo Roots

Several factors, including the number of roots, root geometry, the spatial distribution of the root system, and soil properties, influence the pullout resistance of plant roots. Pachi bamboos typically grow in clusters of culms with dense root systems. Accurately estimating the number of roots in a single bamboo is challenging; however, the number of culms, which are typically uniform in diameter, can be easily identified in the field. In this study, the culm diameters of Pachi bamboo ranged from 7.5 mm to 8.7 mm. This research aimed to establish a correlation between the pullout resistance of bamboo root systems and the number of bamboo culms.

Figure 8 illustrates the relationship between the maximum pullout force (P_{ult}) of Pachi bamboo root systems and the number of culms. The results indicate that the maximum pullout force (P_{ult}) increases with the number of culms. However, one test result (test P4 in Table 1) showed significantly higher pullout resistance compared to the other data points, and the number of culms of the P4 specimen was 9. This unusually high pullout resistance may be attributed to the lower soil water content in the corresponding soil specimen. To enhance the reliability of the regression analysis, the data from Test P4 were excluded. The equation representing the fitted relationship in Figure 8 is as follows:

$$P_{ult} = 46.5(SN) \quad R^2 = 0.56 \tag{3}$$

where P_{ult} is the maximum pullout force (N), and SN represents the number of bamboo culms in a single bamboo cluster. Additionally, Figure 9 demonstrates the positive relationship between the maximum pullout force of the Pachi bamboo root system and the total cross-sectional area of the culms in a cluster. The mathematical equation that fits the data in Figure 9 is as follows:

$$P_{ult} = 0.75(SA) + 100 \quad R^2 = 0.25 \tag{4}$$

where SA is the total cross-sectional area (mm^2) of culms in a Pachi bamboo cluster.

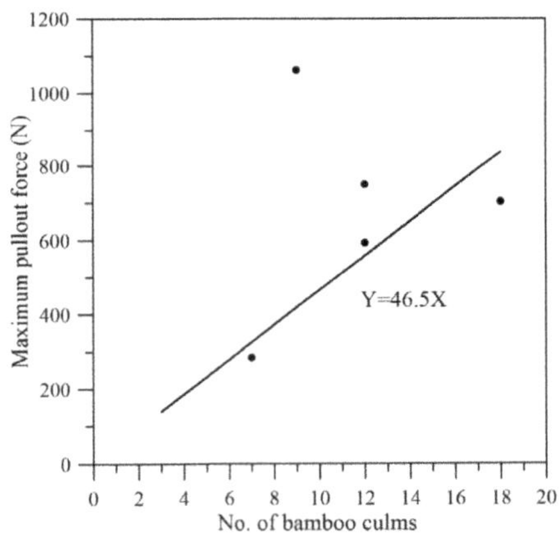

Figure 8. Maximum pullout force vs. number of bamboo culms in a bamboo cluster.

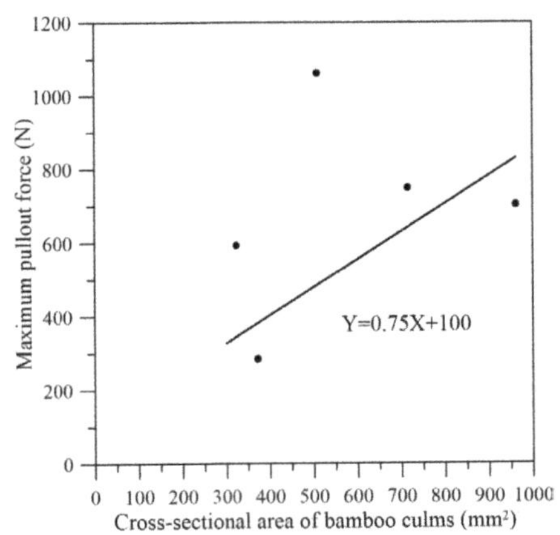

Figure 9. Maximum pullout force vs. cross-sectional area of bamboo culms in a bamboo cluster.

3.5.2. Shear Resistance of Soils Permeated with Pachi Bamboo Root System

Similar to the pullout resistance of bamboo roots discussed earlier, the number of bamboo culms and their cross-sectional area can also be used to characterize the shear resistance of bamboo root systems in soil. Figure 10 illustrates the relationship between the maximum shear resistance of the Pachi bamboo root system and the number of culms. The results show that the maximum shear resistance increases with the number of culms. However, one test result (test S1R in Table 2) exhibited significantly higher shear resistance compared to the other data points due to a low soil water content. The data from test S1R were excluded to refine the regression. The equation describing the fitted relationship between maximum shear resistance and the number of culms is as follows:

$$S_{ult} = 0.53(SN) + 5 \quad R^2 = 0.17 \tag{5}$$

where S_{ult} (kPa) represents the maximum shear resistance. Additionally, Figure 11 shows the relationship between the maximum shear stress of the Pachi bamboo root system and the cross-sectional area (SA) of the culms in a cluster. The mathematical equation for fitting S_{ult} (kPa) and SA (mm^2) is as follows:

$$S_{ult} = 0.01(SA) + 1.5 \quad R^2 = 0.26 \tag{6}$$

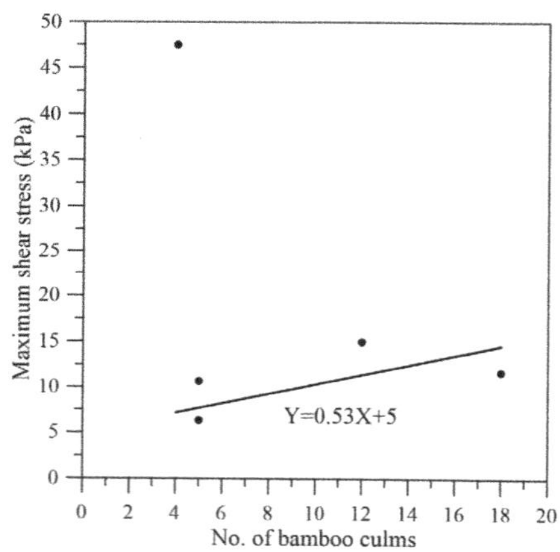

Figure 10. Maximum shear stress vs. number of bamboo culms in a bamboo cluster.

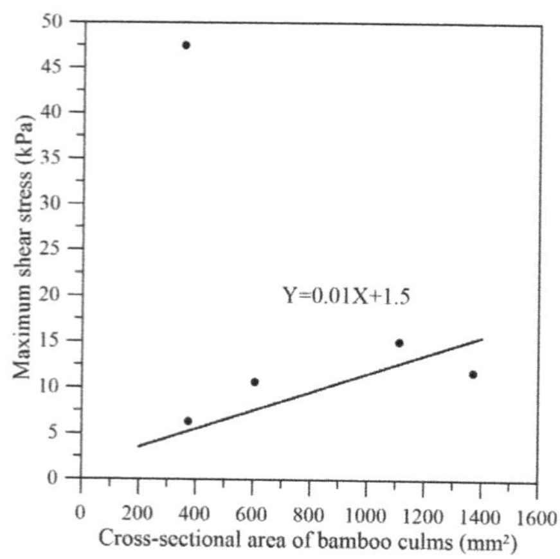

Figure 11. Maximum shear stress vs. cross-sectional area of bamboo culms in a bamboo cluster.

The characteristics of bamboo culms, such as the number of culms and cross-sectional area, can be easily determined in the field. Using the equations developed in this study, the shear resistance of soils with a Pachi bamboo cluster can be effectively estimated.

4. Discussion

4.1. Pachi Bamboo Root System Distribution

Monocot plant species, which are characterized by fibrous root systems—such as grass, rice, maize, corn, barley, millet, and bamboo—typically develop lateral roots concentrated in the topsoil [44]. As a result, their root reinforcement is confined mainly to very shallow depths. In contrast, the Pachi bamboo exhibits a deep-rooting system, with roots extending to depths on the order of meters. Consequently, the root reinforcement provided by the Pachi bamboo is considered significantly deeper than that of the aforementioned species.

The Pachi bamboo root system grows in clumps, forming a dense network of thin roots within each cluster. The bamboo roots intertwine intricately with the surrounding soil, creating cohesive soil–root blocks that serve as natural reinforcements, enhancing the soil's structural stability. This unique root system plays a significant role in preventing soil displacement and mitigating erosion. Researchers have highlighted the importance of bamboo roots in soil stabilization and erosion control due to their distinctive characteristics and adaptability to various environments [19,20]. Zhou et al. (2005) [19] noted that bamboo is equipped with an extensive fibrous root system that integrates seamlessly with its rhizomes. This root–rhizome network not only binds the soil but also effectively protects shallow layers of soil from erosion caused by water and wind forces [19]. The combination of its clumping growth pattern and fibrous root structure, shown in Figure 3, makes Pachi bamboo particularly valuable for soil bioengineering techniques, including slope stabilization, riverbank reinforcement, and degraded land rehabilitation.

The Pachi bamboo root system penetrates deeply into the soil, demonstrating its capacity for soil anchorage and nutrient acquisition. Field observations, shown in Figure 1, showed that the roots of Pachi bamboo extended to depths exceeding 1.6 m within just four years of growth, highlighting its rapid development. Each bamboo cluster, composed of 7–18 culms, contains over 100 thin roots, which together form a dense and complex root network. This dense rooting system differs significantly from that of broadleaf and conifer trees, as the Pachi bamboo's root distribution is highly specialized. Approximately 45–55% of the roots within a cluster grow in a near-vertical orientation, allowing them to penetrate deeply and anchor the plant securely in the soil. Additionally, investigations of 2.2-year-old bamboo root systems revealed that a large number of roots grow in a single cluster and are exceptionally fine, with diameters of less than 4 mm. This fine root structure facilitates effective soil penetration and enhances soil–root cohesion, making it a favorable candidate for applications in erosion control and slope stabilization. The findings of this study provide quantitative insights into the geometric and functional characteristics of the Pachi bamboo root system, offering valuable data for both ecological and engineering applications.

4.2. Tensile Property and Root Reinforcement

The root properties of bamboo have historically received limited attention, particularly concerning the performance of bamboo root systems in soil reinforcement. Few studies have explored the root reinforcement of natural bamboo root systems in detail. For instance, Ma'ruf (2012) [45] conducted laboratory shear tests on soil samples permeated with Apus bamboo roots, finding that the increase in soil shear strength was positively correlated with the soil-root volume ratio. Remarkably, soil reinforced with Apus bamboo roots exhibited a shear strength increase of up to 55% of the peak soil shear strength, even when the soil-root volume ratio was as low as 5%. Additionally, Mulyanti et al. (2024) [46] reported that the tensile strength of bamboo roots ranged between 23 MPa and 45 MPa, while their water content was approximately 85%–90%. Beyond soil reinforcement, bamboo root fibers have also been utilized as reinforcements in composite materials by combining bamboo roots with resins and catalysts [46].

The maximum tensile force of Pachi bamboo roots varies with root diameter, ranging from 10 N to 60 N for root diameters between 1 mm and 3 mm. Figure 12 illustrates the relationship between the maximum tensile force and root diameter for Pachi bamboo and other fibrous root species, including maize, vetiver, ruzi grass, Italian ryegrass, perennial ryegrass, and tall fescue. The results show that the maximum tensile force of Pachi bamboo roots is generally slightly higher than that of the grass species and is comparable to that of vetiver. Notably, several data points for Pachi bamboo exhibit higher tensile forces than those of the other species.

In addition, the relationship of the elastic modulus vs. root diameter of tree roots showed more intraspecific differences compared to the interspecific ones [47]. The elastic modulus of Pachi bamboo roots is 800, 271, and 144 MPa at the root diameter of 1, 2, and 3 mm, respectively. A study by Cislaghi (2021) [47] focused on tree species. The Pachi bamboo root's elastic modulus is greater than that of most of the tree species presented by Cislaghi (2021) [47] at the corresponding root diameter.

This study established relationships between the shear and pullout resistance of Pachi bamboo root systems in the soil and the characteristics of bamboo culms. However, determining the root area ratio (RAR) of the Pachi bamboo root system in the field remains challenging due to the complexity of its root network. The RAR, often used in simple root reinforcement models [1,2,41], is difficult to measure accurately for Pachi bamboo. Conversely, the number and cross-sectional area of bamboo culms in a single cluster are straightforward to count and estimate in the field, and the geometries of individual culms are relatively uniform within a cluster. Based on this, the research presents novel equations for estimating the root reinforcement of the Pachi bamboo root system using the number or total cross-sectional area of bamboo culms in a cluster.

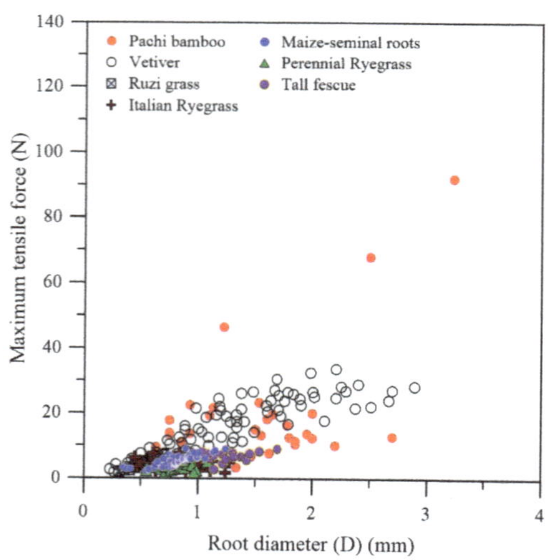

Figure 12. Comparison of root's maximum tensile force vs. root diameter relationship between Pachi bamboos and other fibrous root species [48–51].

4.3. Exploring the Pachi Bamboo in Engineering Practice

Pachi bamboos grow densely in clusters within fields. After several years of growth, a naturally planted bamboo cluster can develop a large number of culms. We investigated the growth characteristics of a single Pachi bamboo cluster planted in 2015 with an initial count of five to six culms. After ten years, this cluster expanded to approximately 550 closely spaced culms. Each culm maintained a consistent diameter of about 20 mm. At ground level, the cluster now spans a diameter of 1.88 m and reaches a height of 4.8 m, with minimal spacing between individual culms. Based on these observations, planting Pachi bamboo clusters—initially containing five to six culms per cluster—at intervals of 2 to 3 m within bamboo clusters on sloped terrain could, over time, develop extensive root systems that significantly reinforce the soil. This characteristic makes Pachi bamboo particularly well-suited for preventing shallow soil failures on slopes.

4.4. Shortcomings and Outlook of the Study

This study presents findings on the root reinforcement characteristics of the Pachi bamboo root system, along with the mechanical properties of its roots. While bamboo is widely distributed globally—particularly throughout Asia—research on the mechanical behavior of bamboo root systems remains limited. In this study, the number of in situ shear and pullout tests conducted on Pachi bamboo roots was constrained by the labor-intensive nature of specimen preparation. Nevertheless, the results provide a practical and straightforward approach for evaluating root reinforcement by Pachi bamboo under field conditions.

The methodologies developed in this study are applicable for assessing soil stability on slopes and along riverbanks where Pachi bamboo is present, underscoring its potential use in engineering applications related to erosion control and slope stabilization. Given that bamboo frequently grows on sloped and rural lands, further research is warranted to investigate the root reinforcement properties of Pachi bamboo at later stages of growth. Such knowledge is essential for the sustainable management of bamboo forests and for maximizing their ecological and engineering benefits.

5. Conclusions

This study examined the reinforcement properties of the Pachi bamboo root system using in situ shear and pullout tests. The relationships between root diameter and both the maximum tensile force and elastic modulus were quantified. Pachi bamboos grow densely in clusters in the field. The root structure of Pachi bamboo is clumping roots, and it grows deeply in the ground, reaching to 1.6 m in four years of growth. The numerous roots of Pachi bamboo make it favorable in protecting against soil failure at shallow depths on slopes.

This study confirmed the hypothesis that the root reinforcement of Pachi bamboo systems is correlated with the characteristics, quantity, and cross-sectional area of bamboo culms in a cluster. Simple formulas were developed to estimate the shear and pullout resistance of Pachi bamboo root systems, based on the characteristics, quantity, and cross-sectional area of bamboo culms within a single cluster. The root system of Pachi bamboo is characterized by its lateral limitations and deep soil penetration, making it potentially effective for reinforcing slopelands.

Moreover, over a 10-year period, the number of culms within a single Pachi bamboo cluster increases significantly, even within a relatively confined area. The root system of Pachi bamboo demonstrates a strong capacity for reinforcing sloped terrain, offering effective stabilization per unit area. These findings support the potential of Pachi bamboo in enhancing soil stabilization and slope stability in engineering practice as well as in bamboo forest management.

Author Contributions: Conceptualization, C.-C.F.; methodology, C.-C.F.; validation, C.-C.F. and C.-H.C.; formal analysis, C.-C.F. and C.-H.C.; investigation, C.-C.F. and C.-H.C.; resources, C.-C.F.; data curation, C.C. and C.-H.C.; writing—original draft preparation, C.-C.F.; writing—review and editing, C.-C.F., C.C., and C.-H.C.; visualization, C.-H.C.; supervision, C.-C.F.; project administration, C.-C.F.; funding acquisition, C.-C.F. All authors have read and agreed to the published version of the manuscript.

Funding: This research was funded by the Taiwan National Science and Technology Council, Taiwan, ROC, grant number MOST 103-2221-E-327-043-MY2.

Data Availability Statement: The data supporting the findings of this research are available upon reasonable request.

Acknowledgments: The authors also would like to thank Jian-Zhi Huang, Ke-Jie Wei, Zhi-Wei Wang, Wen-Quan Xu, Sheng-Fu Lin, and Yen-Wen Chen for their contribution to the fieldwork and data procession.

Conflicts of Interest: Author Chung-Hao Chen was employed by the company CECI Engineering Consultants, Inc., Author Chunhsiung Chen was employed by the company Kuocheng Construction Co. The remaining authors declare that the research was conducted in the absence of any commercial or financial relationships that could be construed as a potential conflict of interest.

References

1. Waldron, L.J. The shear resistance of root-permeated homogeneous and stratified soil. *Soil Sci. Soc. Am. J.* **1977**, *41*, 843–849. [CrossRef]
2. Gray, D.H.; Leiser, A.T. *Biotechnical Slope Protection and Erosion Control*; Van Nostrand Reinhold Co.: New York, NY, USA, 1982; 267p.
3. Endo, T.; Tsuruta, T. Effects of tree root upon the shearing strengths of soils. *Annu. Rep. Hokkaido Branch For. Place Exp. Stn.* **1969**, *18*, 168–179.
4. Hubble, T.C.T.; Docker, B.B.; Rutherfurd, I.D. The role of riparian trees in maintaining riverbank stability: A review of Australian experience and practice. *Ecol. Eng.* **2010**, *36*, 292–304. [CrossRef]
5. Fan, C.-C.; Chen, Y.-W. The effect of root architecture on the shearing resistance of root-permeated soils. *Ecol. Eng.* **2010**, *36*, 813–826. [CrossRef]
6. Waldron, L.J.; Dakessian, S. Soil reinforcement by roots: Calculation of increased soil shear strength from root properties. *Soil Sci.* **1981**, *132*, 427–435. [CrossRef]
7. Schwarz, M.; Cohen, D.; Or, D. Root-soil mechanical interactions during pullout and failure of root bundles. *J. Geophys. Res.* **2010**, *115*, F04035. [CrossRef]
8. O'Loughlin, C.L. An Investigation of the Stability of the Steepland Forest Soils in the Coast Mountains, Southwest British Columbia. Ph.D. Thesis, University of British Columbia, Vancouver, BC, Canada, 1972.
9. Greenwood, J.R.; Norris, J.E.; Wint, J. Assessing the contribution of vegetation to slope stability. *J. Geotech. Eng.* **2004**, *157*, 199–208. [CrossRef]
10. Genet, M.; Stokes, A.; Fourcaud, T.; Norris, J.E. The influence of plant diversity on slope stability in a moist evergreen deciduous forest. *Ecol. Eng.* **2010**, *36*, 265–275. [CrossRef]
11. Fan, C.-C.; Lai, Y.-F. Influence of the spatial layout of vegetation on the stability of slopes. *Plant Soil* **2014**, *377*, 83–95. [CrossRef]
12. Mao, Z.; Yang, M.; Bourrier, F.; Fourcaud, T. Evaluation of root reinforcement models using numerical modelling approaches. *Plant Soil* **2014**, *381*, 249–270. [CrossRef]
13. Masi, E.B.; Tofani, V.; Rossi, G.; Cuomo, S.; Wu, W.; Salciarini, D.; Caporali, E.; Catani, F. Effects of roots cohesion on regional distributed slope stability modelling. *Catena* **2023**, *222*, 106853. [CrossRef]
14. Dupuy, L.; Fourcaud, T.; Stokes, A. A numerical investigation into the influence of soil type and root architecture on tree anchorage. *Plant Soil* **2015**, *278*, 119–134. [CrossRef]
15. Reubens, B.; Poesen, J.; Danjon, F.; Geudens, G.; Muys, B. The role of fine and coarse roots in shallow slope stability and soil erosion control with a focus on root system architecture: A review. *Trees* **2007**, *21*, 385–402. [CrossRef]
16. Stokes, A.; Atger, C.; Bengough, A.G.; Fourcaud, T.; Sidle, R.C. Desirable plant root traits for protecting natural and engineered slopes against landslides. *Plant Soil* **2009**, *324*, 1–30. [CrossRef]
17. Ennos, A.R. The anchorage of leek seedlings: The effect of root length and soil strength. *Ann. Bot.* **1990**, *65*, 409–416. [CrossRef]

18. Bischetti, G.B.; De Cesare, G.; Mickovski, S.B.; Rauch, H.P.; Schwarz, M.; Stangl, R. Design and temporal issues in soil bioengineering structures for the stabilisation of shallow soil movements. *Ecol. Eng.* **2021**, *169*, 106309. [CrossRef]
19. Zhou, B.; Fu, M.; Xie, J.; Yang, X.; Li, Z. Ecological functions of bamboo forest: Research and Application. *J. For. Res.* **2005**, *16*, 143–147.
20. Mishra, G.; Giri, K.; Panday, S.; Kumar, R.; Bisht, N.S. Bamboo: Potential resource for eco-restoration of degraded lands. *J. Biol. Earth Sci.* **2014**, *4*, B130–B136.
21. Tardio, G.; Mickovski, S.B.; Rauch, H.P.; Fernandes, J.P.; Acharya, M.S. The use of bamboo for erosion control and slope stabilization: Soil bioengineering works. In *Bamboo: Current and Future Prospects*; IntechOpen: London, UK, 2018; p. 105.
22. Hairiah, K.; Widianto, W.; Suprayogo, D.; Van Noordwijk, M. Tree roots anchoring and binding soil: Reducing landslide risk in Indonesian agroforestry. *Land* **2020**, *9*, 256. [CrossRef]
23. Lin, D.-G.; Liu, W.-T.; Lin, S.-H. Estimating the Effect of Shear Strength increment due to root on the stability of Makino bamboo forest slopeland. *J. Geoeng.* **2011**, *6*, 73–88.
24. Wang, X.; Keplinger, T.; Gierlinger, N.; Burgert, I. Plant material features responsible for bamboo's excellent mechanical performance: A comparison of tensile properties of bamboo and spruce at the tissue, fibre, and cell wall levels. *Ann. Bot.* **2014**, *8*, 1627–1635. [CrossRef] [PubMed]
25. Li, Z.; Chen, C.; Mi, R.; Gan, W.; Hu, L. A strong, tough, and scalable structural material from fast-rowing bamboo. *Adv. Mater.* **2020**, *32*, 1906308. [CrossRef] [PubMed]
26. Wang, Y.Y.; Wang, X.Q.; Li, Y.Q.; Huang, P.; Fu, S.Y. High-performance bamboo steel derived from natural bamboo. *ACS Appl. Mater. Interfaces* **2020**, *13*, 1431–1440. [CrossRef] [PubMed]
27. Gao, X.; Zhu, D.; Fan, S.; Rahman, M.Z.; Guo, S.; Chen, F. A review of the structural and mechanical properties of bamboo fiber bundle and fiber/bundle reinforced composites. *J. Mater. Res. Technol.* **2022**, *19*, 1162–1190. [CrossRef]
28. Zong, Y.; Chen, X.; Luo, X.; Su, Q.; Zhang, X.; Yan, Y.; Liu, H. Effect of Bamboo Culm Grading on the Properties of Flattened Bamboo Boards. *Forests* **2023**, *14*, 1120. [CrossRef]
29. Rana, A.K.; Mandal, A.; Bandyopadhyay, S. Short jute fiber reinforced polypropylene composites: Effect of compatibiliser, impact modifier and fiber loading. *Compos. Sci. Technol.* **2003**, *63*, 801–806. [CrossRef]
30. Shah, D.U. Developing plant fibre composites for structural applications by optimising composite parameters: A critical review. *J. Mater. Sci.* **2013**, *48*, 6083–6107. [CrossRef]
31. Perremans, D.; Trujillo, E.; Ivens, J.; Van Vuure, A.W. Effect of discontinuities in bamboo fibre reinforced epoxy composites. *Compos. Sci. Technol.* **2018**, *155*, 50–57. [CrossRef]
32. Ahmad, Z.; Upadhyay, A.; Ding, Y.; Emamverdian, A.; Shahzad, A. Bamboo: Origin, habitat, distributions and global perspective. In *Biotechnological Advances in Bamboo*; Ahmad, Z., Ding, Y., Shahzad, A., Eds.; Springer: Singapore, 2021. [CrossRef]
33. Chen, C. Herbarium of Taiwan Forestry Research Institute. Version 1.8. Taiwan Forestry Research Institute. Occurrence Dataset 2021. Available online: https://www.gbif.org/occurrence/1821909071 (accessed on 5 March 2024).
34. POWO Plants of the World Online. Facilitated by the Royal Botanic Gardens, Kew. Published on the Internet. 2024. Available online: https://powo.science.kew.org/taxon/urn:lsid:ipni.org:names:392450-1 (accessed on 16 July 2024).
35. Canadell, J.; Jackson, R.B.; Ehleringer, J.R.; Mooney, H.A.; Sala, O.E.; Schulze, E.D. Maximum rooting depth of vegetation types at the global scale. *Oecologia* **1996**, *108*, 583–595. [CrossRef]
36. Loades, K.W.; Bengough, A.G.; Bransby, M.F.; Hallett, P.D. Planting density influence on fibrous root reinforcement of soils. *Ecol. Eng.* **2010**, *36*, 276–284. [CrossRef]
37. Cohen, D.; Schwarz, M. Tree-root control of shallow landslides. *Earth Surf. Dynam.* **2017**, *5*, 451–477. [CrossRef]
38. Cislaghi, A.; Cohen, D.; Gasser, E.; Bischetti, G.B.; Schwarz, M. Field measurements of passive earth forces in steep, shallow, landslide-prone areas. *J. Geophys. Res. Earth Surf.* **2019**, *124*, 838–866. [CrossRef]
39. Giadrossich, F.; Schwarz, M.; Cohen, D.; Cislaghi, A.; Vergani, C.; Hubble, T.; Phillips, C.; Stokes, A. Methods to measure the mechanical behaviour of tree roots: A review. *Ecol. Eng.* **2017**, *109*, 256–271. [CrossRef]
40. Phillips, C.J.; Marden, M.; Lambie, S.; Watson, A.; Ross, C.; Fraser, S. Observations of below-ground characteristics of young redwood trees (*Sequoia sempervirens*) from two sites in New Zealand—Implications for erosion control. *Plant Soil* **2013**, *363*, 33–48. [CrossRef]
41. Wu, T.H.; McKinell, W.P.; Swanston, D.N. Strength of tree roots and landslides on Prince of Wales Island, Alaska. *Can. Geotech. J.* **1979**, *16*, 19–33. [CrossRef]
42. Deljouei, A.; Cislaghi, A.; Abdi, E.; Borz, S.A.; Majnounian, B.; Hales, T.C. Implications of hornbeam and beech root systems on slope stability: From field and laboratory measurements to modelling methods. *Plant Soil* **2023**, *483*, 547–572. [CrossRef]
43. Vergani, C.; Chiaradia, E.A.; Bischetti, G.B. Variability in the tensile resistance of roots in Alpine forest tree species. *Ecol. Eng.* **2012**, *46*, 43–56. [CrossRef]
44. Topp, C.N.; Benfey, P.N. Growth control of root architecture. In *Plant Biotechnology and Agriculture*; Altman, A., Hasegawa, P.M., Eds.; Prospects for the 21st Century; Academic Press: Amsterdam, The Netherlands, 2012. [CrossRef]

45. Ma'ruf, M.F. Shear strength of Apus bamboo root reinforced soil. *Ecol. Eng.* **2012**, *41*, 84–86. [CrossRef]
46. Mulyanti, J.; Hendra, D.; Megaprastio, B. Utilization of Bamboo Roots as Reinforcing Composite Materials for Light Vehicle Dashboards. *J. Phys.* **2024**, *2739*, 012031. [CrossRef]
47. Cislaghi, A. Exploring the variability in elastic properties of roots in Alpine tree species. *J. For. Sci.* **2021**, *67*, 338–356. [CrossRef]
48. Chimungu, J.G.; Loades, K.W.; Lynch, J.P. Root anatomical phenes predict root penetration ability and biomechanical properties in maize (*Zea mays*). *J. Exp. Bot.* **2015**, *66*, 3151–3162. [CrossRef] [PubMed]
49. Teerawattanasuk, C.; Maneecharoen, J.; Bergado, D.T.; Voottipruex, P.; Le, G.L. Root strength measurements of Vetiver and Ruzi grasses. *Lowl. Technol. Int.* **2014**, *16*, 71–80. [CrossRef] [PubMed]
50. Boldrin, D.; Bengough, A.G.; Lin, Z.; Loades, K.W. Root age influences failure location in grass species during mechanical testing. *Plant Soil* **2021**, *461*, 457–469. [CrossRef]
51. Ekeoma, E.C.; Boldrin, D.; Loades, K.W.; Bengough, A.G. Drying of fibrous roots strengthens the negative power relation between biomechanical properties and diameter. *Plant Soil* **2021**, *469*, 321–334. [CrossRef]

forests

MDPI

Article

Effects of Thinning of the Infected Trees and Cultivating of the Resistant Pines on Soil Microbial Diversity and Function

Xiaorui Zhang [1], Zhuo Liu [1], Mu Cao [1] and Tingting Dai [1,2,*]

1 Co-Innovation Center for the Sustainable Forestry in Southern China, Nanjing Forestry University, Nanjing 210037, China; 17761709897@163.com (X.Z.); 15591861881@163.com (Z.L.); cm129mc@njfu.edu.cn (M.C.)
2 Advanced Analysis and Testing Center, Nanjing Forestry University, Nanjing 210037, China
* Correspondence: 13770647123@163.com

Abstract: Pine wilt disease (PWD) poses a significant threat to pine forest health, making sanitation thinning of infected trees and cultivation of disease-resistant pine stands crucial measures for forest ecosystem restoration. To date, limited studies have systematically investigated how post-sanitation planting of pine-wilt-disease-resistant Pinus species affects soil microbiome, especially regarding bacterial and fungal diversity characteristics, functional succession patterns, and community assembly processes. In this study, we performed a comparative analysis of soil microbial community characteristics and biochemical properties between experimental plots subjected to sanitation thinning and those replanted with disease-resistant pine species. The results indicated that compared to the sanitation-thinned experimental plot, the disease-resistant experimental plots (*Pinus taeda* experimental plot and *Pinus thunbergii* experimental plot) exhibited significantly higher activities of β-glucosidase (S-β-GC), N-acetyl-β-D-glucosidase (S-NAG), and soil arylsulfatase (S-ASF). Compared with the sanitation logging stands, our analysis revealed that the *Pinus taeda* experimental plot and *Pinus thunbergii* experimental plot exhibited significantly higher fungal community evenness (OTUs), greater species abundance (OTUs), and more unique fungal taxa. Furthermore, the edaphic properties—specifically soil moisture content (SMC), pH levels, and total potassium (TK)—significantly influenced the structures of soil bacterial and fungal communities. Compared to the sanitation-thinned experimental plot, wood saprotrophic fungi and ectomycorrhizal fungi exhibited increased abundance in both the *P. taeda* experimental plot and *Pinus thunbergii* experimental plot. Furthermore, the null models indicated that both the *P. taeda* experimental plot and *P. thunbergii* experimental plot enhanced the undominated processes of bacteria and fungi. In summary, our data elucidate the differences in bacterial and fungal responses between pine forests undergoing thinning due to infected trees and those cultivated for disease resistance. This deepens our understanding of microbial functions and community assembly processes within these ecosystems.

Keywords: PWD; soil microorganisms; soil properties; sanitation cutting; *Pinus taeda*; *Pinus thunbergii*

1. Introduction

Pine forests are important ecological treasures on the planet, providing a variety of ecological functions such as biodiversity conservation, carbon storage, soil and water conservation, and timber resources [1,2]. Pine wilt disease (PWD) caused by *Bursaphelenchus xylophilus* Steiner et Buhrer is devastating to pine trees, with infected specimens typically exhibiting rapid wilting followed by death within months [3]. The disease spreads with alarming efficiency through its primary vector (*Monochamus alternatus* Hope), leading to explosive outbreaks that can decimate entire pine forests [4]. Sanitation cutting, as a critical management measure, effectively disrupts disease transmission by removing weakened, dead, and *B. xylophilus*-infested host trees, significantly reducing the risk of epidemic spread [5]. Timely sanitation felling can significantly reduce pine wilt disease transmission risk by 69.28% [6,7]. Cultivating pine resistant to *B. xylophilus* enhances forest resilience through genetic improvement. By selectively breeding resistant species such as *Pinus thunbergii* Parlatore and *Pinus taeda* L.—including resistant cultivars like *P. thunbergii* cv. Asuka—and integrating marker-assisted breeding techniques, sustainable nematode-resistant pine ecosystems can be established [8,9]. To improve the ecological benefits of pine forests, it is essential to rationally cultivate disease-resistant pine species and implement scientific thinning practices to enhance biodiversity. These measures will promote the formation of more stable forest communities and optimize their ecological functions [10].

Soil microorganisms serve as pivotal regulators in terrestrial ecosystems, playing indispensable roles in key ecological processes such as nutrient cycling, pathogen suppression, and plant productivity enhancement [11]. As essential drivers of ecosystem functioning, microbial communities sustain vital ecological services by facilitating nutrient cycling, supporting primary production, and participating in climate regulation [12]. Research demonstrates that alterations in soil biodiversity, particularly microbial diversity, profoundly influence ecosystem functioning [13]. This positive correlation primarily stems from niche complementarity, mutualistic interactions, and the suppression of pathogenic microorganisms [14]. These findings underscore the central importance of soil microorganisms in enhancing forest quality and restoring ecosystem vitality.

Generally, alterations in the soil microbial community structure reflect the ongoing ecological dynamics between the organisms and their environmental context. Key factors that significantly impact microbial communities encompass soil pH [15], soil moisture content [16], organic carbon [17], nutrient availability (including carbon, nitrogen, phosphorus, and sulfur) [18], and biotic elements like plant diversity and specific cultivars [19]. In addition, β-glucosidase (S-β-GC), N-acetyl-β-D-glucosidase (S-NAG), and arylsulfatase (S-ASF) serve as key functional enzymes in soil carbon, nitrogen, and sulfur cycling, whose activities directly regulate soil microbial community functions and plant nutrient availability [20]. The soil itself exhibits high diversity and heterogeneity (with variations in composition and properties across different regions), and its various characteristics (such as pH, nutrient content, moisture, organic matter, etc.) can directly or indirectly influence the composition and activity of microbial communities [21,22]. Due to the intricate and interconnected nature of these influencing factors, gaining a comprehensive understanding of microbial community structure and function remains a significant challenge [23].

Microbial community assembly or community structure is shaped by a combination of deterministic and stochastic ecological processes [24]. Deterministic processes (niche-based theory) result from the predictable filtering effects on species of ecological selection imposed by biotic and abiotic factors that affect the fitness of organisms and thus determine species composition and relative abundance [25]. In contrast, stochastic processes (neutral theory) involve ecological drift (birth, death, immigration, speciation, and limited dispersal) and are not the result of environmentally determined fitness [26,27]. However, current research has predominantly focused on characterizing microbial communities in specific ecosystems (e.g., marine, freshwater, and agricultural systems), while investigations into the evolutionary assembly processes of microbial communities within forest disease management regimes remain notably limited [28–31]. Therefore, gaining a more profound understanding of how the thinning of pine trees infested with pine wood nematode and cultivation of disease-resistant pine trees impact the assembly and diversity of soil microbial communities could significantly contribute to the preservation and restoration of ecological biodiversity, particularly in the temperate forests of China.

The impact of current PWD control measures (selective thinning of infected trees and planting of disease-resistant pines) on soil microbial communities in this region remains poorly understood. To address this knowledge gap, this study collected soil profile samples from these forest stands and employed high-throughput gene sequencing to characterize the microbial community composition. This study sought to: (1) assess the effects of management measures for PWD on the diversity of soil bacterial and fungal communities; (2) identify the main environmental factors responsible for the diversity of soil bacteria and fungi in pine forests after the implementation of the management measures; (3) track the potential functions of soil bacterial and fungal communities after the implementation of the management measures; and (4) elucidate the processes of assembling soil bacterial and fungal communities after the implementation of the management measures.

2. Materials and Methods

2.1. Experimental Protocol and Sample Collection

The study was conducted in Jurong city, Jiangsu Province, located at 119°13′ E, 32°7′ N, with a forest area of 314.14 ha. The climate of the study area is characterized by a subtropical humid monsoon climate, featuring abundant sunlight, four distinct seasons, and obvious mountain climate characteristics. The average annual temperature is 15.2 °C, and the average annual precipitation is 1055.6 mm. The native forest type in this area was a mixed *Pinus massoniana* Lamb and *Larix principis-rupprechtii* Mayr forest. Following the invasion of *B. xylophilus* into the native forest, three experimental plots with similar site conditions were established. The elevation difference between plots was less than 50 m, the slope gradient difference was less than 10 degrees, and the soil type was yellow-brown earth. The following integrated management measures were implemented over the past 20 years: (1) sanitation felling to remove infected and weakened trees; and (2) replanting disease-resistant species separately—*P. taeda* and *P. thunbergii*—in the experimental plots after sanitation felling (detailed characteristics of the three plots are provided in Supplementary Table S1). Since the implementation of these control measures, a comprehensive protection system has been established in the study area, including disease monitoring and stand management practices.

2.2. Soil Sampling

Based on plot accessibility, soil samples were collected in April 2024 using a five-point sampling method. Specifically, within each of the three treatment plots—the sanitation cutting treatment group (PMLG), *P. taeda* replanting group (PTL), and *P. thunbergii* replanting group (PTP)—five biological replicates were established. At each sampling point, the surface cover (including litter, humus, and living ground vegetation) was removed before collecting soil samples at a 10 cm depth [32]. The obtained samples were evenly divided into two portions: one for soil microbial community analysis and the other for soil physicochemical property determination. For preservation, all samples were stored in an ultra-low-temperature freezer at −80 °C. High-throughput sequencing analysis was performed by Shanghai Majorbio Bio-Pharm Technology Co. (Shanghai, China) (www.majorbio.com (accessed on 7 May 2025)).

2.3. Determination of Enzymatic Activity and Physicochemical Properties of the Soil

The activities of aryl sulfatase (S-ASF), acid phosphatase (S-ACP), N-acetyl-β-D-glucosidase (S-NAG), and β-glucosidase (S-β-GC), which are associated with the cycling of sulfur, phosphorus, and carbon, were assessed using kits supplied by Suzhou Comin Biotechnology ((www.cominbio.com (accessed on 7 May 2025)) [33]. For comprehensive details regarding the methodology, please refer to the instructions available on the Suzhou Keming Biotechnology Co. website. The soil water content (SWC, %) was determined by drying at 105 °C, while the soil pH was measured with a pH meter after mixing the soil and water (2.5:1) [34]. The soil organic carbon (SOC) and total nitrogen (TN) contents were determined via an elemental analyzer (EA 3000, Vector, Redavalle, Italy), whereas the soil total phosphorus (TP), total potassium (TK), and total sulfur (TS) contents were quantified via an OLYMPUS XRF analyzer (made in Center Valley, PA, USA) [35,36]. All soil enzyme activity data reported in this study were measured using air-dried soil samples. Soil properties from the same sampling site (analyzed in the same project; see Supplementary Material, Table S1) were included solely as supplementary variables for correlation analyses with microbial communities and are not discussed in the main text.

2.4. DNA Extraction, PCR Amplification, and Illumina MiSeq Sequencing

A QIAamp kit was used for DNA extraction from different soil samples according to the manufacturer's instructions. The concentration and quality of the extracted DNA were determined via a Thermo Scientific NanoDrop2000 spectrophotometer (Thermo Fisher Scientific, Wilmington, DE, USA), and the integrity of the extracted DNA was tested via 1% agarose gel electrophoresis [37]. Using amplicon sequencing to identify fungi and bacteria in soil, the primer pairs 338F (5′-ACTCCTACGGGAGGCAGCAG-3′)/806R (5′-GGACTACHVGGGTWTCTAAT-3′) were employed to amplify the V3-V4 region of the 16S rRNA gene for bacterial community analysis, while the primer pair ITS1F (5′-CTTGGTCATTTAGAGAGGAAGTAA-3′)/ITS2R (5′-GCTGCGTTCTTCATCGATGC-3′) was used to amplify the ITS1 region (Internal Transcribed Spacer 1) for fungal community identification and analysis [38]. The PCR conditions consisted of denaturation at 95 °C for 2 min, followed by 25 cycles of 95 °C for 30 s, 55 °C for 30 s, and 72 °C for 60 s; a 10 min extension at 72 °C for 10 min; a 10 min extension at 4 °C for 10 min; and an extension at 72 °C for 10 min, followed by holding at 4 °C. The PCR products were detected via electrophoresis on a 2% agarose gel and recovered by cutting the gel via an AxyPrepDNA Gel Recovery Kit (AXYGEN, Union City, CA, USA) [38]. The PCR products were quantified via the QuantiFluor™-ST Blue Fluorescence Quantification System (Promega, Madison, WI, USA), with reference to the preliminary quantification results of electrophoresis. PCR products were prepared with the NEXTFLEX Rapid DNA-Seq Kit (Bioo Scientific, Austin, TX, USA) and sequenced in paired-end mode (2 × 300 bp) on the Illumina MiSeq platform (Illumina, San Diego, CA, USA). The original sequences were submitted and stored in the NCBI Sequence Read Archive (SRA), with accession number SUB14777552.

2.5. Bioinformatics Analysis

First, we filtered fast for quality via FASTQ software (version 0.19.6) and merged it via Flash (version 1.2.11) to merge with the following criteria: (i) filter out bases with quality values below 20, set a window of 50 bp, and remove reads for bases containing N; (ii) merge sequences with overlap times greater than 10 bp and allow a maximum mismatch of 2 bp; and (iii) qiime (version 1.9.1) was used for denoising, removing sequences annotated to chloroplasts and mitochondria in the samples and thinning the resulting data via a minimum sample sequence levelling process. We subsequently used Uparse (version 11) to manipulate the classification units (OTUs) for annotation and finally used the RDP classifier (version 2.13) to compare the measured microbial data with the Silicon Valley data. The measured microbial data were compared to the Silva 138 database (bacteria) and the Unite 9.0/its_fungi database (fungi), with a confidence threshold of 70% and an OTU sequence similarity of 97%.

2.6. Statistical Analyses

First, we employed one-way analysis of variance (ANOVA) combined with false discovery rate (FDR) correction to assess the significance of intergroup differences in soil enzyme activities, physicochemical properties, and microbial α-diversity indices (Coverage, Simpson, Shannoneven, Chao). Next, based on the Bray–Curtis distance matrix, we analyzed β-diversity differences at the OTU level using analysis of similarities (ANOSIM) and principal coordinate analysis (PCoA). To identify differentially abundant taxa, we applied linear discriminant analysis effect size (LEfSe) (LDA threshold > 2.5) and an all-against-all multi-group comparison strategy to screen for significantly different bacterial/fungal taxa (genus level) across the three sample types. To elucidate the influence of environmental factors on soil microbial communities, we first screened environmental parameters using variance inflation factor (VIF < 20). Subsequently, redundancy analysis (RDA/CCA) (default standardization) was performed in conjunction with the Bray–Curtis distance matrix to reveal microbial community–environmental factor relationships. The relative contributions of these factors were quantified using variance partitioning analysis (VPA). Based on the abundance data of the top 500 OTUs, we constructed a Spearman rank correlation network ($|R| > 0.8$, $p < 0.05$; implemented with the igraph package and visualized using Gephi 0.9.1). Functional annotation was conducted using FUNGuild (for fungi) and FAPROTAX (for bacteria) to identify indicator taxa significantly associated with composition/abundance [39,40]. We evaluated the roles of selection and dispersal in community turnover by employing null model approaches. In doing so, we determined phylogenetic and taxonomic beta diversity indices (β-nearest taxon index, βNTI) and utilized the Bray–Curtis-based Raup–Crick metric (RCbray) to assess variations in both phylogenetic and taxonomic diversity [41].

3. Results

3.1. Soil Enzyme Activities in Three Pine Plots

PMLG, PTL, and PTP showed significant differences in soil enzyme activities related to carbon, nitrogen, and sulfur cycling. Notably, S-β-GC and S-NAG esterase were significantly greater in PTL's soils and PTP's soils than in PMLG's soils. S-β-GC was higher in PTL's soils than in PTP's soils, and interestingly, the activity of S-NAG was the opposite in both. In addition, compared to PMLG's soils, S-ACP and S-ASF esterase activities in PTL's soil changed but not significantly (Table 1).

Table 1. Study of enzyme activities related to soil nutrient cycling in different *Pinus* plantations.

Foresty-Type	S-β-GC (umol/d/g)	S-NAG (umol/d/g)	S-ACP (umol/d/g)	S-ASF (umol/d/g)
Mixed pine forests	2.71 ± 1.87^{b}	7.96 ± 1.23^{c}	17.48 ± 1.70^{a}	0.66 ± 0.07^{b}
Pinus thunbergii forests	17.06 ± 3.36^{a}	18.83 ± 1.67^{b}	21.72 ± 8.99^{a}	1.99 ± 0.56^{a}
Pinus tada forests	12.12 ± 1.08^{b}	27.72 ± 7.26^{a}	21.52 ± 3.66^{a}	1.22 ± 0.166^{b}

Note: The difference identification symbols (such as a, b, c) are used to indicate significant differences between different treatment groups or samples. S-β-GC: β-glucosidase; S-NAG: N-acetyl-β-D-glucosidase; S-ACP: acid phosphatase; S-ASF: aryl sulfatase.

3.2. Bacterial and Fungal Community Composition and Diversity

A total of 8255 bacterial OTUs and 3913 fungal OTUs were obtained from the 15 soil samples, and the *P. taeda* replanting group (PTL) soil microorganisms had the highest abundance of fungi (61.03% of the total fungal OTUs) and bacteria (65.18% of the total bacterial OTUs) (Figure S1a,b). The minimum coverage index for each sample exceeded 99.80%, indicating sufficient sequencing depth and high coverage (Figure 1c). The fungal communities in the soil samples comprised Basidiomycota, Ascomycota, and Mortierellomycota (total abundance > 87%) (Figure 1a). Furthermore, Proteobacteria, Acidobacteria, Actinobacteria, and Chloroflexi (total abundance > 76%) were dominant in all samples (Figure 1b).

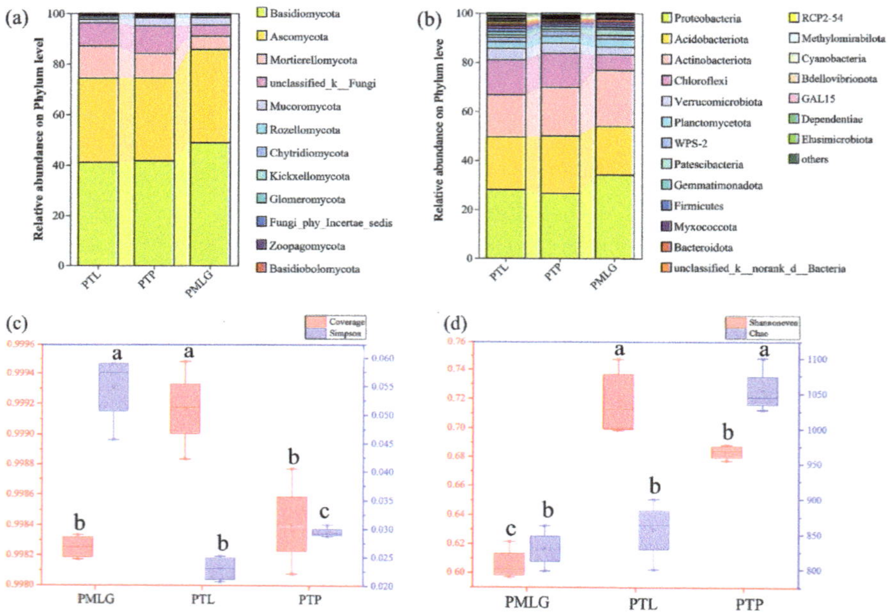

Figure 1. Phylum-level taxonomic structure and α-diversity (OTUs) of soil microbial communities across the three plantation plots. (**a**) Relative abundance of major taxa at the fungal level. (**b**) Relative abundance of major taxa at the bacterial level. (**c**) Simpson and Coverage indices in fungi. (**d**) Shannoneven and Chao indices in fungi. Notes: PMLG: sanitation cutting treatment group; PTL: *Pinus taeda* replanting group; PTP: *Pinus thunbergii* replanting group. Lowercase letters indicate significant differences ($p < 0.05$) between different plantations.

Replanting with *P. taeda* or *P. thunbergii* after sanitation cutting significantly enhances the richness and diversity of soil fungal communities. α-diversity analysis revealed that both the *P. taeda* replanting group (PTL) and *P. thunbergii* replanting group (PTP) exhibited significantly higher soil microbial Chao index (richness) and Shannoneven index (diversity) compared to the sanitation cutting treatment group (PMLG) (Figure 1d). Notably, however, the sanitation cutting treatment group (PMLG) displayed the highest Simpson index (dominance), with statistically significant differences ($p < 0.05$) from the other plots in soil. Although the Shannon index, Chao index, and Simpson index showed differences in bacterial abundance in the three plots, none reached statistical significance (Figure 1c,d) ($p < 0.05$).

3.3. Beta Diversity

The results of principal component analysis (PCA) (Figure 2a) showed that the fungal community structure differed significantly among the three habitats (R = 0.9902, p = 0.001), with axis 1 and axis 2 explaining 36.72% and 18.07% of the variance in the community structure, respectively. Bacterial community structure differed significantly among the three habitats (R = 0.4941, p = 0.001), with axis 1 and axis 2 explaining 43.84% and 13.55% of the variance in community structure (Figure 2b).

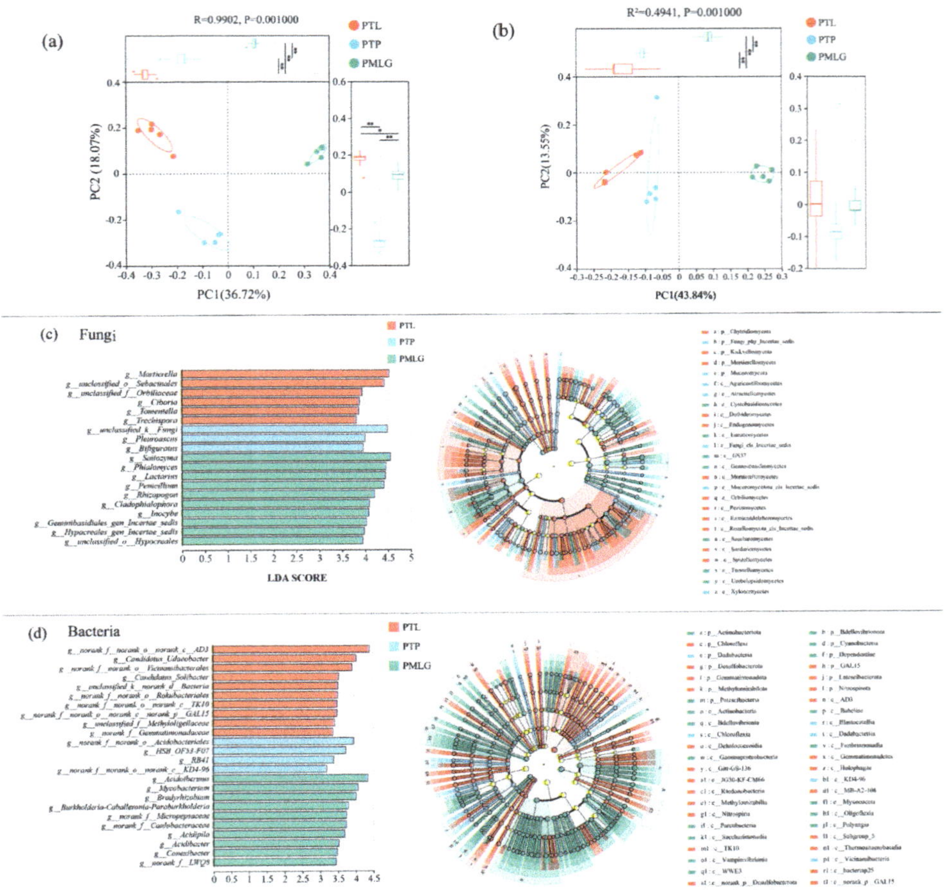

Figure 2. Changes in the soil microbial community structure in three pine plots. (**a**) PCoA of differences in fungal community structure. (**b**) PCoA of differences in bacterial community structure. Analysis of LEfSe species differences in fungal (**c**) and bacterial (**d**) communities. Note: * indicates a significance level of $p < 0.05$, ** indicates a significance level of $p < 0.01$, *** indicates a significance level of $p < 0.001$. PMLG: sanitation cutting treatment group; PTL: *Pinus taeda* replanting group; PTP: *Pinus thunbergii* replanting group.

The soil microbial community data were analyzed from the phylum to species level using LEfSe method to find the species that differed significantly in abundance between groups, of which there were 19 fungal genera with LDA scores greater than 2.5, as shown in Figure 2c. Among them, 6 genera were significantly enriched in the *P. taeda* replanting group (PTL), 3 genera were significantly enriched in the *P. thunbergii* replanting group (PTP), and 10 genera were significantly enriched in the sanitation cutting treatment group (PMLG). From the species evolution analysis, the genera significantly enriched in PTL were mainly from Chytridiomycota; the genera significantly enriched in PTP were mainly from Kickxellomycota and Mucoromycota. In terms of soil bacteria, there were 10, 4, and 10 genera significantly enriched in PMLG, PTP, and PTL respectively. From the species evolution analysis, the genera significantly enriched in PTL were mainly from Chloroflexi, Desulfobacterota, Gemmatimonadata, and Methylomirabilota; the genera significantly enriched in PTP were mainly from Dadabacteria; and the genera significantly enriched in PMLG were mainly from Actinobacteriota (Figure 2d). This suggests that the soil-enriched species-specific taxa differed significantly among the three habitats.

3.4. Correlations of Soil Microbial Communities with Environmental Variables

To investigate the effects of replanting disease-resistant *P. thunbergii* and *P. taeda* after sanitation cutting on soil microbial community structure, we performed redundancy analysis (RDA) to examine the relationships between microbial community composition and nine key environmental variables (including five soil physicochemical properties and four enzyme activity indicators) that were pre-selected through variance inflation factor (VIF) analysis (Tables S2 and S3).

Among the environmental variables, S-ASF exhibited the longest vector length in the RDA biplot, followed by SMC ($r^2 = 0.7213$, $p < 0.002$) and pH ($r^2 = 0.7638$, $p < 0.001$), indicating their most significant influence on soil fungal communities. In contrast, TP and S-β-GC showed the shortest vectors, demonstrating minimal impact on bacterial communities (Figure 3a). Variance partitioning analysis (VPA) revealed that soil enzyme activity was the primary driver of fungal community variation, accounting for 5.21% of the explained variance, while soil nutrients contributed 23.55% (Figure 3b). Notably, the three treatment groups formed distinct clusters in the ordination space, suggesting that replanting with disease-resistant pines significantly altered soil fungal communities, resulting in unique community characteristics for each group. In bacterial communities, SMC ($r^2 = 0.6525$, $p < 0.004$) displayed the longest vector, followed by TK ($r^2 = 0.5333$, $p < 0.009$), highlighting their strong influence. S-β-GC showed the shortest vector, indicating negligible effects (Figure 3c). Interestingly, in contrast to fungal communities, soil physicochemical properties emerged as the dominant factor explaining bacterial community variation, with enzyme activity and soil nutrients explaining 10.90% and 23.52% of the variance, respectively (Figure 3d).

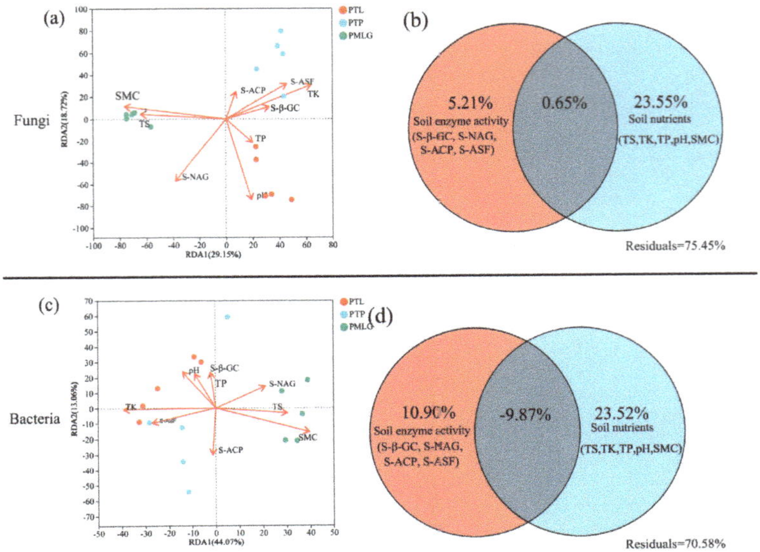

Figure 3. Associations between microbial communities and environmental factors. (**a**) Redundancy analysis of fungal community structure and its relationship with soil properties. (**b**) Variance partitioning analysis (VPA) of fungal communities showing the contributions of environmental factors to fungal communities. (**c**) Redundancy analysis of the bacterial community structure and its relationship with the soil properties. (**d**) Variance partitioning analysis (VPA) of the bacterial community showing the contributions of environmental factors to the bacterial community. PMLG: mixed pine forests; PTL: *Pinus tada* forests; PTP: *Pinus thunbergii* forests.

3.5. Potential Functional Structure of Different Microbial Communities

The three most abundant functional characteristics in the sanitation cutting treatment group (PMLG), *P. taeda* replanting group (PTL), and *P. thunbergii* replanting group (PTP) are undefined saprotroph, ectomycorrhizal, and endophyte–litter saprotroph–soil saprotroph–undefined saprotroph (Figure 4a). Compared to the PMLG group, the PTL and PTP groups showed a significant increase in the relative abundance of endophyte–litter saprotroph–soil saprotroph–undefined saprotroph (ELSSUS), wood saprotroph (WS), and ectomycorrhizal (EC) fungi. While the top five functions of soil bacterial relative abundance were not significantly different in the three experimental plots., we observed significant differences in the functions of the other bacterial communities (Figure 4b). Compared to PMLG, PTP had significantly higher functional taxa of bacterial denitrification, nitrous oxide denitrification, and nitrite denitrification in soil bacterial community function (Figure 4d). Only the anoxygenic_photoautotrophic were significantly increased in the bacterial community functions of PTL compared to PMLG (Figure 4c).

3.6. Deterministic and Stochastic Assembly of Soil Microbial Communities

The β-NTI and RCbray values, as determined between the samples, elucidate the ecological processes governing microbial communities.

Although the βNITs of the sanitation cutting treatment group (PMLG) soil microbial communities were not significantly different from those of the *P. taeda* replanting group (PTL) and *P. thunbergii* replanting group (PTP) ($p > 0.05$), their medians were all less than 2, suggesting that stochastic processes mainly control the formation of soil fungal communities (Figure 5a). Significant changes in soil bacterial community assembly occurred in PTP and PTL compared to PMLG. Although homogenizing dispersal contributed to the bacterial community composition in PMLG, the bacterial community composition that determined PMLG remained dominated by stochastic processes (homogenizing dispersal and undominated) (Figure 5b). The magnitude of undominated soil increased in the order PTL > PTP > PMLG, and it played a significant role in the composition of bacterial communities.

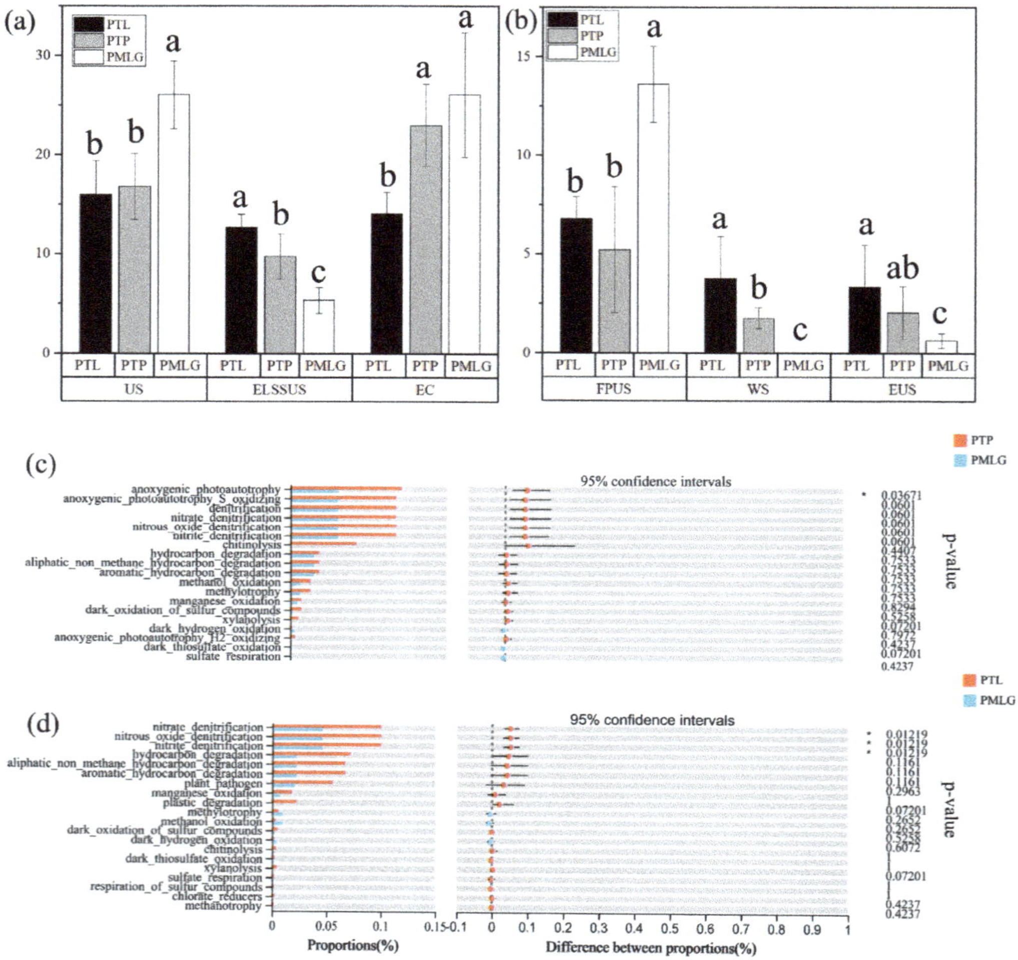

Figure 4. Functional predictive analyses of bacterial communities and bacterial communities. (**a**,**b**) FUNGuild functional prediction component difference tests for fungal communities. (**c**,**d**) FAPROTAX functional prediction component difference tests for bacterial communities. Notes: The difference notation symbols (such as a, b, c) are used to indicate significant differences among different treatment groups. * indicates a significance level of $p < 0.05$, ** indicates a significance level of $p < 0.01$, *** indicates a significance level of $p < 0.001$. US: undefined saprotroph; ELSSUS: endophyte–litter saprotroph–soil saprotroph–undefined saprotroph; EC: ectomycorrhizal; FPUS: fungal parasite–undefined saprotroph; WS: wood saprotroph; EUS: ectomycorrhizal–undefined saprotroph. PMLG: sanitation cutting treatment group; PTL: *Pinus taeda* replanting group; PTP: *Pinus thunbergii* replanting group.

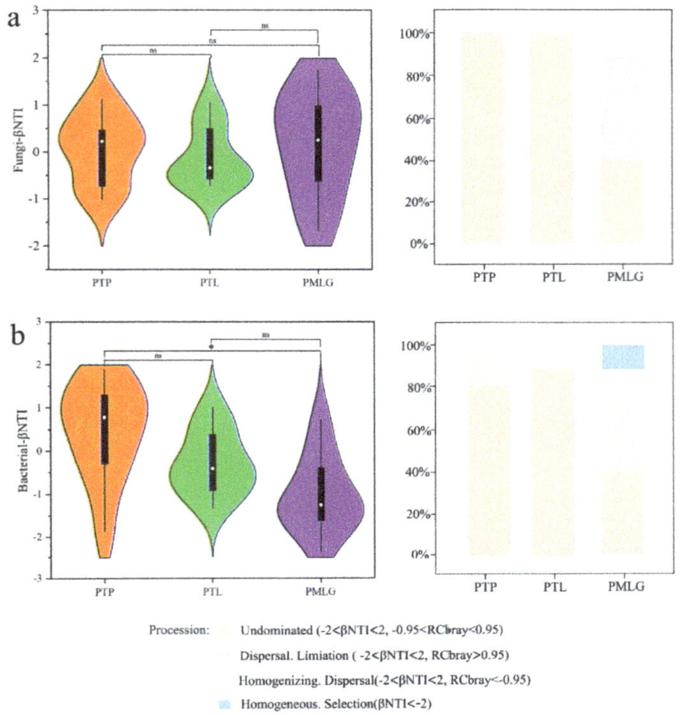

Figure 5. Distribution of β-NTI values for fungal (**a**) and bacterial (**b**) community comparisons. Notes: * indicates a significance level of $p < 0.05$. ns indicates a significance level of $p > 0.05$. PMLG: sanitation cutting treatment group; PTL: *Pinus taeda* replanting group; PTP: *Pinus thunbergii* replanting group.

4. Discussion

4.1. Effects of Infected Wood Harvesting and Planting of Disease-Resistant Pine Trees on Soil Enzyme Activity

Planting pines such as *P. thunbergii* and *P. taeda* has been shown to increase soil microbial activity and soil organic matter decomposition, which leads to a raise in soil enzyme activities [42,43]. This largely explains the increase in soil enzyme activities commonly observed after planting disease-resistant pine trees. β-Glucosidase catalyzes the hydrolysis and biodegradation of various β-glucosides present in plant residues and is a major source of carbon for the growth and activity of soil microorganisms [44]. The soil β-glucosidase (S-β-GC) activity was significantly greater in *P. thunbergii* plots and *P. taeda* plots than in sanitation cutting plots (Table 1), and planting *P. thunbergii* resulted in a 459% increase in the soil SOC content. This phenomenon may be attributed to the active participation of soil microorganisms in the decomposition and transformation of organic matter through a variety of metabolic pathways that contribute to the stabilization of organic carbon, thereby affecting soil carbon storage and turnover [45]. Soil S-NAG catalyzes the hydrolysis of N-acetyl-β-D-glucosamine to release nitrogen from the soil; however, high levels of SOC promote nitrogen release. Therefore, the activity of soil S-NAG in *P. taeda* plots increased [46]. The soil arylsulfatase activity decreased with increasing TS content after replanting pine trees after sanitation cutting. A lower TS content stimulates microbial communities to produce high-quality S-ASF to increase the soil S content, resulting in lower S-ASF activity in sanitation cutting plots than in *P. thunbergii* plots and *P. taeda* plots [47].

4.2. Microbial Communities and Their Diversity

Research indicates that planting *P. taeda* or *P. thunbergii* after sanitation logging can effectively restore the richness and diversity of soil fungal communities. This aligns with the findings of Lin et al., who observed that while thinning temporarily suppresses fungal diversity, reforestation facilitates its recovery [48]—a trend consistent with the observations in the *P. taeda* and *P. thunbergii* sample plots of this study. The soil fungal community was dominated by Basidiomycota, Ascomycota, and Mortierellomycota, while the bacterial community was primarily composed of Proteobacteria, Acidobacteria, Actinobacteria, and Chloroflexi. These findings align with global studies on forest soil microbiomes [49], where Basidiomycota and Ascomycota are recognized as core fungal phyla in forest ecosystems, playing key roles in lignin degradation and symbiotic relationships [50]. Meanwhile, the enrichment of Proteobacteria and Acidobacteria is typically associated with organic matter decomposition and acidic soil conditions [51]. Notably, the high abundance of Mortierellomycota may be linked to root exudates from *P. taeda* in our study sites, as this phylum is known to form mutualistic relationships with plant roots [52]. Similarly, the enrichment of Chloroflexi in the *P. thunbergii* replanting group (PTP) likely reflects their adaptation to pine litter decomposition.

4.3. Relationships Between Soil Properties and Microorganisms

Microbial and soil properties are intricately linked, and changes in soil nutrients and soil enzyme activities affect the abundance of microbial communities [53]. Compared with the sanitation cutting treatment group (PMLG), the soil enzyme activities of S-NAG and S-β-GC in the *Pinus taeda* replanting group (PTL) and *Pinus thunbergii* replanting group (PTP) were significantly increased, which was probably due to the relative increase in difficult-to-degrade organic carbon as a result of mesquite logging [54]. However, the higher moisture content in the sanitation cutting treatment group (PMLG) compared to the *Pinus taeda* replanting group (PTL) and *Pinus thunbergii* replanting group (PTP) indicates greater plant diversity and unfragmented litter content, which contributes to its ability to retain higher water levels [55].

The results of RDA suggested that the microbial community showed a significant correlation with SMC, pH, and TK. Changes in soil water content directly affect the composition and structure of microbial communities, and drought increases the abundance and diversity of fungal communities. SMC affects the community structure of soil microorganisms, and fungi gradually dominate the microbial community under drought conditions [56]. The study proved that SMC was the most important driver of soil fungal community changes in the region. The effect of TK on the bacterial community may be due to the ability of soil bacteria to promote the solubilization of insoluble potassium through the decomposition of minerals, so that soil bacteria have a more adequate supply of potassium, which in turn promotes the growth and reproduction of soil bacteria and enhances the biological activity of the soil [57]. Previous studies have established that soil pH is the most significant factor influencing the composition of the soil microbial community in the region [58]. In our findings, compared with the sanitation cutting treatment group (PMLG), the *Pinus taeda* replanting group (PTL) and *Pinus thunbergii* replanting group (PTP) showed an important change in pH. As a general rule, pH is a crucial factor influencing the structure of the soil fungal community; however, it exerts no significant influence on the bacterial community [59]. This may be due to the fact that the Acidobacteria gates in the soil affected the results of the experiment.

4.4. Functional Changes in Soil Microbial Communities

The findings of this study demonstrate a significant increase in the relative abundance of ectomycorrhizal fungi in both the *P. taeda* replanting group (PTL) and *P. thunbergii* replanting group (PTP). This phenomenon suggests that the replanting of disease-resistant pine trees may exert a positive influence on the structure of soil microbial communities. Ectomycorrhizal fungi form mutualistic symbiotic relationships with plant roots, substantially enhancing the host plant's efficiency in nutrient and water uptake [60]. Previous studies have confirmed a significant positive correlation between ectomycorrhizal fungi abundance and plant health status as well as growth indicators [61]. Notably, the abundance of the endophyte–litter saprotroph–soil saprotroph–undefined saprotroph functional guild also exhibited an increasing trend. This guild plays a pivotal role in litter decomposition and organic matter mineralization, and its elevated abundance may indicate that the replanting measures have enhanced soil organic matter breakdown and nutrient cycling efficiency [62]. Concurrently, the observed variation in wood saprotroph abundance could be linked to increased woody debris input following replanting. These taxa contribute to carbon cycling through lignin and cellulose decomposition [63]. Denitrifiers exhibited significantly higher abundance in the *P. thunbergii* replanting group (PTP). As crucial nitrogen-cycling agents, this increase likely reflects enhanced soil nitrogen transformation following replanting, potentially driven by elevated soil organic matter and improved microenvironments [64]. Anoxygenic phototrophs showed distinct enrichment in the *P. thunbergii* replanting group (PTP), suggesting potential adaptation to altered light availability and organic matter composition in surface soils post-replanting [65].

4.5. Soil Microbial Community Assembly

The processes influencing the composition of soil microbial communities are pivotal to the study of microbial ecology [24]. Microbial community assembly is governed by the processes affecting constituent species, such as selection, drift, speciation, and dispersal [66]. Dispersal and species formation are influenced by both stochastic and deterministic factors [67]. Our findings indicate that although undominated processes in microbial community assembly were attenuated in the *Pinus taeda* replanting group (PTL) and *Pinus thunbergii* replanting group (PTP) compared to the sanitation cutting treatment group (PMLG), bacterial homogenizing dispersal was significantly increased in the sanitation cutting treatment group (PMLG). Undominated processes played a significant role in the assembly of the microbial community compared with homogenizing dispersal. Moreover, bacterial communities, which tend to have broader niches, were more significantly influenced by homogenizing dispersal than were fungal communities [68]. Compared to fungi, bacterial community composition is relatively increased by homogenizing dispersal, which may be the result of decomposing difficult-to-degrade organic matter to provide substrate for symbiotic utilization by bacteria [69]. In addition, higher SMC suppresses fungal abundance and diversity, which may be why diffusion limitation in the sanitation cutting treatment group (PMLG) is important for microbial community assembly [70]. The stochastic nature and unpredictability of microbial composition are heightened by reduced resource competition, ecological niche selection, and amplified preferential attachment effects [71]. To better mechanistically understand the role of environmental factors in driving microbial communities in the intercropping of infected trees and planting of disease-resistant pine forests, future studies need to consider incorporating local factors into temporal sampling designs.

5. Conclusions

Our results showed that after pine forests were infested by pine wood nematode, by either inter-planting infected wood (sanitation-thinned experimental plot) or cultivating new pine forests resistant to the disease (*P. thunbergii* and *P. taeda*), there were significant differences in the structure and function of the soil microbial communities, especially the 468 fungi. The activities of enzymes such as β-glucosidase (S-β-GC), N-acetyl-β-D-gluco-sidase (S-NAG), and aryl sulfate (S-ASF) were significantly higher in the *Pinus taeda* experimental plot and *Pinus thunbergii* experimental plot compared to the sanitation-thinned experimental plot. In addition, the bacterial and fungal communities in the *Pinus taeda* experimental plot and *Pinus thunbergii* experimental plot significantly differed from the soil bacterial and fungal communities in sanitation-thinned experimental plot because of their significant β-diversity, different indicator groups, and unique functional properties. Soil moisture content, pH, and total potassium were the most important factors affecting the process of bacterial and fungal community assembly in soil. Wood saprotrophic fungi and ectomycorrhizal fungi exhibited increased abundance in both the *P. taeda* experimental plot and *Pinus thunbergii* experimental plot. Zero-modeling analyses indicated that undominated processes were among the major processes involved in microbial community assembly. Dispersal limitation and homogenizing dispersal played important roles in fungal and bacterial community assembly in the sanitation-thinned experimental plot, respectively, but their roles were diminished in the *P. taeda* experimental plot and *Pinus thunbergii* experimental plot. In the context of management measures taken after the occurrence of pine wood nematode disease, these results provide new insights into the understanding of soil microbial changes and contribute to the continued development of sustainable management of pine forest ecosystems.

Supplementary Materials: The following supporting information can be downloaded at: https://www.mdpi.com/article/10.3390/f16050813/s1, Table S1. Basic properties of the three different pinus plots. Significance indicators: a, ab, b. Table S2. Soil physicochemical properties of three pinus plots. Significance indicators: a, ab, b. Table S3. Soil factors of three pinus plots screened via variance inflation factor (VIF) analysis. Figure S1. Wayne analysis of microbial OTUs andα-diversity (OTUs) of soil microbial communities across the three plantation plots: (a) fungi; (b) bacteria. (c) Simpson and coverage indices in bacteria. (d) Shannoneven and Chao indices in bacteria. PMLG: sanitation cutting treatment group; PTL: *Pinus taeda* replanting group; PTP: *Pinus thunbergii* replanting group.

Author Contributions: X.Z.: writing—review and editing, writing—original draft, software, methodology, investigation, formal analysis, data curation. Z.L.: software, methodology, data curation. M.C.: software, methodology, data curation. T.D.: writing—review and editing, supervision, project administration, conceptualization. All authors have read and agreed to the published version of the manuscript.

Funding: This work was supported by the National Natural Science Foundation of China (32471873); the STI 2030—Major Projects (2023ZD0405605); National Key R&D Program of China (2023YFD1401304); Natural Science Foundation of Jiangsu Province, China (BK20231291); and China Postdoctoral Science Foundation Project (2024M751426).

Data Availability Statement: Data will be made available on request.

Conflicts of Interest: The authors declare that they have no known competing financial interests or personal relationships that could have appeared to influence the work reported in this paper.

References

1. Torres, I.; Moreno, J.M.; Morales-Molino, C.; Arianoutsou, M. Ecosystem services provided by pine forests. In *Pines and Their Mixed Forest Ecosystems in the Mediterranean Basin*; Springer: Berlin/Heidelberg, Germany, 2021; pp. 617–629.
2. Dering, M.; Baranowska, M.; Beridze, B.; Chybicki, I.J.; Danelia, I.; Iszkuo, G.; Kvartskhava, G.; Kosiński, P.; Rczka, G.; Thomas, P.A. The evolutionary heritage and ecological uniqueness of Scots pine in the Caucasus ecoregion is at risk of climate changes. *Sci. Rep.* **2021**, *11*, 22845. [CrossRef] [PubMed]
3. Aierken, N.; Wang, G.; Chen, M.Y.; Chai, G.Q.; Han, X.Y.; Qian, Z.H.; Zhang, X.L. Assessing global pine wilt disease risk based on ensemble species distribution models. *Ecol. Indic.* **2024**, *167*, 112691. [CrossRef]
4. Qi, J.; Nan, J.; Zhao, X.; Liang, C.; Fan, J.; He, H. Genetic Structure of *Monochamus alternatus* (Hope) in Qinling-Daba Mountains and Expansion Trend: Implications for Pest Prevention and Management. *Ecol. Evol.* **2024**, *14*, e70373. [CrossRef] [PubMed]
5. Zhang, H.F.; Chen, S.Y.; Liu, G.; Wang, Y.X. Effects of sanitation cutting pine wilt diseased trees on the stand structure of pure Pinus massoniana plantatio. *J. Zhejiang AF Univ.* **2020**, *37*, 745–751.
6. Huang, Y.A. Control Effects of Pine Wood Nematode Disease Transmitted by Monochamus alternatus Hope by Sanitation Cuttings. *Biol. Disaster Sci.* **2015**, *38*, 31–34.
7. Roberts, M.; Gilligan, C.A.; Kleczkowski, A.; Hanley, N.; Whalley, A.E.; Healey, J.R. The effect of forest management options on forest resilience to pathogens. *Front. For. Glob. Change* **2020**, *3*, 7. [CrossRef]
8. Wang, X.Y.; Wu, X.Q.; Wen, T.Y.; Feng, Y.Q.; Zhang, Y. Transcriptomic analysis reveals differentially expressed genes associated with pine wood nematode resistance in resistant *Pinus thunbergii*. *Tree Physiol.* **2023**, *43*, 995–1008. [CrossRef]
9. Nose, M.; Shiraishi, S. *Breeding for Resistance to Pine Wilt Disease*; Springer: Tokyo, Japan, 2008.
10. Zhu, S.W.; Liu, L.X.; Hu, X.F.; Dai, W.; Wang, Y.R. The effects of different thinning intensities on the understory vegetation characteristics of mixed forests of *Larix principis-rupprechtii*. *For. Eng.* **2024**, *40*, 47–55.
11. Han, S.; Tan, S.; Wang, A.; Chen, W.; Huang, Q. Bacterial rather than fungal diversity and community assembly drive soil multifunctionality in a subtropical forest ecosystem. *Environ. Microbiol. Rep.* **2022**, *14*, 85–95. [CrossRef]
12. Singh Rawat, V.; Kaur, J.; Bhagwat, S.; Arora Pandit, M.; Dogra Rawat, C. Deploying microbes as drivers and indicators in ecological restoration. *Restor. Ecol.* **2023**, *31*, e13688. [CrossRef]
13. Yang, G.; Ryo, M.; Roy, J.; Lammel, D.R.; Ballhausen, M.B.; Jing, X.; Zhu, X.F.; Rillig, M.C. Multiple anthropogenic pressures eliminate the effects of soil microbial diversity on ecosystem functions in experimental microcosms. *Nat. Commun.* **2022**, *13*, 4260. [CrossRef] [PubMed]
14. Zhang, N.; Nunan, N.; Hirsch, P.R.; Sun, B.; Zhou, J.; Liang, Y. Theory of microbial coexistence in promoting soil–plant ecosystem health. *Biol. Fertil. Soils* **2021**, *57*, 897–911. [CrossRef]
15. She, J.R.; Zheng, Q.; Zhang, X.; Yu, N.H.; Li, H. Study on rhizosphere microbial community structure and soil physical and chemical properties of rare trees in Xiangxi Region. *J. Cent. South Univ. For. Technol.* **2023**, *43*, 104–115.
16. Wang, N.; Gao, J.; Wei, J.; Liu, Y.; Zhuang, X.L.; Zhuang, G.Q. Effects of Wetland Reclamation on Soil Microbial Community Structure in the Sanjiang Plain. *Environ. Sci.* **2019**, *40*, 2375–2381.
17. Chen, X.L.; Hu, M.Y.; Zheng, G.C.; Chen, H.Y.H. Persistent soil organic carbon deficits from converting primary forests to plantations and secondary forests in subtropical China. *Glob. Ecol. Conserv.* **2023**, *45*, e02530. [CrossRef]
18. Yang, X.M.; Feng, Q.; Zhu, M.; Zhang, J.T.; Yang, L.S.; Li, R.L. The Impact of Artificial Restoration of Alpine Grasslands in the Qilian Mountains on Vegetation, Soil Bacteria, and Soil Fungal Community Diversity. *Microorganisms* **2024**, *12*, 854. [CrossRef]
19. Meng, W.J.; Li, Y.L.; Qu, Z.L.; Zhang, Y.M.; Liu, B.; Liu, K.; Gao, Z.W.; Dong, L.N.; Sun, H. Fungal community structure shifts in litter degradation along forest succession induced by pine wilt disease. *Microbiol. Res.* **2024**, *280*, 127588. [CrossRef]
20. Qiao, H.; Mo, X.Q.; Luo, Y.H.; Liu, X.Y.; Hu, Y.J.; Chen, X.B.; Su, Y.R. Patterns of soil ecoenzymatic stoichiometry and its influencing factors during stand development in *Camellia oleifera* plantations. *Acta Ecol. Sin.* **2019**, *39*, 1887–1896.
21. Xun, W.B.; Li, W.; Xiong, W.; Ren, Y.; Liu, Y.P.; Miao, Y.Z.; Xu, Z.H.; Zhang, N.; Shen, Q.R.; Zhang, R.F. Diversity-triggered deterministic bacterial assembly constrains community functions. *Nat. Commun.* **2019**, *10*, 3833. [CrossRef]
22. Yan, Y.; Klinkhamer, P.G.L.; Veen, J.A.V.; Kuramae, E.E. Environmental filtering: A case of bacterial community assembly in soil. *Soil Biol. Biochem.* **2019**, *136*, 107531. [CrossRef]
23. Lan, G.Y.; Wu, Z.X.; Yang, C.; Sun, R.; Chen, B.Q.; Zhang, X.C. Forest conversion alters the structure and functional processes of tropical forest soil microbial communities. *Land Degrad. Dev.* **2021**, *32*, 613–627. [CrossRef]
24. Nemergut, D.R.; Schmidt, S.K.; Fukami, T.; O'Neill, S.P.; Bilinski, T.M.; Stanish, L.F.; Knelman, J.E.; Darcy, J.L.; Lynch, R.C.; Wickey, P.; et al. Patterns and processes of microbial community assembly. *Microbiol. Mol. Biol. Rev.* **2013**, *77*, 342–356. [CrossRef] [PubMed]
25. Zhao, P.; Liu, J.; Jia, T.; Luo, Z.; Li, C.; Chai, B. Assembly mechanisms of soil bacterial communities in subalpine coniferous forests on the Loess Plateau, China. *J. Microbiol.* **2019**, *57*, 461–469. [CrossRef]
26. Zhou, J.Z.; Liu, W.Z.; Deng, Y.; Jiang, Y.H.; Wang, A.J. Stochastic Assembly Leads to Alternative Communities with Distinct Functions in a Bioreactor Microbial Community. *Mbio* **2013**, *4*, 10–1128. [CrossRef] [PubMed]

27. Harris, K.; Parsons, T.L.; Ijaz, U.Z.; Lahti, L.; Holmes, I.; Quince, C. Linking statistical and ecological theory: Hubbell's unified neutral theory of biodiversity as a hierarchical Dirichlet process. *Proc. IEEE* **2015**, *105*, 516–529. [CrossRef]
28. Zhao, Z.; Li, H.; Sun, Y.; Yang, Q.; Fan, J. Contrasting the assembly of phytoplankton and zooplankton communities in a polluted semi-closed sea: Effects of marine compartments and environmental selection. *Environ. Pollut. (Barking Essex 1987)* **2021**, *285*, 117256. [CrossRef]
29. Vignola, M.; Werner, D.; Wade, M.J.; Meynet, P.; Davenport, R.J. Medium shapes the microbial community of water filters with implications for effluent quality. *Water Res.* **2018**, *129*, 499–508. [CrossRef]
30. Yu, C.F.; Zhu, Z.Y.; Zhang, M.X. Unveiling the impact and mechanisms of Cd-driven ecological assembly and coexistence of bacterial communities in coastal sediments of Yellow Sea. *J. Hazard. Mater.* **2023**, *460*, 132309. [CrossRef]
31. Goss-Souza, D.; Mendes, L.W.; Rodrigues, J.L.M.; Tsai, S.M. Ecological processes shaping bulk soil and rhizosphere microbiome assembly in a long-term Amazon forest-to-agriculture conversion. *Microb. Ecol.* **2020**, *79*, 110–122. [CrossRef]
32. He, H.R.; Xu, M.Z.; Li, W.T.; Chen, L.; Chen, Y.A.; Moorhead, D.L.; Brangari, A.C.; Liu, J.; Cui, Y.X.; Zeng, Y.; et al. Linking soil depth to aridity effects on soil microbial community composition, diversity and resource limitation. *Catena* **2023**, *232*, 107393. [CrossRef]
33. Ghazouani, S.; Béjaoui, Z.; Michael, P.; Spiers, G.; Beckett, P.; Gtari, M.; Nkongolo, K. Rhizobioaugmentation of *Casuarina glauca* with N-fixing actinobacteria Frankia decreases enzymatic activities in wastewater irrigated soil: Effects of Frankia on *C. glauca* growth. *Ecotoxicology* **2020**, *29*, 417–428. [CrossRef]
34. Shang, H.B. *Methods of Conventional Analysis of Resources and Environment*; Northwest Agriculture and Forestry University Press: Yangling, China, 2010.
35. Sun, Y.; Shi, Y.; Tang, Y.; Tian, J.; Wu, X. Correlation between plant diversity and the physicochemical properties of soil microbes. *Appl. Ecol. Environ. Res.* **2019**, *17*, 10371–10388. [CrossRef]
36. Afonin, F.V.P. X-ray fluorescence analysis. *Z. Vsesoyuznogo Khimicheskogo Obs. Im. DI Mendeleeva* **1980**, *25*, 610–615.
37. Wang, J.P.; Chen, G.R.; Ji, S.H.; Zhong, Y.Q.; Zhao, Q.; He, Q.Q.; Wu, Y.H.; Bing, H.J. Close relationship between the gene abundance and activity of soil extracellular enzyme: Evidence from a vegetation restoration chronosequence. *Soil Biol. Biochem.* **2023**, *177*, 108929. [CrossRef]
38. Zhou, X.; Wang, J.T.; Liu, F.; Liang, J.M.; Zhao, P.; Tsui, C.K.M.; Cai, L. Cross-kingdom synthetic microbiota supports tomato suppression of Fusarium wilt disease. *Nat. Commun.* **2022**, *13*, 7890. [CrossRef] [PubMed]
39. Tanunchai, B.; Ji, L.; Schroeter, S.A.; Wahdan, S.F.M.; Hossen, S.; Buscot, Y.F.; Lehnert, A.S.; Alves, E.G.; Hilke, I.; Gleixner, G.; et al. FungalTraits vs. FUNGuild: Comparison of Ecological Functional Assignments of Leaf- and Needle-Associated Fungi Across 12 Temperate Tree Species. *Microb. Ecol.* **2023**, *85*, 411–428. [CrossRef]
40. Yang, Z.C.; Peng, C.Y.; Cao, H.M.; Song, J.J.; Gong, B.; Li, L.; Wang, L.; He, Y.; Liang, M.; Lin, J.C. Microbial functional assemblages predicted by the FAPROTAX analysis are impacted by physicochemical properties, but C, N and S cycling genes are not in mangrove soil in the Beibu Gulf, China. *Ecol. Indic.* **2022**, *139*, 108887. [CrossRef]
41. Stegen, J.C.; Lin, X.J.; Fredrickson, J.K.; Chen, X.Y.; Kennedy, D.W.; Murray, C.J.; Rockhold, M.L.; Konopka, A. Quantifying community assembly processes and identifying features that impose them. *Isme J.* **2013**, *7*, 2069–2079. [CrossRef]
42. Hou, Y.P.; Liu, L.; Chu, H.; Ma, S.J.; Zhao, D.; Liang, R.R. Effects of exotic plant *Rhus typhina* invasion on soil properties in different forest types. *Acta Ecol. Sin.* **2015**, *35*, 5324–5330.
43. Guo, J.Y.; Wang, L.; Yang, J.J.; Ma, F.Y.; Ma, X.S.; Wang, W.B.; Dong, Y.F. Understory vegetation patterns and soil characteristics of a *Pinus thunbergii* plantation in mountainous land of Shandong Province. *J. Zhejiang AF Univ.* **2018**, *35*, 209–218.
44. Mól, P.C.G.; Júnior, J.C.Q.; Veríssimo, L.A.A.; Boscolo, M.; Gomes, E.; Minim, L.A.; Da Silva, R. β-glucosidase: An overview on immobilization and some aspects of structure, function, applications and cost. *Process Biochem.* **2023**, *130*, 26–39. [CrossRef]
45. Wu, H.W.; Cui, H.L.; Fu, C.X.; Li, R.; Qi, F.Y.; Liu, Z.L.; Yang, G.; Xiao, K.Q.; Qiao, M. Unveiling the crucial role of soil microorganisms in carbon cycling: A review. *Sci. Total Environ.* **2024**, *909*, 168627. [CrossRef]
46. Yang, M.; Yang, D.; Yu, X. Soil microbial communities and enzyme activities in sea-buckthorn (*Hippophae rhamnoides*) plantation at different ages. *PLoS ONE* **2018**, *13*, e0190959. [CrossRef] [PubMed]
47. Chu, H.; Wang, C.; Wang, H.; Chen, H.; Tang, M. Pine wilt disease alters soil properties and root-associated fungal communities in *Pinus tabulaeformis* forest. *Plant Soil* **2016**, *404*, 237–249. [CrossRef]
48. Lin, Y.T.; Whitman, W.B.; Coleman, D.C.; Chiu, C.Y. Effects of reforestation on the structure and diversity of bacterial communities in subtropical low mountain forest soils. *Front. Microbiol.* **2018**, *9*, 1968. [CrossRef]
49. Tedersoo, L.; Mikryukov, V.; Anslan, S.; Bahram, M.; Khalid, A.N.; Corrales, A.; Abarenkov, K. The Global Soil Mycobiome consortium dataset for boosting fungal diversity research. *Fungal Divers.* **2021**, *111*, 573–588. [CrossRef]
50. Tian, H.M.; Li, L.Z.; Zhu, Y.P.; Wang, C.C.; Wu, M.X.; Shen, W.X.; Li, C.R.; Li, K. Soil fungal community and co-occurrence network patterns at different successional stages of black locust coppice stands. *Front. Microbiol.* **2025**, *16*, 1528028. [CrossRef]

51. Kim, H.S.; Lee, S.H.; Jo, H.Y.; Finneran, K.T.; Kwon, M.J.; Man, J.K. Diversity and composition of soil Acidobacteria and Proteobacteria communities as a bacterial indicator of past land-use change from forest to farmland. *Sci. Total Environ.* **2021**, *797*, 148944. [CrossRef]
52. Rosling, A.; Eshghi Sahraei, S.; Kalsoom Khan, F.; Desirò, A.; Bryson, A.E.; Mondo, S.J.; Grigoriev, L.V.; Bonito, G.; Sánchez-García, M. Evolutionary history of arbuscular mycorrhizal fungi and genomic signatures of obligate symbiosis. *BMC Genom.* **2024**, *25*, 529. [CrossRef]
53. Li, X.; Liu, Y.; Wu, G.P.; Lie, Z.Y.; Sheng, H.; Aguila, L.C.R.; Khan, M.Q.; Liu, X.J.; Zhou, S.Y.D.; Wu, T. Mixed plantations do not necessarily provide higher ecosystem multifunctionality than monoculture plantations. *Sci. Total Environ.* **2024**, *914*, 170156. [CrossRef]
54. Tan, R.; Yu, S.Q.; LI, Y.; Wang, X.F.; Xu, X.Y.; Li, Y.H.; Wang, W.F. Effect of thinning restoration on enzyme activity and enzyme stoichiometry in the topsoil of oak-pine mixed forest. *J. Zhejiang AF Univ.* **2024**, *41*, 1201–1210.
55. Wei, X.; Liang, W.J. Regulation of stand density alters forest structure and soil moisture during afforestation with *Robinia pseudoacacia* L. and *Pinus tabulaeformis* Carr on the Loess Plateau. *For. Ecol. Manag.* **2021**, *491*, 119196. [CrossRef]
56. Guo, L.; Chen, X.H.; Sheng, Y.Z.; Yang, N.; Hou, A.K.; Fang, H.S. Impact of soil fissure status on microbial community in mining-disturbed area, the northern Shaanxi province. *Front. Microbiol.* **2024**, *15*, 1463665. [CrossRef]
57. Jing, J.Y.; Cong, F.W.; Bezemer, T.M. Legacies at work: Plant–soil–microbiome interactions underpinning agricultural sustainability. *Trends Plant Sci.* **2022**, *27*, 781–792. [CrossRef] [PubMed]
58. Zhou, X.Q.; Guo, Z.Y.; Chen, C.G.; Jia, Z.J. Soil microbial community structure and diversity are largely influenced by soil pH and nutrient quality in 78-year-old tree plantations. *Biogeosciences* **2017**, *14*, 2101–2111. [CrossRef]
59. Wang, X.J.; Zhang, Z.C.; Yu, Z.Q.; Shen, G.F.; Cheng, H.F.; Tao, S. Composition and diversity of soil microbial communities in the alpine wetland and alpine forest ecosystems on the Tibetan Plateau. *Sci. Total Environ.* **2020**, *747*, 141358. [CrossRef]
60. Lehto, T.; Zwiazek, J.J. Ectomycorrhizas and water relations of trees: A review. *Mycorrhiza* **2011**, *21*, 71–90. [CrossRef]
61. Anthony, M.A. Does ectomycorrhizal fungal biodiversity affect tree growth? *Fungal Ecol.* **2025**, *74*, 101413. [CrossRef]
62. Clemmensen, K.E.; Bahr, A.; Ovaskainen, O.; Dahlberg, A.; Ekblad, A.; Wallander, H.; Stenlid, J.; Finlay, R.D.; Wardle, D.A.; Lindahl, B.D. Roots and associated fungi drive long-term carbon sequestration in boreal forest. *Science* **2013**, *339*, 1615–1618. [CrossRef]
63. Maraón-Jiménez, S.; Radujkovi, D.; Verbruggen, E.; Grau, O.; Cuntz, M.; Peuelas, J.; Richter, A.; Schrumpf, M.; Rebmann, C. Shifts in the abundances of saprotrophic and ectomycorrhizal fungi with altered leaf litter inputs. *Front. Plant Sci.* **2021**, *12*, 682142.
64. Du, L.; Zhong, H.; Guo, X.; Li, H.; Xia, J.; Chen, Q. Nitrogen fertilization and soil nitrogen cycling: Unraveling the links among multiple environmental factors, functional genes, and transformation rates. *Sci. Total Environ.* **2024**, *951*, 175561. [CrossRef] [PubMed]
65. George, D.M.; Vincent, A.S.; Mackey, H.R. An overview of anoxygenic phototrophic bacteria and their applications in environmental biotechnology for sustainable Resource recovery. *Biotechnol. Rep.* **2020**, *28*, e00563. [CrossRef]
66. Ma, J.M.; Ma, K.; Liu, J.L.; Chen, N.N. Rhizosphere soil microbial community under ice in a high-latitude wetland: Different community assembly processes shape patterns of rare and abundant microbes. *Front. Microbiol.* **2022**, *13*, 783371. [CrossRef]
67. Willig, M.R.; Moulton, M.P. The role of stochastic and deterministic processes in structuring Neotropical bat communities. *J. Mammal.* **1989**, *70*, 323–329. [CrossRef]
68. Schmidt, S.K.; Nemergut, D.R.; Darcy, J.L.; Lynch, R. *Do Bacterial and Fungal Communities Assemble Differently During Primary Succession?* Wiley Online Library: Hoboken, NJ, USA, 2014; pp. 254–258.
69. Dong, L.; Li, M.X.; Li, S.; Yue, L.Y.; Ali, M.; Han, J.; Lian, W.H.; Hu, C.J.; Lin, Z.L.; Shi, G.Y. Aridity drives the variability of desert soil microbiomes across north-western China. *Sci. Total Environ.* **2024**, *907*, 168048. [CrossRef]
70. Lu, M.; Tian, K.; Sun, X.Y.; Ren, Y.L.; Wang, H.; Peng, S.X. Variation of soil fungal community characteristics of typical wetland in Napahai between dry wet seasons under different waterlogging conditions. *Sci. Silvae Sin.* **2018**, *54*, 98–109.
71. Zhou, J.Z.; Ye, D.; Zhang, P.; Xue, K.; Liang, Y.T.; Nostrand, J.D.V.; Yang, Y.F.; He, Z.L.; Wu, L.Y.; Stahl, D.A. Stochasticity, succession, and environmental perturbations in a fluidic ecosystem. *Proc. Natl. Acad. Sci. USA* **2014**, *111*, E836–E845. [CrossRef] [PubMed]

forests

MDPI

Article

The Effect of Acid Rain and Understory Vegetation Removal on the Biological Activity of the Soils of the *Cinnamomum camphora* (Linn) Presl Plantation

Zaihua He [1,†], Yini Liu [1,†], Yonghui Lin [1], Xiangshi Kong [2], Hong Lin [3] and Xingbing He [1,*]

[1] College of Biology and Environmental Sciences, Jishou University, Jishou 416000, China; 17775865595@163.com (Y.L.)
[2] Key Laboratory for Ecotourism of Hunan Province, School of Tourism, Jishou University, Jishou 416000, China; kongxiangshi@126.com
[3] Institute of Applied Ecology, School of Food Science, Nanjing Xiaozhuang University, 3601 Hongjing Avenue, Nanjing 211171, China; linhong@njxzc.edu.cn
* Correspondence: hexb@jsu.edu.cn
† These authors contributed equally to this work.

Abstract: Acid rain and understory vegetation removal are critical drivers altering soil ecosystem alterations. However, the mechanisms by which these factors influence soil moisture dynamics, nutrient availability, and microbially mediated enzyme activities remain insufficiently elucidated. This study investigated the impacts of simulated acid rain and understory vegetation removal on soil properties, enzyme activities, and microbial community in a subtropical *Cinnamomum camphor* (Linn) Presl plantation. The results indicated that acid rain and understory vegetation removal significantly decreased the soil organic carbon (SOC) while concurrently elevating the C-acquiring enzyme activities and microbial C limitation. Understory vegetation removal markedly reduced the soil moisture, nutrient availability, and N- and P-acquiring enzyme activities. Additionally, acid rain increased the bacterial diversity, but the understory vegetation removal increased the fungal diversity. Moreover, both acid rain and understory vegetation removal enhanced the bacterial community deterministic processes and destabilized the community by shifting generalists toward specialists, but had no significant effect on the fungal community structure. Partial least squares path modeling revealed that the bacterial stability loss intensified the C limitation, while the fungal stability regulated the P limitation. Collectively, the findings highlighted the critical role of understory vegetation in buffering the soil microclimate and nutrient cycling, and demonstrated that bacterial communities are more responsive to acid rain and understory vegetation removal than fungal communities. This study provides insights into the mechanisms by which anthropogenic disturbances alter soil ecological functions in subtropical plantations, emphasizing the need for integrated forest management strategies to conserve and manage soil ecosystems in subtropical plantations.

Keywords: anthropogenic disturbances; microbial community; enzyme stoichiometry; nutrient availability; subtropical plantations

1. Introduction

Acid rain, one of the most pervasive anthropogenic environmental stressors, has attracted widespread concerns due to its negative impact on ecosystems [1–3]. Soil, as an essential component of forest ecosystems, is highly susceptible to external environmental disturbances [4,5]. Numerous studies have shown that acid rain can accelerate soil acidification, reduce nutrients availability, and elevate toxic aluminum (Al^{3+}) levels, thus destabilizing soil structure and functions [1,3,6]. Soil microorganisms and enzymes play central roles in soil ecological processes, such as carbon (C) sequestration and nutrient cycling, and their activities are sensitive to environmental changes, such as acid rain [7,8]. In general, soil acidification caused by acid rain can inhibit microbial activity, reduce microbial biomass, and, subsequently, alter microbial communities and enzyme activities [9,10]. Existing studies have also suggested that soil microbial activity and enzyme activities can be directly suppressed by acid rain, thus influencing soil ecological functions [6,10,11]. For instance, Wang et al. reported that acid rain decreased soil hydrolytic enzyme activities and then slowed the rate of litter decomposition [6]. A meta-analysis also indicated that acid rain inhibited the growth of soil microbes and reduced soil enzyme activities [10,12]. In addition, acid rain can indirectly influence soil microorganisms and enzyme activities by influencing plant diversity, as well as the aboveground and underground biomass [1,13,14]. Meanwhile, for plants, as the main interceptors and receptors of acid rain, the understory vegetation modulates acid rain's impacts on soil microorganisms and enzymatic processes to some extent [14,15].

Understory vegetation, as an essential components of forest ecosystem, is crucial to the processes and functions of forest ecosystems, like ecosystem productivity, nutrient cycling, and water conservation [14,16]. To decrease competition between canopy trees and understory vegetation, removing the latter has been a traditional forest management practice, particularly in plantations [17,18]. However, the removal of understory vegetation can alter the soil microenvironment and nutrient availability, which greatly impact the soil microbial community and enzyme activities [14,19]. For instance, several studies have reported that root exudates and litter from understory vegetation exert substantial impacts on soil nutrient dynamics (the content and bioavailability) [20,21], and, subsequently, result in significantly negative, positive, or no effects on soil enzymatic activities and microbial biomass [19,22–24]. These studies indicated that the effects of understory vegetation removal on microorganisms and soil enzyme activities exhibit significant variability across ecosystem types and environmental conditions, and needs further study.

Soil microorganisms, particularly bacteria and fungi, are recognized as important regulators of nutrient cycling and sensitive to external environmental disturbances. However, bacterial and fungal communities responded differently to acid rain or understory vegetation removal in forest ecosystems [25,26]. Microbial communities typically derive their energy from soil organic matter (e.g., organic matter mineralization) [27,28]; yet, bacteria and fungi have different metabolic preferences, where bacteria are normally characterized by using labile C resources, while fungi exhibit greater metabolic efficiency in processing recalcitrant C compounds [29,30]. Moreover, fungi generally have a considerably higher osmotic stress tolerance capabilities than bacteria [30,31]. Consequently, bacterial communities may be more responsive to acid rain or understory vegetation removal than fungal communities, due to the fact that bacteria are more susceptible to the soil microenvironment and nutrient availability [25,32]. Extracellular enzymes, secreted by soil microorganisms, are protein catalysts that drive essential ecological processes, including the decomposition of soil organic matter and the biogeochemical cycling of nutrients, such as C, nitrogen (N), and phosphorus (P) [7,33]. Consequently, the changes in the soil microbial community, combined with modifications to the nutrient availability, can alter extracellular enzyme secretion and their stoichiometry (e.g., the vector length and vector angle). Furthermore, we speculate that soil enzyme stoichiometry is closely related to the microbial community. However, in subtropical *Cinnamomum camphor* (Linn) Presl (*C. camphor*) plantations, the underlying mechanism by which the alterations in the soil moisture and nutrient availability, driven by distinct environmental stressors like acid rain and understory vegetation removal, regulate microbial community dynamics and enzyme activities remain poorly understood, necessitating further investigation.

In this study, we carried out a simulated acid rain experiment combined with understory vegetation removal in a subtropical *C. camphor* plantation forest located in Jishou, a region within southwestern China severely impacted by acid rain [34], to assess the effects of acid rain and understory vegetation removal on the soil microbial community and enzyme activities and stoichiometry, while elucidating their relevance. Specifically, we hypothesized that (1) the soil bacterial community would be more responsive to acid rain and understory vegetation removal than the soil fungal community; (2) soil enzyme activities and their stoichiometry would be significantly affected by the soil microbial community. By elucidating these mechanisms, this work aims to provide sustainable management practices that conserve soil ecological functions in subtropical plantations.

2. Materials and Methods

2.1. Study Site Description

The study was implemented in a subtropical *C. camphor* plantation located at 28°17′ N, 109°43′ E, with an elevation of 258 m above sea level. The study site spanned 1000 m^2 adjacent to Jishou University in Jishou City, China. The mean annual temperature is about 16.5 °C, the mean annual precipitation is about 1400 mm–1800 mm, primarily falling during the rainy reason from April to June, and the mean annual humidity is about 82%. The soil is commonly classified as Ultisol, which is primarily developed from limestone [35]. The soil moisture content, pH, total organic C, and ammonium nitrogen is about 25.81%, 6.26, 40.93 g/kg, and 35.05 mg/kg, respectively.

The *C. camphor* plantation, established 40 years ago on land cleared from a *Pinus massoniana* forest, has an understory vegetation coverage of approximately 85%. The current understory plants are dominated by five fern species (*Cyclosorus acuminatus*, *Arachniodes rhomboidea*, *Polystichum tsus-simense*, *Cyrtomium fortunei*, and *Microlepia marginata*), a palm (*Trachycarpus fortunei*), and a grass (*Veronicastrum stenostachyum*). These species play critical roles in maintaining ecosystem functions, including nutrient cycling, soil and water conservation, and so on [14,35]. Some other plantation characteristics were recorded in prior studies [35,36].

2.2. Experimental Design and Soil Sampling

In January 2023, four 10 m × 10 m plots, characterized by similar site conditions, were established within the *C. camphor* plantation, with each plot separated by >20 m. Each plot has four 1 m × 1 m subplots, separated by more than 2 m buffer zones to minimize cross-treatment interference. Briefly, we conducted a complete factorial combination experiment of simulated acid rain and understory vegetation removal to give a total of four treatments. The treatments included (1) the control (CK, no acid rain and understory vegetation removal), (2) acid rain only (AR), (3) understory vegetation removal only (UR), and combined acid rain and understory vegetation removal (UA), with four replicates for each treatment in the *C. camphor* plantation.

The mother solution (0.1 mol/L) of acid rain was formed using H_2SO_4 (98, wt%) to avoid nitrogen fertilizer effects. The mother solution was then diluted with deionized water to obtain a pH = 4.0 for the simulated acid rain. For the AR and UA, 2 L simulated acid rain was sprayed on each subplot every 15 days [26], while the UR and CK received an equivalent volume of deionized water. For the UR and UA, the shoots and visible roots on the surface of all understory vegetation were manually excised using machetes. Each month, germinating understory vegetation were removed manually during the experiment. The experiment lasted for 18 months (from January 2023 to July 2024).

In July 2024, the topsoil (0–10 cm) was sampled via a five-point method [2], put into plastic bags, mixed thoroughly, and then the 16 soil samples were immediately transported back to the laboratory in an icebox. After taking out visible stones and plant residues, the soil samples were sieved (2 mm mesh) and then split into three parts. The first part was air-dried for soil physicochemical properties analysis, the second part was kept fresh for soil enzymatic activity determination, and the third part was stored at −80 °C to examine the soil microbial community.

2.3. Soil Physicochemical Property, CO_2 Release, and Enzyme Analysis

The soil moisture content (MC) was assessed by drying a fresh soil sample at 105 °C for 48 h. The soil pH was determined using a pH meter after shaking soil/deionized water (w/v) suspensions at a ratio of 1:2.5 for 30 min. The soil ammonium nitrogen (AN) was measured using a continuous flow autoanalyzer (AA3, iFIA, Beijing, China) following extraction with 2 mol/L KCl. Dissolved organic carbon (DOC) was extracted with a 0.5 mol/L K_2SO_4, passed through a 0.45 μm filter, and then measured via a TOC analyzer (TOC-L, Shimadzu, Kyoto, Japan). The soil organic carbon (SOC) was analyzed using dichromate–sulfuric acid oxidation method [37]. The soil total nitrogen (STN) was determined through the Kjeldahl method [37]. The soil total phosphorus (STP) was analyzed colorimetrically after the soil was digested with H_2SO_4-$HClO_4$ solution [37].

Approximately 0.5 g of fresh soil was placed in a sealed flask and then incubated in 25 °C darkness for 2 days. The released CO_2 was absorbed using a 0.5 mol/L NaOH solution and titrated by using 0.05 mol/L HCl. The released CO_2 was expressed as $\mu mol \cdot g^{-1}$dry soil$\cdot d^{-1}$.

The potential activity of the extracellular enzymes was quantified in soil suspensions using a spectrophotometry protocol with appropriate substrates [38]. Briefly, we homogenized 20 g of soil in 100 mL of acetic acid buffer (pH = 5.5). The resulting slurry was used to perform spectrophotometry for the potential enzyme activity. These included β-1,4-glucosidase (BG) and cellobiohydrolase (CBH) using a 4-nitrophenyl-β-D-linked substrates (cellobioside and glucopyranoside); leucine aminopeptidase (LAP) using L-leucine-p-nitroanilide; β-1,4-N-acetyl-glucosaminnidase (NAG) using 4-nitrophenyl N-acetyl-β-D-glucosaminide; acid phosphatase (AP) using p-nitrophenyl disodium phosphate substrates. The units of enzyme activity were calculated as units of substrate hydrolyzed product per gram per hour ($nmol \cdot g^{-1}$ dry soil$\cdot h^{-1}$).

2.4. Soil Microbial Community Analysis

Total genomic DNA was extracted from well-homogenized soil sample using the E.Z.N.A.® soil DNA Kit (Omega Bio-tek, Norcross, GA, USA) following the manufacturer's instructions. The soil bacterial V4–V5 regions of the 16S rRNA were amplified with the primers 515F (5′-GTGCCAGCMGCCGCGG-3′) and 907R (5′-CCGTCAATTCMTTTRAGTTT-3′). Fungal communities were targeted through the amplification of the ITS1 region using the primers ITS1F (5′-CTTGGTCATTTAGAGGAAGTAA-3′) and ITS2R (5′-GCTGCGTTCTTCATCGATGC-3′), with adapter and barcode sequences appended. PCR reactions were performed in triplicate with a 20 μL mixture containing 4 μL of 5 × FastPfu Buffer, 2 μL of 2.5 mM dNTPs, 0.8 μL of each primer (5 μM), 0.4 μL of FastPfu Polymerase, and 10 ng of template DNA. Amplicons were extracted from 2% agarose gels and purified using the AxyPrep DNA Gel Extraction Kit (Axygen Biosciences, Union City, CA, USA) according to the manufacturer's instructions. Purified PCR products were quantified by Qubit®3.0 (Life Invitrogen, Waltham, MA, USA) and the pooled DNA product was used to construct the Illumina pair-end library following Illumina's genomic DNA library preparation procedure. Then, the amplicon library was paired-end sequenced (2 × 300) on an NGS platform (Shanghai BIOZERON Biotech. Co., Ltd., Shanghai, China) following the standard protocols. Raw sequences were deposited in the NCBI SRA database under accession number PRJNA1224727.

Raw sequences were first demultiplexed–filtered using Trimmomatic [39] complemented by custom perl scripts, implementing the following stringent criteria: (i) the 300 bp reads were truncated at positions with an average quality score < 20 within a 10 bp sliding window, with the subsequent removal of any truncated reads shorter than 50 bp; (ii) strict sequence-matching requirements were enforced—exact barcode matching with ≤2 nucleotide mismatches allowed in the primer regions, coupled with the elimination of reads containing ambiguous base calls; (iii) overlap-based assembly required minimum 10-bp overlaps between paired reads, discarding unassembled fragments. The curated sequences were then analyzed in QIIME2 (2020.11) [40] using the DADA2 algorithm to obtain the biological reads (i.e., amplicon sequence variants, ASVs) [41].

The bacterial and fungal sequences were classified with the SILVA (v.138) and UNITE databases, respectively. After the sequence classification, we retained only the classified bacterial and fungal ASVs. We then rarefied the 16S and ITS dataset to 11,579 and 3774 ASVs per sample, respectively.

2.5. Statistical Analysis

The microbial observed ASVs, Chao1, and phylogenetic diversity (PD) were calculated using the "microeco" package [42]. The microbial community niche breadth index was determined according to Levins [43], and the generalist species and specialist species were classified by calculating the occurrences of ASVs from 1000 permutations simulated via the "EcolUtils" package [44], and an ASV was classified generalist or specialist depending on whether its observed occurrence exceeded the upper 95% confidence interval or was below the lower 95% confidence interval [45,46]. The average variation degree (AVD) was calculated to estimate the soil microbial community stability among different treatments, where a lower AVD indicates higher microbiome stability [47]. Phylogenetic normalized stochasticity ratio (pNST) calculations were performed with the "NST" package to characterize the bacterial and fungal community assembly processes [48,49]. The pNST index had a cutoff at 0.5 to delineate the assembly processes that were more deterministic (<0.5) and more stochastic (>0.5).

To quantify soil microbial metabolic limitations in C, N, and P, we employed vector analysis based on extracellular enzyme stoichiometry. This method interprets enzyme activity ratios as proxies for microbial resource allocation. The vector length reflects the relative investment of the microbial metabolism in C versus nutrient acquisition, while the vector angle indicates the preferential limitation between P and N [50]. Specifically, vector length quantifies the degree of C limitation, that is, a longer vector length corresponds to a stronger microbial C restriction. The vector angle distinguishes N and P limitation, that is, an angle below 45° indicates N limitation; otherwise, it is the P limitation. These metrics were calculated using the following equations:

$$Vector\ length = \mathrm{sqrt}\left(x^2 + y^2\right) \tag{1}$$

$$Vector\ angle = \mathrm{degrees}(\mathrm{atan2}(x, y)) \tag{2}$$

where x represents ln(CBH + BG)/ln(AP) and y represents ln(CBH + BG)/ln(NAG + LAP).

Additionally, the effects of acid rain and the understory vegetation removal on soil physicochemical properties, enzyme activities, and the microbial community indices were quantified and compared as follows:

$$R = ln\left(\frac{T}{C}\right) \quad (3)$$

where T and C represent the specific values of physicochemical properties, enzyme activities, or microbial community indices for the treatments and control, respectively. Values of $R > 0$ indicate a positive effect (i.e., acid rain increased the variable value relative to the control), while $R < 0$ indicates a negative effect. The vector length, vector angle, and pNST was analyzed using the ANOVA test, and a post hoc Tukey test was performed to detect differences between individual treatments, with Holm correction applied to adjust the p-values for multiple comparisons, ensuring a robust statistical analysis. Statistical significance was determined at $p < 0.05$.

Redundancy analysis (RDA) was performed to examine the relationships between enzyme activities or their stoichiometry, soil physicochemical properties, and biotic factors using the "vegan" package. Key discriminating variables were identified through a 'stepwise selection' process. Variable importance quantification was implemented using the "rdacca.hp" package [51], which calculates the relative contribution of each explanatory variable to dependent variables. Finally, the "plspm" package was employed to develop a partial least squares path modeling (PLS-PM) to study the primary mechanisms on which acid rain and understory vegetation removal mediate the microbial metabolism limitation through soil physicochemical properties and the soil microbial community. The PLS-PM model fitness was assessed using the goodness-of-fit (GoF) index [52], with interpretation thresholds defined as weak (GoF > 0.1), moderate (GoF > 0.25), and strong (GoF > 0.36) [53]. All of the analyses were performed in R 4.4.0.

3. Results

3.1. Soil Physicochemical Properties and CO_2 Release

Regardless of the acid rain treatment, understory vegetation removal significantly decreased the soil MC compared to the control (Figure 1), and the interaction of the understory vegetation and acid rain also had significant negative effect on the MC ($p < 0.001$; Figure 1). The soil pH value in the CK, UR, AR, and UA treatments was 6.28, 6.28, 6.19, and 6.22, respectively, indicating that acid rain slightly reduced the soil pH, but there were no significant difference among all of the treatments in the *C. camphor* plantation (Figure 1; Table S1). However, the DOC in the UR, AR, and UA treatments decreased by 26.32%, 3.83%, and 16.18%, respectively. The AN in UR, AR, and UA treatments decreased by 16.83%, 15.32%, and 25.99%, respectively. Moreover, understory vegetation removal and the interaction of understory removal with acid rain had an obviously negative effect on the DOC and AN (Figure 1). In addition, understory vegetation removal and acid rain had significantly negative main and interactive effects on the SOC (Figure 1). Compared with the CK, the UR, AR, and UA treatments decreased the SOC by 23.38%, 16.24%, and 18.34%, respectively. Only the understory vegetation had a significantly negative effect on the STN, and neither the understory vegetation removal nor acid rain had a marked effect on the STP in the *C. camphor* plantation (Figure 1). For the CO_2 release, compared to the CK, there were significant positive effects in the AR and UA treatments (Figure 1).

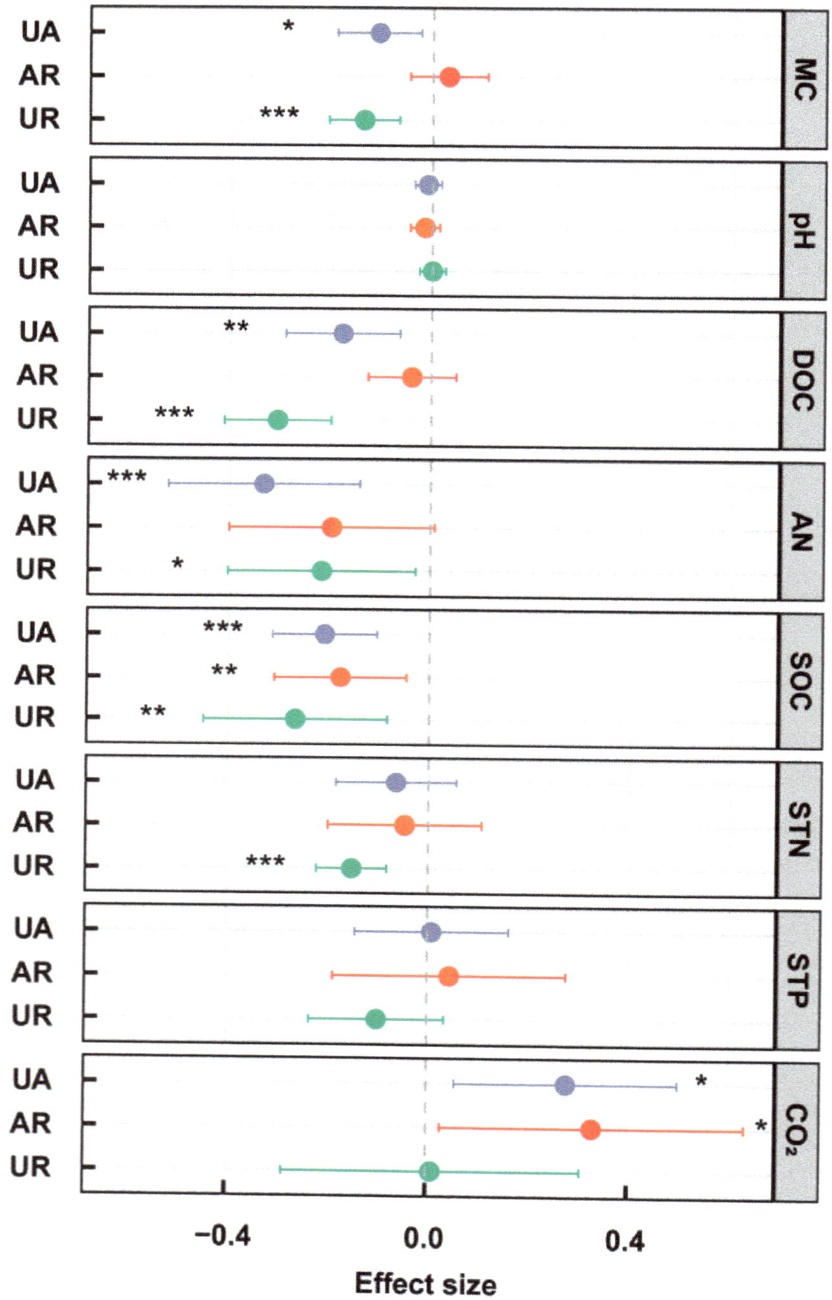

Figure 1. The effect sizes of simulated acid rain and understory vegetation removal on the soil physicochemical properties and CO_2 release in the *C. camphor* plantation. Effect sizes are presented as mean values ± 95% confidence intervals ($n = 4$). *** $p < 0.001$, ** $p < 0.01$, and * $p < 0.05$. UR stands for understory vegetation removal, AR stands for simulated acid rain, and UA stands for understory vegetation removal and simulated acid rain.

3.2. Soil Enzyme Activities

In general, soil potential enzyme activities were significantly affected by either acid rain or understory vegetation removal in the *C. camphor* plantation (Figure 2). Specifically, acid rain had a significantly positive effect on the CBH and BG activities, understory vegetation removal had a markedly positive effect on the BG and had a significantly negative effect on the NAG and AP activities, while the understory vegetation and acid rain had a significantly positive interactive effect on the BG activity and a negative effect on the LAP activity in the *C. camphor* plantation (Figure 2). Compared to the CK, the enzyme vector lengths of the UR, AR, and UA increased by 9.08%, 10.35%, and 12.43%, respectively, and understory vegetation removal and acid rain had significantly positive main and interactive effects on the vector length (Figure 2). However, understory vegetation removal and acid rain did not have any marked effects on the vector angle (Figure 2). Additionally, the enzyme vector lengths were > 1 and the vector angles were > 45° in all of the treatments, indicating that the microbial metabolism was likely limited by both C and P, but P restriction was more obvious in the *C. camphor* plantation (Figure 2).

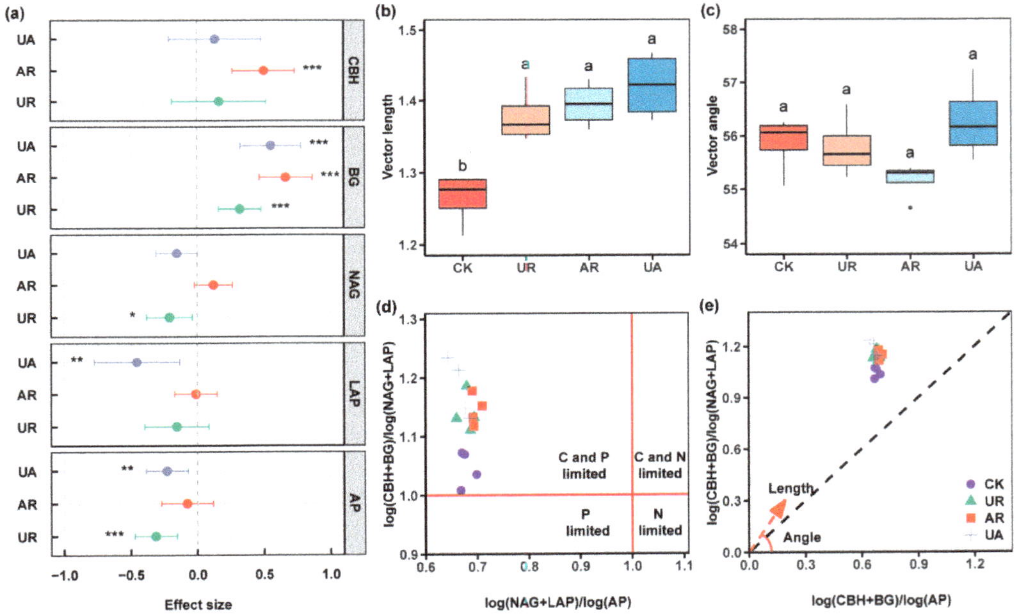

Figure 2. The effect sizes of simulated acid rain and understory vegetation removal on the soil enzyme activities (**a**), the variation in the vector lengths and angles across the treatments (**b**,**c**), potential C and nutrient limitations across the treatments, (**d**) and the enzyme vector model of the extracellular enzyme stoichiometry (**e**) in the *C. camphor* plantation. Effect sizes are presented as mean values ± 95% confidence intervals ($n = 4$). *** $p < 0.001$, ** $p < 0.01$, and * $p < 0.05$. Distinct lowercase letters indicate significant differences between treatments ($p < 0.05$). UR stands for understory vegetation removal, AR stands for simulated acid rain, and UA stands for understory vegetation removal and simulated acid rain.

3.3. Soil Microbial Community

Acid rain and the interaction of understory vegetation removal and acid rain had significantly positive effects on the bacterial-observed ASVs, Chao1, and PD; however, only the understory vegetation removal had markedly positive effects on the fungal-observed species, Chao1, and PD in the *C. camphor* plantation (Figure 3).

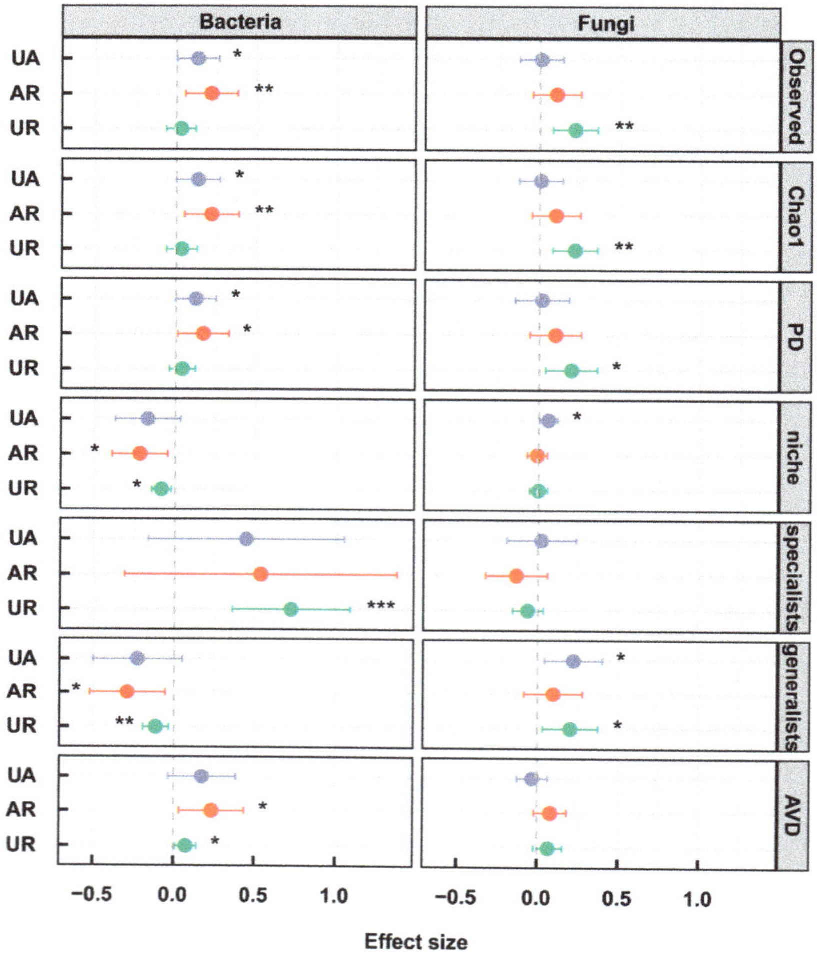

Figure 3. The effect sizes of simulated acid rain and understory vegetation removal on the soil microbial alpha diversity, niche breadth, specialists, generalists, and AVD in the *C. camphor* plantation. Effect sizes are presented as mean values ± 95% confidence intervals ($n = 4$). *** $p < 0.001$, ** $p < 0.01$, and * $p < 0.05$. UR stands for understory vegetation removal, AR stands for simulated acid rain, and UA stands for understory vegetation removal and simulated acid rain.

To study the differences in the microbial community structure, we calculated the community-level habitat niche breadth indices, the AVD (average variation degree), and the phylogenetic normalized stochasticity ratio (pNST) in all of the treatments. For the bacterial community structure, compared to the CK, the niche breadth indices of the UR, AR, and UA decreased by 7.41%, 18.92%, and 15.19%, respectively, but the understory vegetation and acid rain had no significant interactive effect on niche breadth (Figure 3). Simultaneously, the UR had a significantly positive and negative effect on the specialist ratio and generalist ratio, respectively. Acid rain only had a markedly negative effect on the generalist ratio. Conversely, the AVD values for the bacterial community were 8.64%, 27.40%, and 19.50% higher in the UR, AR, and UA treatments than in the CK treatment. Moreover, the increase in the AR treatment was significant (Figure 3), and the lower AVD value indicates higher microbiome stability. The results indicated that the bacterial community in the CK was more stable than in the other three treatments, that is, understory vegetation removal and acid rain would make the bacterial community more vulnerable in the *C. camphor* plantation. However, for the fungal community, the understory vegetation removal had an obviously positive effect on the generalist ratio, and the understory vegetation and acid rain had significantly positive effect on the niche breadth indices and generalist ratio (Figure 3). Moreover, there was no significant difference in the fungal community AVD values across all of the treatments in the *C. camphor* plantation (Figure 3).

The pNST based on the null model was used to quantitatively assess the microbial community assembly process of the four treatments in the *C. camphor* plantation. For the bacterial community, the pNST values for all of the treatments were lower than 50%, indicating that the bacterial community assembly process was more deterministic in all of the treatments. Despite that, the pNST values were distinctly higher in the CK treatment, at 30.43%, 22.54%, and 24.01%, than in the UR, AR, and UA treatments (Figure 4), thus demonstrating that understory vegetation removal, acid rain, and its interaction strengthened the deterministic assembly process. Different from the bacterial community, an opposite trend was observed for the fungal community assembly. Specifically, the pNST values were 0.48, 0.60, 0.62, and 0.58 for the CK, UR, AR, and UA treatments, respectively, indicating that the fungal community assembly process changed from more deterministic to more stochastic (Figure 4). However, there was no significant difference between the treatments.

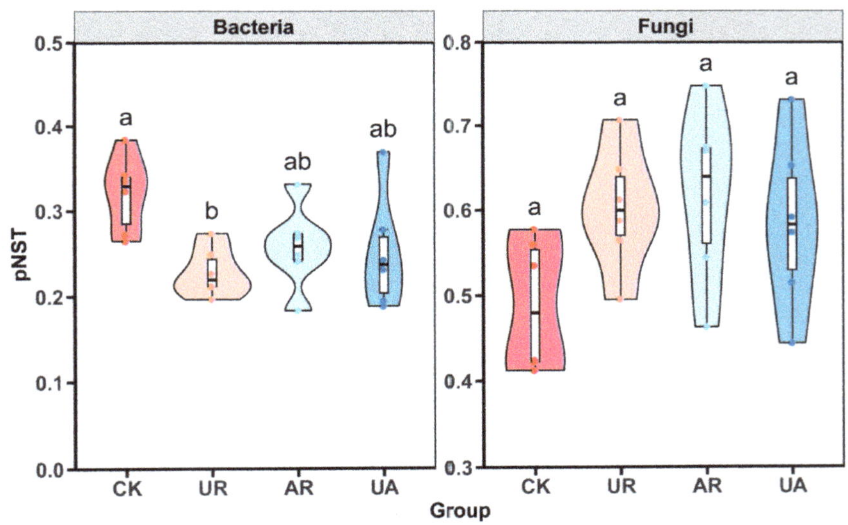

Figure 4. The effect of simulated acid rain and understory vegetation removal on the soil microbial community assembly process in the *C. camphor* plantation. Distinct lowercase letters indicate significant differences between the treatments ($p < 0.05$). UR stands for understory vegetation removal, AR stands for simulated acid rain, and UA stands for understory vegetation removal and simulated acid rain.

3.4. The Relationship of Soil Physicochemical Properties, Enzyme Activities, and Microbial Community

The RDA results indicated that the soil enzyme activities were significantly affected by the MC, bacterial AVD values, and fungal diversity (Figure 5), and their explained variations were 22.57%, 13.93%, and 4.19%, respectively, in this study. The NAG, LAP, and AP activities were strongly correlated with the MC and fungal diversity, while the BG and CBH activities were strongly related to the bacterial AVD values and microbial activities in the *C. camphor* plantation. However, the enzyme stoichiometries were significantly affected by the AN, MC, AVD_fungi, and pH, and their explained variations were 17.04%, 15.18%,

12.49%, and 11.08%, respectively. Moreover, the vector length was closely related to the AN and pH, while the vector angle was strongly correlated with the fungal AVD values in this study.

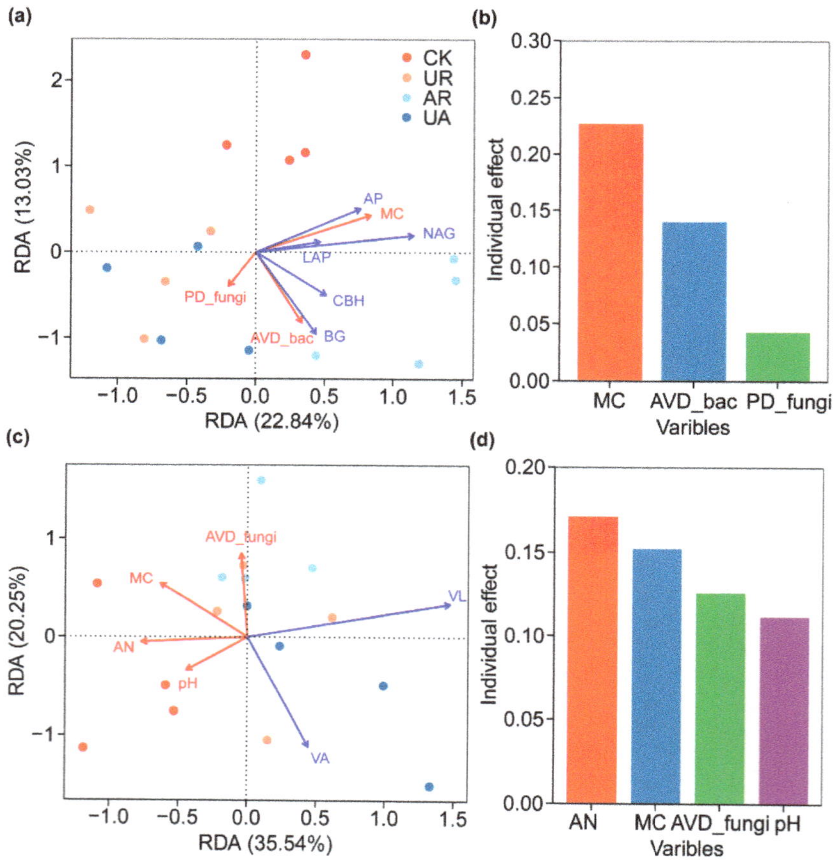

Figure 5. Effects of soil physicochemical properties and microbial community on soil enzyme activities (**a**) and their stoichiometry (**c**), and individual explanation ratios of predictors (**b**,**d**) calculated via hierarchical variation partitioning in the *C. camphor* plantation. UR stands for understory vegetation removal, AR stands for simulated acid rain, and UA stands for understory vegetation removal and simulated acid rain.

The PLS-PM showed the effects of the soil physicochemical properties, bacterial diversity, and bacterial stability on the microbial C limitation in the *C. camphor* plantation. The soil properties (−0.684), pH (−0.300), soil elements (−0.218), bacterial niche (−0.361), and bacterial stability (−0.364) had a negative total effect on the microbial C restriction, while acid rain (0.311), understory vegetation removal (0.554), and C-acquiring enzyme activities (0.701) showed a positive total effect on it (Figure 6). Meanwhile, the soil properties (−0.477) had a negative effect on the microbial P restriction, but understory vegetation removal (0.391), soil elements (0.111), fungal generalists ratio (0.338), fungal niche breadth (0.366), and fungal stability (0.537) had a positive effect on the microbial P limitation in the *C. camphor* plantation.

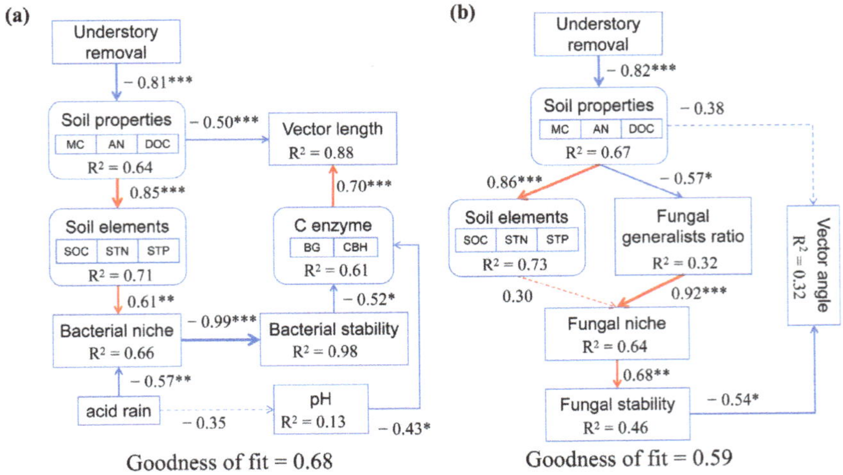

Figure 6. The primary mechanism of acid rain and understory vegetation removal mediating the microbial metabolism C limitation (**a**) and P limitation (**b**) through soil physicochemical properties and microbial community using partial least squares path modeling. Red and blue solid arrows indicate significant positive and negative paths, respectively. Soild and dash line indicate significant and non-significant paths. The thickness represents the strength of the path. *** $p < 0.001$, ** $p < 0.01$, and * $p < 0.05$. The microbial metabolism C limitation is represented by the vector length, while the microbial metabolism P limitation is represented by the vector angle.

4. Discussion

4.1. Responses of Soil Physicochemical Properties, CO_2 Release, and Enzyme Activities

In the present study, simulated acid rain had a noticeable negative effect on the SOC, which was similar to the results published by previous studies [3,5]. There are several reasons to explain the findings. One possible explanation is that the decreased SOC could be ascribed to the solubility of the soil organic C promoted by acid rain and leached out [3,15,25], particularly in subtropical plantations with a high rainfall intensity. Another contributing factor involved microbial metabolic trade-offs [54,55], in which, under acid rain stress, soil microorganisms might reallocate C and energy from growth and maintenance toward stress tolerance and resource acquisition, thereby reducing the SOC stabilization [30,56]. This interpretation aligns with the elevated CO_2 release and extracellular enzyme stoichiometry patterns observed in our study, which reflect intensified microbial C limitation and organic matter mineralization. It is commonly anticipated that, once organic matter decomposition increases, the microbial-derived CO_2 release from soil will also increase [57]. Moreover, the increased C-acquiring enzyme CBH and BG activities under acid rain also support this interpretation. These results showed that soil microorganisms secreted more extracellular enzymes to metabolize organic matter in order to resist adverse environmental conditions. Furthermore, acid rain might indirectly reduce SOC inputs by impairing the plant photosynthetic efficiency and root exudate production, thereby reducing the amount and quality of C entering the soil [3,12,15]. Collectively, these mechanisms—leaching, microbial metabolic shifts, and reduced carbon inputs—synergistically drive SOC depletion in acid rain-affected subtropical plantations.

Similar with acid rain, there was a strong negative effect of understory vegetation removal on the SOC, but its mechanism was different. It is generally accepted that understory vegetation has a substantial impact on the soil microclimate and nutrient cycling in forest ecosystems [14,16]. Understory vegetation removal can change the amount and quality of C input to the soil [14,23,24]. Therefore, on the one hand, the decreased SOC could be ascribed to directly reducing the aboveground and belowground C quantity inputs [23,24]. On the other hand, understory vegetation removal reduced the soil nutrient availability, such as soil DOC and AN, which could intensify the microbial C restriction and subsequently accelerate soil organic matter mineralization, ultimately reducing the SOC [19,24,58]. The latter interpretation is also supported by the results of the increased C-acquiring enzyme CBH and BG activities in the present study. Additionally, the presence of understory vegetation can alleviate soil surface runoff and nutrient loss by modifying the amount of throughfall interception [14,59], which was supported by the reduced soil MC, DOC, and AN in understory vegetation removal in this study. Consequently, considering the significant influences from understory vegetation on soil properties, the above explanations probably account for the patterns observed here. Collectively, our results underscored the crucial role of understory vegetation in regulating soil C cycling in the *C. camphor* plantations.

4.2. Responses of Microbial Community

In our study, the bacterial diversity was notably positively affected by acid rain and its interaction with understory vegetation removal, but the fungal diversity was only significantly positively affected by understory vegetation removal. In general, acid rain is expected to reduce soil bacterial diversity due to its toxic effects on bacterial communities [14,25]. However, we observed an increase in the bacterial diversity under the AR, which was consistent with prior studies reporting similar trends [3,60,61]. One possible reason for these findings is that soil acidification selectively suppresses acid-sensitive bacterial taxa while favoring acid-tolerant oligotrophic species through niche filtering [62,63]. Another possible reason is that different nutrient availability might drive shifts in the microbial diversity [26,64]; the soil low C availability induced by acid rain was likely to be mutually disadvantaged to bacteria with rather copiotrophic lifestyles, but was likely to be advantaged to oligotrophic bacteria through interspecific competition [25,65]. Previous studies also reported that the increase in C and P limitation significantly increased bacterial diversity [65], which was consistent with our study. In addition, these explanations were supported by the decrease in bacterial generalists and the increase in bacterial specialists in this study. Simultaneously, the change in bacterial generalists and specialists resulted in the notable reduction in the bacterial community niche breadth. The results indicated that the bacterial community was significantly affected by acid rain. In contrast, fungal diversity was not significantly affected by acid rain, but instead showed a pronounced response to understory vegetation removal, likely because fungi are more resistant to acidity [30,66]. The possible reasons for the significant increase in fungal diversity in the UR are that understory vegetation removal reduced the input labile organic compounds, such as the soil DOC, and fungi are often associated with recalcitrant organic compounds [30,32], which was supported by the increase in generalists.

In contrast to the microbial diversity, both acid rain and understory vegetation removal significantly destabilized the bacterial communities. This aligned with previous findings that canopy nitrogen addition and understory removal reduced bacterial and fungal community stability [67]. However, the cause of this phenomenon was not exactly the same. Although high microbial diversity is generally associated with enhanced community stability [47,67], in contrast to this, in the present study, bacterial communities with a high diversity had low community stability, which was presumably due to the high diversity of bacteria caused by the increase in specialists and the decrease in generalists. As a result, these specialists could lead to intense competition and less cooperation with each other [3], which could impede the exchange and efficient use of soil nutrients and resources, ultimately destabilizing the bacterial community structure [67]. Simultaneously, fewer generalists were present, leaving the community more vulnerable to environmental change [68], which could also destabilize the community. Alternatively, another possible reason was that acid rain and understory vegetation removal led to more deterministic processes of bacterial community assembly through niche partitioning and species interactions [67,69], which will eventually destabilize the community. However, the response of the fungal community to acid rain and understory vegetation removal was relatively stable. The fungal community stability and assembly processes were not notably influenced by acid rain and understory vegetation removal. The following reason might be used to explain the findings. On the one hand, fungi are known to be more adaptable to complex environmental variations, such as acid rain and drought [30]. Moreover, previous studies also reported that the fungal community did not respond sensitively to soil acidification and understory vegetation removal [25,70]. On the other hand, fungi can make more effective use of recalcitrant organic matter, whereas acid rain and understory vegetation removal might mainly reduce labile organic matter and eventually exacerbate the utilization of recalcitrant C [30,32], which was supported by the increase in the enzyme vector length. Therefore, the stability of fungal communities contributed to the absence of significant differences in fungal community assembly processes across the treatments.

However, the interactive effects of acid rain and understory vegetation removal did not significantly affect bacterial and fungal community assembly processes and stability, likely because the specialist and generalist ratios in the UA were not significantly affected, which consequently maintained the bacterial community niche breadth and stabilized the bacterial community. However, the stability of the fungal community structure might be attributed mainly to the fungal diversity stability. Overall, this work revealed that the response of the soil bacterial and fungal communities to acid rain and understory vegetation removal was different in the *C. camphor* plantations, that is, the soil bacterial community was generally more responsive to acid rain and understory vegetation removal in the *C. camphor* plantations, which was supported our first hypothesis.

4.3. Relationship Among Measured Variables

Prior studies have reported that shifts in microbial C source utilization resulted from changes in the microbial community structure [65,71], which was consistent with our study, showing that microbial community stability significantly affected the microbial metabolic nutrition restriction. Specifically, in the present study, the RDA and PLS-PM indicated that acid rain and understory vegetation removal affected soil enzyme activities and their stoichiometry by altering soil physicochemical properties and microbial community stability. These results supported our second hypothesis. C-acquiring enzymes and the vector length were mainly regulated by the soil bacterial community, while N- and P-acquiring enzymes and the vector angle were mainly affected by the soil fungal community. It is possible to attribute this to microbial resource usage preferences, with bacteria being able to use labile C more efficiently and fungi being able to use recalcitrant C more effectively [30,32], while N and P are more often contained in complex organic matter. Therefore, the shift in the nutrient availability and microclimate under acid rain and understory vegetation removal would amplify the impact of environmental filtering on the bacterial community and reduce its stability [67]. However, the unstable bacterial community structure was mainly due to a decrease in the community niche breadth, which led to increased competition and decreased cooperation, requiring more energy to maintain, and ultimately to stronger microbial C limitation, which is in agreement with a previous study [65]. However, probably because fungi are more resistant to external disturbances, the fungal community stability was not significantly affected by acid rain and understory vegetation removal, allowing the vector angle to remain relatively stable. These interpretations were supported by the results of our PLS-PM. Collectively, our results elucidated the linkages between microbial stability and microbial nutrient limitation under acid rain and understory vegetation removal in the *C. camphor* plantations. While the bacterial stability was closely tied to the C limitation, the fungal stability correlated with the P limitation. However, given that this study is only a short-term experiment and lacks continuous samples, the underlying mechanisms between microbial community stability and diversity and microbial metabolism require further investigation in the future.

5. Conclusions

This study explored the impacts of simulated acid rain and understory vegetation removal on the biological activity of the soils in a *C. camphor* plantation. The findings revealed the distinct impacts of these external disturbances. Acid rain significantly altered the SOC, CO_2 release, enzyme activities, and their stoichiometry, while also restructuring the microbial community. Understory vegetation removal markedly reduced the MC and nutrient availability, affected the enzyme stoichiometry balances, and also shifted the microbial community. The elevated bacterial diversity led to reduced bacterial community stability, likely due to the increase in specialists. In contrast, the fungal community demonstrated greater resistance to external perturbations, attributed to their metabolic preference and tolerance to osmotic stress. Mechanistically, the bacterial community instability was firmly closely related to the microbial C limitation (i.e., the vector length), whereas the fungal community stability was more closely associated with the microbial P limitation (i.e., the vector angle). Overall, our study advanced the understanding of the mechanisms by which acid rain and understory vegetation removal destabilize microbial communities and increase microbial metabolic limitations. Critically, they highlight the indispensable role of understory vegetation in buffering soil microclimates, sustaining nutrient cycling, and maintaining community stability in plantations. Finally, these finding emphasized the need for integrated forest management strategies to conserve and manage soil ecosystems in subtropical plantations.

Supplementary Materials: The following supporting information can be downloaded at: https://www.mdpi.com/article/10.3390/f16030525/s1, Table S1: Soil physicochemical properties in treatments in the *Cinnamomum. camphor* (Linn) Presl plantation.

Author Contributions: Conceptualization, X.H. and Z.H.; methodology, Z.H. and Y.L. (Yonghui Lin); software, Z.H. and Y.L. (Yini Liu); validation, Y.L. (Yonghui Lin); formal analysis, X.H. and Z.H.; investigation, Y.L. (Yini Liu), Z.H. and X.K.; data curation, Z.H. and X.K.; writing—original draft preparation, Z.H. and Y.L. (Yini Liu); writing—review and editing, X.H., Z.H. and H.L.; visualization, Z.H. and Y.L. (Yini Liu); projection administration, X.H. and Z.H. All authors have read and agreed to the published version of the manuscript.

Funding: This work was supported by the General Program of Scientific Research Foundation of Hunan Provincial Education Department (22C0278), the Jishou University Doctoral Talent Introduction Project (1122007), the National Natural Science Foundation of China (grant numbers 32060332, 31670624, and 32160356), the Youth Program of Scientific Research Foundation of Hunan Provincial Education Department (24B0500), and the Natural Science Foundation of Hunan Province (2025JJ60205 and 2025JJ50112).

Data Availability Statement: The datasets in this study are available in Supplementary Material Table S1.

Conflicts of Interest: The authors declare no conflicts of interest.

References

1. Meng, C.; Tian, D.; Zeng, H.; Li, Z.; Yi, C.; Niu, S. Global soil acidification impacts on belowground processes. *Environ. Res. Lett.* **2019**, *14*, 074003. [CrossRef]
2. Zheng, Y.; Wang, Y.; Zheng, Y.; Li, Y. Effects of simulated acid rain on soil enzyme activity and related chemical indexes in woodlands. *Forests* **2022**, *13*, 860. [CrossRef]
3. Liu, Z.; Chen, J.; Su, Z.; Liu, Z.; Li, Y.; Wang, J.; Wu, L.; Wei, H.; Zhang, J. Acid rain reduces plant-photosynthesized carbon sequestration and soil microbial network complexity. *Sci. Total Environ.* **2023**, *873*, 162030. [CrossRef] [PubMed]
4. Liu, X.; Li, C.; Meng, M.; Zhai, L.; Zhang, B.; Jia, Z.; Gu, Z.; Liu, Q.; Zhang, Y.; Zhang, J. Comparative effects of the recovery from sulfuric and nitric acid rain on the soil enzyme activities and metabolic functions of soil microbial communities. *Sci. Total Environ.* **2020**, *714*, 136788. [CrossRef]
5. Ding, Y.; Sun, L.; Li, C.; Chen, M.; Zhou, Y.; Meng, M.; Li, Z.; Zhang, J.; Liu, X. Effects of short-term simulated acid rain and nitrogen deposition on soil nutrients and enzyme activities in *Cunninghamia lanceolata* plantation. *Front. Ecol. Evol.* **2024**, *12*, 1365954. [CrossRef]
6. Wang, C.; Guo, P.; Han, G.; Feng, X.; Zhang, P.; Tian, X. Effect of simulated acid rain on the litter decomposition of *Quercus acutissima* and *Pinus massoniana* in forest soil microcosms and the relationship with soil enzyme activities. *Sci. Total Environ.* **2010**, *408*, 2706–2713. [CrossRef]
7. Burns, R.G.; DeForest, J.L.; Marxsen, J.; Sinsabaugh, R.L.; Stromberger, M.E.; Wallenstein, M.D.; Weintraub, M.N.; Zoppini, A. Soil enzymes in a changing environment: Current knowledge and future directions. *Soil Biol. Biochem.* **2013**, *58*, 216–234. [CrossRef]
8. Jansson, J.K.; Hofmockel, K.S. Soil microbiomes and climate change. *Nat. Rev. Microbiol.* **2020**, *18*, 35–46. [CrossRef]
9. Lv, Y.; Wang, C.; Jia, Y.; Wang, W.; Ma, X.; Du, J.; Pu, G.; Tian, X. Effects of sulfuric, nitric, and mixed acid rain on litter decomposition, soil microbial biomass, and enzyme activities in subtropical forests of China. *Appl. Soil Ecol.* **2014**, *79*, 1–9. [CrossRef]
10. Shi, Z.; Zhang, J.; Zhang, H.; Wei, H.; Lu, T.; Chen, X.; Li, H.; Yang, J.; Liu, Z. Response and driving factors of soil enzyme activity related to acid rain: A meta-analysis. *Environ. Sci. Pollut. Res.* **2023**, *30*, 105072–105083. [CrossRef]
11. Tang, L.; Lin, Y.; He, X.; Han, G. Acid rain decelerates the decomposition of *Cunninghamia lanceolata* needle and *Cinnamomum camphora* leaf litters in a karst region in China. *Ecol. Res.* **2019**, *34*, 193–200. [CrossRef]
12. Liu, Z.; Shi, Z.; Wei, H.; Zhang, J. Acid rain reduces soil CO_2 emission and promotes soil organic carbon accumulation in association with decreasing the biomass and biological activity of ecosystems: A meta-analysis. *Catena* **2022**, *208*, 105714. [CrossRef]
13. Liu, X.; Zhao, W.; Meng, M.; Fu, Z.; Xu, L.; Zha, Y.; Yue, J.; Zhang, S.; Zhang, J. Comparative effects of simulated acid rain of different ratios of SO_4^{2-} to NO_3^- on fine root in subtropical plantation of China. *Sci. Total Environ.* **2018**, *618*, 336–346. [CrossRef] [PubMed]
14. Deng, J.; Fang, S.; Fang, X.; Jin, Y.; Kuang, Y.; Lin, F.; Liu, J.; Ma, J.; Nie, Y.; Ouyang, S.; et al. Forest understory vegetation study: Current status and future trends. *For. Res.* **2023**, *3*, 6. [CrossRef]
15. Zhang, Y.; Li, J.; Tan, J.; Li, W.; Singh, B.P.; Yang, X.; Bolan, N.; Chen, X.; Xu, S.; Bao, Y.; et al. An overview of the direct and indirect effects of acid rain on plants: Relationships among acid rain, soil, microorganisms, and plants. *Sci. Total Environ.* **2023**, *873*, 162388. [CrossRef]
16. Landuyt, D.; De Lombaerde, E.; Perring, M.P.; Hertzog, L.R.; Ampoorter, E.; Maes, S.L.; De Frenne, P.; Ma, S.; Proesmans, W.; Blondeel, H.; et al. The functional role of temperate forest understorey vegetation in a changing world. *Glob. Change Biol.* **2019**, *25*, 3625–3641. [CrossRef]
17. Wagner, R.G.; Little, K.M.; Richardson, B.; Mcnabb, K. The role of vegetation management for enhancing productivity of the world's forests. *Forestry* **2006**, *79*, 57–79. [CrossRef]
18. Giuggiola, A.; Zweifel, R.; Feichtinger, L.M.; Vollenweider, P.; Bugmann, H.; Haeni, M.; Rigling, A. Competition for water in a xeric forest ecosystem—Effects of understory removal on soil micro-climate, growth and physiology of dominant Scots pine trees. *For. Ecol. Manag.* **2018**, *409*, 241–249. [CrossRef]
19. Yang, Y.; Zhang, X.; Zhang, C.; Wang, H.; Fu, X.; Chen, F.; Wan, S.; Sun, X.; Wen, X.; Wang, J. Understory vegetation plays the key role in sustaining soil microbial biomass and extracellular enzyme activities. *Biogeosciences* **2018**, *15*, 4481–4494. [CrossRef]
20. Wang, F.; Zou, B.; Li, H.; Li, Z. The effect of understory removal on microclimate and soil properties in two subtropical lumber plantations. *J. For. Res.* **2014**, *19*, 238–243. [CrossRef]
21. Fang, X.M.; Wang, G.G.; Xu, Z.J.; Zong, Y.Y.; Zhang, X.L.; Li, J.J.; Wang, H.; Chen, F.S. Litter addition and understory removal influenced soil organic carbon quality and mineral nitrogen supply in a subtropical plantation forest. *Plant Soil* **2021**, *460*, 527–540. [CrossRef]
22. Zhao, J.; Wang, X.; Shao, Y.; Xu, G.; Fu, S. Effects of vegetation removal on soil properties and decomposer organisms. *Soil Biol. Biochem.* **2011**, *43*, 954–960. [CrossRef]
23. Zhang, S.; Yang, X.; Li, D.; Li, S.; Chen, Z.; Wu, J. A meta-analysis of understory plant removal impacts on soil properties in forest ecosystems. *Geoderma* **2022**, *426*, 116116. [CrossRef]

24. Deng, C.; Lyu, M.; Xiong, X.; Peñuelas, J.; Sardans, J.; Li, X.; Lin, W.; Yang, Y.; Xie, J. Understory ferns removal downregulates microbial carbon use efficiency and carbon accrual in previously degraded lands. *Agr. For. Meteorol.* **2023**, *340*, 109631. [CrossRef]
25. Choma, M.; Tahovská, K.; Kaštovská, E.; Bárta, J.; Růžek, M.; Oulehle, F. Bacteria but not fungi respond to soil acidification rapidly and consistently in both a spruce and beech forest. *FEMS Microbiol. Ecol.* **2020**, *96*, fiaa174. [CrossRef]
26. Li, Y.; Wang, Y.; Zhang, W. Impact of simulated acid rain on the composition of soil microbial communities and soil respiration in typical subtropical forests in Southwest China. *Ecotox. Environ. Saf.* **2021**, *215*, 112152. [CrossRef]
27. Hooper, D.U.; Bignell, D.E.; Brown, V.K.; Brussard, L.; Mark Dangerfield, J.; Wall, D.H.; Wardle, D.A.; Coleman, D.C.; Giller, K.E.; Lavelle, P.; et al. Interactions between aboveground and belowground biodiversity in terrestrial ecosystems: Patterns, mechanisms, and feedbacks. *Bioscience* **2000**, *50*, 1049. [CrossRef]
28. Delgado-Baquerizo, M.; Reich, P.B.; Khachane, A.N.; Campbell, C.D.; Thomas, N.; Freitag, T.E.; Abu Al-Soud, W.; Sørensen, S.; Bardgett, R.D.; Singh, B.K. It is elemental: Soil nutrient stoichiometry drives bacterial diversity. *Environ. Microbiol.* **2017**, *19*, 1176–1188. [CrossRef]
29. Reischke, S.; Rousk, J.; Bååth, E. The effects of glucose loading rates on bacterial and fungal growth in soil. *Soil Biol. Biochem.* **2014**, *70*, 88–95. [CrossRef]
30. Wang, C.; Kuzyakov, Y. Mechanisms and implications of bacterial-fungal competition for soil resources. *ISME J.* **2024**, *18*, wrae073. [CrossRef]
31. Griffiths, B.S.; Ritz, K.; Ebblewhite, N.; Dobson, G. Soil microbial community structure: Effects of substrate loading rates. *Soil Biol. Biochem.* **1998**, *31*, 145–153. [CrossRef]
32. Deng, W.; Lu, Y.; Lyu, M.; Deng, C.; Li, X.; Jiang, Y.; Zhu, H.; Yang, Y.; Xie, J. Chemical composition of soil carbon is governed by microbial diversity during understory fern removal in subtropical pine forests. *Sci. Total Environ.* **2024**, *914*, 169904. [CrossRef] [PubMed]
33. Sinsabaugh, R.L. Phenol oxidase, peroxidase and organic matter dynamics of soil. *Soil Biol. Biochem.* **2010**, *42*, 391–404. [CrossRef]
34. Xuan, C.; Xiaoran, S.; Zhaoji, S.; Jiaen, Z.; Zhong, Q.; Huimin, X.; Hui, W. Analysis of the spatio-temporal changes in acid rain and their causes in China (1998–2018). *J. Resour. Ecol.* **2021**, *12*, 593–599. [CrossRef]
35. He, X.; Lin, Y.; Han, G.; Ma, T. Litterfall interception by understorey vegetation delayed litter decomposition in *Cinnamomum camphora* plantation forest. *Plant Soil* **2013**, *372*, 207–219. [CrossRef]
36. He, X.; Han, G.; Lin, Y.; Tian, X.; Xiang, C.; Tian, Q.; Wang, F.; He, Z. Diversity and decomposition potential of endophytes in leaves of a *Cinnamomum camphora* plantation in China. *Ecol. Res.* **2011**, *27*, 273–284. [CrossRef]
37. Bao, S. *Soil and Agricultural Chemistry Analysis*; China Agriculture Press: Beijing, China, 2000.
38. Du, J.; Zhang, Y.; Guo, W.; Li, N.; Gao, C.; Cui, M.; Lin, Z.; Wei, M.; Zhang, H. Chronic impacts of TiO_2 nanoparticles on *Populus nigra* L. leaf decomposition in freshwater ecosystem. *J. Hazard. Mater.* **2018**, *350*, 121–127. [CrossRef]
39. Bolger, A.M.; Lohse, M.; Usadel, B. Trimmomatic: A flexible trimmer for Illumina sequence data. *Bioinformatics* **2014**, *30*, 2114–2120. [CrossRef]
40. Bolyen, E.; Rideout, J.R.; Dillon, M.R.; Bokulich, N.A.; Abnet, C.; Al-Ghalith, G.A.; Alexander, H.; Alm, E.J.; Arumugam, M.; Asnicar, F.; et al. Reproducible, interactive, scalable and extensible microbiome data science using QIIME 2. *Nat. Biotechnol.* **2019**, *37*, 852–857. [CrossRef]
41. Callahan, B.J.; McMurdie, P.J.; Rosen, M.J.; Han, A.W.; Johnson, A.J.A.; Holmes, S.P. DADA2: High-resolution sample inference from Illumina amplicon data. *Nat. Methods* **2016**, *13*, 581–583. [CrossRef]
42. Liu, C.; Cui, Y.; Li, X.; Yao, M. microeco: An R package for data mining in microbial community ecology. *FEMS Microbiol. Ecol.* **2020**, *97*, fiaa255. [CrossRef] [PubMed]
43. Levins, R. *Evolution in Changing Environments: Some Theoretical Explorations*; Princeton University Press: Princeton, NJ, USA, 1968.
44. Salazar, G. EcolUtils: Utilities for Community Ecology Analysis. R Package Version 0.1. 2015. Available online: https://github.com/GuillemSalazar/EcolUtils (accessed on 10 May 2023).
45. Wu, W.; Logares, R.; Huang, B.; Hsieh, C. Abundant and rare picoeukaryotic sub-communities present contrasting patterns in the epipelagic waters of marginal seas in the northwestern Pacific Ocean. *Environ. Microbiol.* **2017**, *19*, 287–300. [CrossRef] [PubMed]
46. Zhang, J.; Zhang, B.; Liu, Y.; Guo, Y.; Shi, P.; Wei, G. Distinct large-scale biogeographic patterns of fungal communities in bulk soil and soybean rhizosphere in China. *Sci. Total Environ.* **2018**, *644*, 791–800. [CrossRef]
47. Xun, W.; Liu, Y.; Li, W.; Ren, Y.; Xiong, W.; Xu, Z.; Zhang, N.; Miao, Y.; Shen, Q.; Zhang, R. Specialized metabolic functions of keystone taxa sustain soil microbiome stability. *Microbiome* **2021**, *9*, 35. [CrossRef]
48. Ning, D.; Deng, Y.; Tiedje, J.M.; Zhou, J. A general framework for quantitatively assessing ecological stochasticity. *Proc. Natl. Acad. Sci. USA* **2019**, *116*, 16892–16898. [CrossRef]
49. Ning, D.; Wang, Y.; Fan, Y.; Wang, J.; Van Nostrand, J.D.; Wu, L.; Zhang, P.; Curtis, D.J.; Tian, R.; Lui, L.; et al. Environmental stress mediates groundwater microbial community assembly. *Nat. Microbiol.* **2024**, *9*, 490–501. [CrossRef]
50. Moorhead, D.L.; Sinsabaugh, R.L.; Hill, B.H.; Weintraub, M.N. Vector Analysis of Ecoenzyme Activities Reveal Constraints on Coupled C, N and P Dynamics. *Soil Biol. Biochem.* **2016**, *93*, 1–7. [CrossRef]

51. Lai, J.; Zou, Y.; Zhang, J.; Peres-Neto, P.R. Generalizing hierarchical and variation partitioning in multiple regression and canonical analyses using the rdacca.hp R package. *Methods Ecol. Evol.* **2022**, *13*, 782–788. [CrossRef]
52. Henseler, J.; Sarstedt, M. Goodness-of-fit indices for partial least squares path modeling. *Comput. Stat.* **2012**, *28*, 565–580. [CrossRef]
53. Wetzels, M.; Odekerken-Schröder, G.; Van Oppen, C. Using PLS path modeling for assessing hierarchical construct models: Guidelines and empirical illustration. *Mis Quart.* **2009**, *33*, 177–195. [CrossRef]
54. Liu, Z.; Wei, H.; Zhang, J.; Saleem, M.; He, Y.; Zhong, J.; Ma, R. Higher sensitivity of soil microbial network than community structure under acid rain. *Microorganisms* **2021**, *9*, 118. [CrossRef] [PubMed]
55. Liu, Z.; Wei, H.; Zhang, J.; Saleem, M.; He, Y.; Zhong, J.; Ma, R. Seasonality regulates the effects of acid rain on microbial community in a subtropical agricultural soil of Southern China. *Ecotox. Environ. Saf.* **2021**, *224*, 112681. [CrossRef] [PubMed]
56. Ye, C.; Chen, D.; Hall, S.J.; Pan, S.; Yan, X.; Bai, T.; Guo, H.; Zhang, Y.; Bai, Y.; Hu, S. Reconciling multiple impacts of nitrogen enrichment on soil carbon: Plant, microbial and geochemical controls. *Ecol. Lett.* **2018**, *21*, 1162–1173. [CrossRef] [PubMed]
57. Liu, Z.; Shan, X.; Wei, H.; Zhang, J.; Saleem, M.; Li, D.; Zhang, Y.; Ma, R.; He, Y.; Zhong, J.; et al. Idiosyncratic responses of microbial communities and carbon utilization to acid rain frequency in the agricultural and forest soils. *Glob. Ecol. Conserv.* **2021**, *26*, e01429. [CrossRef]
58. Lyu, M.; Xie, J.; Giardina, C.P.; Vadeboncoeur, M.A.; Feng, X.; Wang, M.; Ukonmaanaho, L.; Lin, T.; Kuzyakov, Y.; Yang, Y. Understory ferns alter soil carbon chemistry and increase carbon storage during reforestation with native pine on previously degraded sites. *Soil Biol. Biochem.* **2019**, *132*, 80–92. [CrossRef]
59. Jiang, M.; Lin, T.; Shaner, P.L.; Lyu, M.; Xu, C.; Xie, J.; Lin, C.; Yang, Z.; Yang, Y. Understory interception contributed to the convergence of surface runoff between a Chinese fir plantation and a secondary broadleaf forest. *J. Hydrol.* **2019**, *574*, 862–871. [CrossRef]
60. Wang, L.; Chen, Z.; Shang, H.; Wang, J.; Zhang, P. Impact of simulated acid rain on soil microbial community function in Masson pine seedlings. *Electron. J. Biotechnol.* **2014**, *17*, 199–203. [CrossRef]
61. Liu, Z.; Li, D.; Zhang, J.; Saleem, M.; Zhang, Y.; Ma, R.; He, Y.; Yang, J.; Xiang, H.; Wei, H. Effect of simulated acid rain on soil CO_2, CH_4 and N_2O emissions and microbial communities in an agricultural soil. *Geoderma* **2020**, *366*, 114222. [CrossRef]
62. Wu, Y.T.; Gutknecht, J.; Nadrowski, K.; Geißler, C.; Kühn, P.; Scholten, T.; Both, S.; Erfmeier, A.; Böhnke, M.; Bruelheide, H.; et al. Relationships between soil microorganisms, plant communities, and soil characteristics in Chinese subtropical forests. *Ecosystems* **2012**, *15*, 624–636. [CrossRef]
63. Wang, Y.; Li, X.; Datta, R.; Chen, J.; Du, Y.; Du, D. Key factors shaping prokaryotic communities in subtropical forest soils. *Appl. Soil Ecol.* **2022**, *169*, 104162. [CrossRef]
64. Cline, L.C.; Hobbie, S.E.; Madritch, M.D.; Buyarski, C.R.; Tilman, D.; Cavender-Bares, J.M. Resource availability underlies the plant-fungal diversity relationship in a grassland ecosystem. *Ecology* **2017**, *99*, 204–216. [CrossRef] [PubMed]
65. Kang, H.; Xue, Y.; Cui, Y.; Moorhead, D.L.; Lambers, H.; Wang, D. Nutrient limitation mediates soil microbial community structure and stability in forest restoration. *Sci. Total Environ.* **2024**, *935*, 173266. [CrossRef] [PubMed]
66. Hou, Z.; Zhang, X.; Chen, W.; Liang, Z.; Wang, K.; Zhang, Y.; Song, Y. Differential responses of bacterial and fungal community structure in soil to nitrogen deposition in two planted forests in southwest China in relation to pH. *Forests* **2024**, *15*, 1112. [CrossRef]
67. Li, D.; Wu, J. Canopy nitrogen addition and understory removal destabilize the microbial community in a subtropical Chinese fir plantation. *J. Environ. Manag.* **2024**, *354*, 120407. [CrossRef]
68. Li, Y.; Wang, J.; Shen, C.; Zhou, G.; Delgado-Baquerizo, M.; Ge, Y. Microbial diversity losses constrain the capacity of soils to mitigate climate change. *Glob. Change Biol.* **2024**, *30*, e17601. [CrossRef]
69. Chase, J.M.; Myers, J.A. Disentangling the importance of ecological niches from stochastic processes across scales. *Philos. T. R. Soc. B.* **2011**, *366*, 2351–2363. [CrossRef]
70. Xi, D.; Jin, S.; Wu, J. Soil bacterial community is more sensitive than fungal community to canopy nitrogen deposition and understory removal in a Chinese fir plantation. *Front. Microbiol.* **2022**, *13*, 1015936. [CrossRef]
71. Cline, L.C.; Zak, D.R. Soil microbial communities are shaped by plant-driven changes in resource availability during secondary succession. *Ecology* **2015**, *96*, 3374–3385. [CrossRef]

forests

MDPI

Article

How Natural Regeneration After Severe Disturbance Affects Ecosystem Services Provision of Andean Forest Soils at Contrasting Timescales

Juan Ortiz [1], Marcelo Panichini [2], Pablo Neira [3], Carlos Henríquez-Castillo [4], Rocio E. Gallardo Jara [1], Rodrigo Rodriguez [5], Ana Mutis [6], Camila Ramos [1,7], Winfred Espejo [1], Ramiro Puc-Kauil [8], Erik Zagal [1], Neal Stolpe [1], Mauricio Schoebitz [1], Marco Sandoval [1] and Francis Dube [9,*]

1 Soil and Natural Resources Department, Faculty of Agronomy, University of Concepcion, Vicente Mendez 595, Chillan 3812120, Ñuble Region, Chile; rgallardo@udec.cl (R.E.G.J.); camiramos@udec.cl (C.R.); ezagal@udec.cl (E.Z.); mschoebitz@udec.cl (M.S.)
2 National Agricultural Research Institute (INIA), Carillanca Station, km 10, Cajón Road, Vilcún 4880000, La Araucania, Chile
3 Institut de Recerca de la Biodiversitat (IRBio), Universitat de Barcelona, 08028 Barcelona, Spain
4 Faculty of Health Sciences, Universidad del Alba, Av. Ejército Libertador 171, Santiago 8320000, Santiago City, Chile
5 Biocontrol Research Laboratory, University of La Frontera, Temuco 4811230, La Araucania, Chile; r.rodriguez09@ufromail.cl
6 Department of Chemical Sciences and Natural Resources, Faculty of Engineering and Science, University of La Frontera, Temuco 4811230, La Araucania, Chile; ana.mutis@ufrontera.cl
7 Doctoral Program in Agronomy, Faculty of Agronomy, University of Concepcion, Vicente Mendez, 595, Chillan 3812120, Ñuble Region, Chile
8 National Technological of Mexico, Venustiano Carranza Superior Technological Institute, Puebla 73049, Mexico; ramiro.selvastropicales@gmail.com
9 Department of Silviculture, Faculty of Forest Sciences, University of Concepción, Victoria 631, Casilla 160-C, Concepción 4030000, Biobio Region, Chile
* Correspondence: fdube@udec.cl

Abstract: Chile holds ~50% of temperate forests in the Southern Hemisphere, thus constituting a genetic–ecological heritage. However, intense anthropogenic pressures have been inducing distinct forest structural-regeneration patterns. Accordingly, we evaluated 22 soil properties at 0–5 and 5–20 cm depths in two protected sites, with similar perturbation records but contrasting post-disturbance regeneration stages: long-term secondary forest (~50 y) (SEC_{FORST}) (dominated by *Chusquea* sp.-understory) and a short-term forest after disturbance (~5 y) (FAD_{IST}) within a *Nothofagus* spp. forest to determine the potential of these soils to promote nutrient availability, water cycling, soil organic carbon (SOC) sequestration ($CO_2 \rightarrow$SOC), and microbiome. Results detected 93 correlations ($r \geq 0.80$); however, no significant differences ($p < 0.05$) in physical or chemical properties, except for infiltration velocity (+27.97%), penetration resistance (−23%), SOC (+5.64%), and % Al saturation (+5.64%) relative to SEC_{FORST}, and a consistent trend of suitable values 0–5 > 5–20 cm were estimated. The SOC$\rightarrow CO_2$ capacity reached 4.2 ± 0.5 (FAD_{IST}) and 2.7 ± 0.2 Mg C y^{-1} (SEC_{FORST}) and only microbial abundance shifts were observed. These findings provide relevant insights on belowground resilience, evidenced by similar ecosystem services provision capacities over time, which may be influenced progressively by opportunistic *Chusquea* sp.

Keywords: *Nothofagus* spp.; south-central Chile; passive management; biosphere reserve; volcanic soils; *Chusquea* sp.; secondary forest

1. Introduction

Terrestrial organic carbon (C) stored in soil (SOC) controls multiple scalar natural processes as ecosystem functionality (water and nutrient cycling), which are primarily mediated by microorganisms [1]. However, historically, ~132 Pg of SOC pool (8.8% of total) have been released into the atmosphere in response to human actions (greenhouse gas emissions) (SOC→CO_2), causing land degradation [2] and climate change [3–6], and currently accounting for a net loss trend of 0.22–0.53 Pg y^{-1} [7]. Notwithstanding the above, soils represent the greatest CO_2→SOC potential globally, and specific capacities are associated with climatic zones, individual native properties, and historical land use records [8], where deep soils (≥1 m) from temperate to cold zones and mineral fractions rich in Fe-oxides and active clays, probably fulfilling the best conditions [8,9]. Regarding most favorable ecosystems facilitating (SOC→CO_2), forests have a great sink potential, storing 861–967 Pg of SOC (0–1 m), encompassing up to 30% of total emissions within the pedosphere [10–13]. Particularly, temperate forests, defined as "highly-seasonal biomes, dominated by freeze-hardy woody species due to remarkable intervals of long growing and a cold winter period" [14], are distributed across 23.5–66.5° N and S latitudes, covering 19–26 % of global forest area, having a presence on all continents (except Africa), usually meet the criteria mentioned above [12,15].

The main aspects associated with the relevant CO_2→SOC capacity in temperate are (a) at ecosystem level: (i) constant inputs of SOC precursors (plant debris), (ii) permanent soil cover, (iii) deep-rooted systems, increasing the uptake-recycling nutrients, and (iv) suitable conditions for biological activity [16]; (b) at pedosphere level: (i) an increase in recalcitrance due to lignin-rich substrates, (ii) lowering soil temperature–moisture reducing SOC mineralization rates, (iii) the formation of stable aggregates, and (iv) mineral-associated C [9,17–19]; and (c) at microscopic level: (i) inputs of fresh organics (e.g., leaf/herbaceous strata litter), (ii) nutrient bioavailability processes such as organic re-synthesis, nutrient mining cycling (e.g., N, S, P, Fe, Cu) for plant development, and (iii) structural and functional microbial pools (e.g., active microorganisms and those able to restore the soil to a pre-disturbance condition), leading to re-aggregation processes mediated by biological byproducts [2,9,17,20,21].

However, forest degradation caused a decline of 3% of the global area (4128 M ha) over the last 35 y, reversing the CO_2→SOC processes and mechanisms previously described [22,23]. One of the most severely affected regions is South America, reaching a forest loss area of 88,803 K ha [10,24]. In addition, it is estimated that only less than 20% of temperate forests are present in the Southern Hemisphere, resulting in one of the highest levels of endemism (85% of woody species and 34% of genera, prevailing evergreen species) [12,15].

The previous scenario is critical for genetic preservation in Chile, which holds 50 and 79.5% of the total temperate forest area in the Southern Hemisphere and continental levels, respectively (33–55° S latitude, 14.41 M ha), dominated by *Nothofagus* spp. (pure or mixed) [12,25–28]. Local forest suppression has been subordinated to agricultural purposes, excessive logging for fuelwood, over-grazing–browsing, and the replacement by commercial plantations (e.g., *Pinus radiata*, *Eucalyptus nitens*) due to the high productivity and timber quality of these forest types (including soils) [28,29]. In response, nowadays, Chile holds 23% of the protected forest area in South America, due to extensive government and institutional efforts that have led to the implementation of land reclamation programs (including a sustainable agro-silvicultural policy framework) [12]. Nonetheless, vast derelict territories consequently undergo natural regeneration processes (e.g., secondary forest), many of them based on Andisols, which are of high agricultural importance, accounting for 50–60 of arable land [30]. The above suggests a need to evaluate the soil ecosystem services status in those areas since their inherent high $CO_2 \rightarrow SOC$ potential is widely reported in the literature [7,31,32].

Accordingly, we aimed to compare different soil properties that are useful for establishing fertility status and the current capacity to provide ecosystem services of two contrasting time-lapse scenarios within an unmanaged post-disturbance *Nothofagus* sp. association forest; both scenarios are highly representative of the aforementioned context and dominant biome type across the south-central macro-zone of Chile.

We hypothesized that soil properties–ecosystem services capacities of long-term passive management have been influenced by the effects of emergent now dominant Chusquea spp. understory, particularly, generating (i) a robust microbiome and particularly decomposer communities; (ii) a greater fertility status; (iii) better physical aptitudes related to hydric conductivity; and iv) a lower annual $CO_2 \rightarrow SOC$ capacity rate.

2. Material and Methods

2.1. Study Area Description

The study was carried out at the "Ranchillo Alto" state-owned property, a 635 ha native forest that was declared "National Heritage" and a "Protected National Asset" "by the Chilean Government; it is located in the Ñuble Region (Figure 1), nearby the town of Yungay, within the biological corridor "Nevados de Chillán—Laguna del Laja", which was designated as a "Biosphere Reserve" by UNESCO [33]. The local climate is classified as temperate Mediterranean, with average temperatures of 13.5–25 °C during the summer and a pronounced winter season from May to September, with frequent snowfall–frost events and a mean annual rainfall of over 3.000 mm [33]. Through a recognition visit and preliminary site examination (e.g., interviews with local landowners, soil sampling, pers. commun.) [34], it was determined that the area possesses a historical record of over-management–utilization of native *Nothofagus* spp. involving diverse detrimental activities (such as those mentioned in the previous section), causing the interruption of biomass–litter production, which is the partial to total detachment of soil cover and even topsoil, evidenced by the presence of gullies, leading to consequent massive losses of soil organic matter and nutrients (Figure 2) [21,35].

However, those processes of land degradation occurred at different spatial–temporal scales that do not always agree with the government–social policies implemented to cease them, and (if this is the case), nor do the actions implemented for its reclamation (e.g., agroforestry systems, conservation silviculture) [29,35], or, in the absence of these, different timeline patterns of natural regeneration (e.g., passive management), leading to a complex mosaic of ecological–vegetational conditions [33].

Accordingly, we evaluated two temporal-contrasting sectors within the derived secondary forest, both meeting natural regeneration/no history of post-disturbance management and a similar land degradation record. A description of each condition is provided as follows (Table 1).

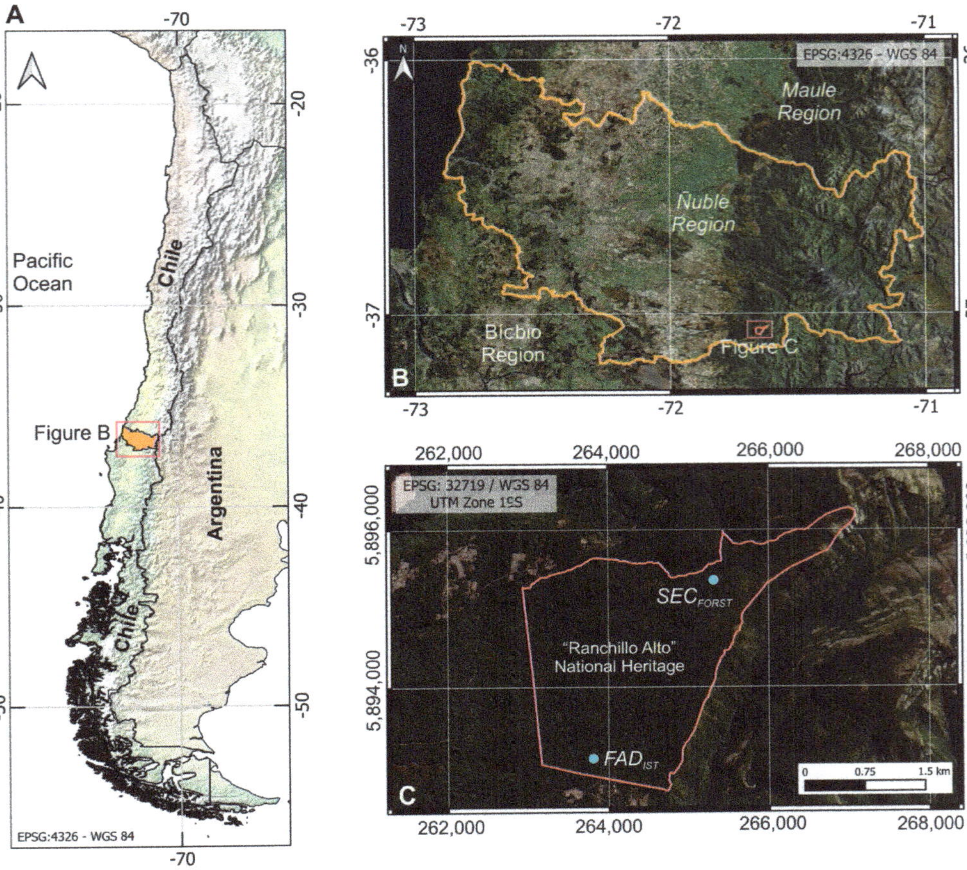

Figure 1. Approaching maps illustrating study site. (**A**) national map of central-south Chile, highlighting Ñuble Region in orange, (**B**) regional map of Ñuble, and the location of the Ranchillo Alto site in southern part of the region, (**C**) localization of the protected area Ranchillo Alto and the position of the FAD_{IST} and SEC_{FORST} analyzed in this study.

Table 1. Description of study cases.

Condition	Location	Assessed Area (ha)	N° Plots and Area (ha)	Tree Density (N° ha^{-1}); Average Tree Height (m) and Diameter at Breast Height (DBH) (cm)	Forest Species	Understory-Herbaceous Strata Composition	Degradation Record
FAD_{IST}	37°03′31.87″ S, 71°38′21.76″ W 1367 m.a.s.l	4	3 × 1.33	60; 15 and 34	Roble (*Nothofagus obliqua* (Mirb.) Oerst. - coihue (*Nothofagus dombeyi*) (Mirb.) Oerst. (5:1)	Vetch (*Fabaceae purpurea*), clover (*Trifolium incarnatum*, *T. subterraneum*, and *T. vesiculosum*), *Lolium multiflorum westerwoldicum*, *Phalaris acuatica*, *Lolium perenne*, oats (*Avena sativa*), *Festuca arundinacea*, *Dactylis glomerata* L.), *Chusquea* sp., and there-sprouting of Radal (*Lomatia hirsuta* (Lam.) Diels)	Evidence of wildfire, agricultural burning, and intense logging. Testimony of use of intensive animal–mechanical loads; cattle grazing–browsing, prolonged commercial, and domestic timber harvesting

Table 1. *Cont.*

Condition	Location	Assessed Area (ha)	N° Plots and Area (ha)	Tree Density (N° ha^{-1}); Average Tree Height (m) and Diameter at Breast Height (DBH) (cm)	Forest Species	Understory-Herbaceous Strata Composition	Degradation Record
SEC_{FORST}	37°4′43.01″ S, 71°39′25.32″ W 1250 m.s.n.m	4	3 × 1.33	296; 33 and 46	Roble (*Nothofagus obliqua*)- coihue (*Nothofagus dombeyi*) (8:3)	Primarily dominated by *Chusquea* sp., a circumstance that has been reported in the literature during regenerative dynamics processes post-disturbance (e.g., Muñoz and Gonzalez, 2009; Donoso et al., 2022 [28,36])	Evidence of fire events. Testimony of use of intensive animal loads; cattle/goat/equine grazing and/or browsing, prolonged commercial and domestic timber harvesting

Figure 2. Photographs of the study area, (**A**) original degraded site overview, (**B**) FAD_{IST}, and (**C**) SEC_{FORST}. Photo credits: F. Dube.

The soils were classified as medial, amorphic, and mesic Typic Haploxerands within the "Santa Barbara" series with an Ap/A/AB/Bw/BC/C sequence, according to the Soil Survey Staff (2022) [37] and Stolpe (2006) [38].

2.2. Soil Sampling and Characterization

To analyze soil properties under different post-disturbance conditions, we collected soil samples from two distinct forest sites: a long-term secondary forest (SEC_{FORST}) and a recently disturbed area (FAD_{IST}). These samples allow us to compare how soil chemistry, microbiology, and physics vary over time. In January 2020, three composite soil samples (2 kg each) were taken from three representative plots (1.33 ha each) at two depths: 0–5 cm in triplicate (n = 36), which represents the most biologically active soil layer, and 5–20 cm, where deeper-rooted microbial communities and chemical interactions occur. This depth selection ensures that we capture key soil processes occurring at different strata.

2.2.1. Physical Trials

To assess how soil structure changes over time, we measured different physical properties that influence water retention, root penetration, and overall soil stability. These properties help determine whether the soil is improving or degrading after disturbance, which is crucial for forest recovery. The following tests were performed: (i) Bulk density (BD) measures how compacted the soil is, which affects root growth and water movement. A denser soil can limit plant development and microbial activity. We determined BD by extracting soil cores (211 cm^3) and drying them at 105 °C until reaching a constant weight; (ii) soil particle density (PD), estimated by the pycnometer method (Pobel, Gay-Lussac pycnometer with thermometer 10–35 °C/0.2 °C; frosted 10/19, 50 mL, Madrid, Spain) [39]; (iii) net pore space (POR), which was calculated from BD and PD values:

$$POR = [1 - (BD/PD)] \times 100; \quad (1)$$

(iv) infiltration rate-velocity ($I_{NF}V$), unsatured (K), which was tested utilizing an infiltrometer model Mini Disk Infiltrometer S (Pullman, WA, USA), where the K value (cm day^{-1}) was estimated according to Zhang [40], through cumulative infiltration measurements; (v) water holding capacity (WHC) was estimated according to Zagal et al. [41], placing samples into cones previously sealed at the bottom at 1:2 soil–water ratio for 12 h. Therefore, the tape was cautiously perforated in order to drain the formed solution while it was collected in plastic bottles and its volume was then measured; (iv) water-stable aggregates (WSA) (%) were estimated by the Kemper and Rosenau [42] method, which consisted of sieving (0.250 mm) each sample, then immersing them into an aluminum capsule containing distilled water for 3 min (at 35 rep min^{-1}) by using an Eijkelkamp Agrisearch Equipment (Giesbeek, The Netherlands), and the resulting dispersed soil was dried at 105°C, while the remaining soil was placed into another aluminum capsule containing sodium hexametaphosphate (2 g L^{-1}) for 15 min (35 rep min^{-1}), and the dispersed soil was dried at 105 °C. Finally, samples for both procedures were weighed to calculate their ratio/percentage in comparison with that of the original sample, and (vii) penetration resistance ($P_{EN}R$) was estimated using a penetrometer model Soil Compaction Tester Dickey–John (Auburn, IL, USA), where a total of 60 measurement points were established by longitudinal transects to ensure representativeness.

2.2.2. Chemical Trials

Chemical analyses were performed according to methods compiled by Sadzawka et al. [43] at the Agricultural Research Institute of Chile (INIA-Quilamapu, Chillán, Chile). After the air drying procedure and extraction of a <2 mm fraction from the samples, particular analytical methods were conducted as follows: pH was determined potentiometrically in water (pH-H_2O) a 1:2.5 ratio (HI 4211, Hanna Instruments, Smithfield, RI, USA); NH_4^+ and NO_3^- were analyzed by KCl extraction (2M) for 1 h, then filtered prior to determination by automated colorimetry by using an autoanalyzer (segmented flux spectrophotometer, Skalar, Breda, The Netherlands); P Olsen P was determined using 0.5 M $NaHCO_3$ at pH 8.5, and therefore estimated by the Murphy and Riley method and turbidimetry; K^+, Ca^{2+}, and Mg^{2+} by an extraction with 1 mol/L ammonium acetate solution at pH 7.0 and determination by spectrophotometry atomic absorption (969, Unicam, Ilminster, UK); effective cation exchange capacity (ECEC), which was determined by the sum of Ca+ Mg + K + Na+ Al; S by an extraction with dihydrogen solution calcium phosphate 0.01 mol/L and turbidimetric determination; and exchangeable aluminum (Al_{EXCH}) was analyzed by extraction with potassium chloride solution 1 mol/L and determination by spectrophotometry atomic absorption (969, Unicam, Ilminster, UK), whereas percent aluminum saturation (% Al_{SAT}) was determined by the following formula:

$$\% Al_{SAT} = (Al_{EXCH}/ECEC)*100 \quad (2)$$

where % Al_{SAT} is aluminum saturation, Al_{EXCH} is exchangeable aluminum and ECEC is effective cation exchange capacity.

The total N and SOC contents were determined by dry combustion (Leco CN-2000, macro-analyzer, LECO Corporation, Saint Joseph, MI, USA) [44] at the Soil and Natural Resources Laboratory (Faculty of Agronomy, University of Concepción).

In addition, SOC stock was estimated by

$$\text{SOC stock} = \text{H} \times \text{BD} \times \text{SOC} \times 100 \quad (3)$$

where H is the soil depth (cm); BD is bulk density (g cm^{-3}); and SOC is total SOC content (g kg^{-1}).

2.2.3. Molecular Microbiome Analysis

Samples consisting of 10 g of soil were taken from each site at 0–5 and 5–20 cm, sealed in plastic bags, and transported within coolers for later analysis. To analyze the microbial communities in the soil, we extracted DNA from each sample. DNA extraction isolates genetic material from bacteria and fungi, allowing us to identify their diversity and functions. We used a Macherey Nagel soil DNA extraction kit (Macherey Nagel Labs, Germany) according to the manufacturer's protocol, ensuring high-quality DNA for sequencing. The DNA was verified by electrophoresis in an agarose gel, according to Brody and Kern [45], and the concentration was evaluated using a Thermo Scientific NanoDrop™ 3300 Spectrophotometer. DNA samples were merged (from triplicate), lyophilized, and stored for further sequencing. To combine multiple primers from a single genomic DNA extraction, a Fluidigm technology technique was performed according Mallot et al. [46], in order to minimice PCR errors and cross. For that purpose, the following primers were selected: Bacteria F515 (5′-GTGYCAGCMGCCGCGGTAA-3′) and 806R (5′-GGACTACNVGGGTWTCTAAT-3′); Eukarya: Euk1391f (5′-GTACACACCGCCCGTC-3′) and EukBr (5′-TGATCCTTCTGCAGGTTCACCTAC-3′); and Fungi: ITS3 (5′-GGACTACVSGGGTATCTAAT-3′)–ITS4 (5′-TCCTCCGCTTATTGATATGC-3′).

Sequencing was performed using a single 2 × 250 Illumina Hiseq2000 run. Every single sample was analyzed with the dada2 package (v1.11.3) in R (v3.4.4). To ensure high-quality sequence data, a rigorous quality control process was performed following the standard Mothur pipeline. Low-quality bases and ambiguous sequences were filtered using the screen.seqs and filter.seqs commands, applying strict thresholds to remove sequencing artifacts. Chimeric sequences were identified and removed using chimera.vsearch, ensuring only high-confidence reads were retained for downstream analysis. Paired-end reads were merged using the make.contigs command, which aligns overlapping sequences based on quality scores to generate accurate consensus sequences. This step minimizes sequencing errors and ensures the reliability of the assembled reads before taxonomic classification and community analysis. Taxonomic identification was carried out using the GTDB database (release 202) for bacteria, the PR2v5 database for Eukarya, and the UNITE ITS database for fungi (v209). For the subsequent analyses, the microeco package in the R environment was employed. Specifically, we calculated the relative abundance of ASVs in each sample and conducted a community analysis, encompassing observed composition and alpha-diversity and beta-diversity indices. Following this, functional profiles were generated using a Functional Annotation of Prokaryotic Taxa (FAPROTAX) [47] for bacterial and fungal traits for fungi [48]. This tool facilitates the prediction of the functional roles of microorganisms based on their taxonomic information and offers insights into the functional potential of the microbial community. As environmental factors and microbial functions can be generally applied to explain microbial community structure and assembly, we correlated the diversity indices and functional potential with the environmental variables using Spearman's correlation method.

2.3. Statistical Analysis

To determine whether soil properties differed significantly between SEC_{FORST} and FAD_{IST}, we used statistical tests ('ARTool' package [49]) to compare their physical, chemical, and microbial characteristics. By applying non-parametric and correlation analyses, we identified patterns in how soil recovers after disturbance. In all cases, significant differences were considered with a $p < 0.05$ significance level. All statistical analyses and graphs were performed using RStudio Statistical Software (V4.1.0, the RCore Team 2021).

To compare microbial taxa, we used non-parametric statistical tests due to the non-normal distribution of abundance data. Specifically, taxa abundances were analyzed using the Kruskal–Wallis test, which allows for comparisons across multiple groups without assuming normality. When significant differences were found, pairwise comparisons were conducted using Dunn's test with Bonferroni correction to control for multiple testing. This approach ensures robust statistical inference while accounting for the distributional characteristics of microbial abundance data.

3. Results and Discussion

3.1. Physical Properties

Physical results are presented in Table 2, showing no significant differences ($p < 0.05$), except for $I_{NF}Vk$ (SEC_{FORST} > FAD_{IST}) and PEN_{RES} (FAD_{IST} > SEC_{FORST}), both presumably related to greater SOC contents and the associated biological processes in SEC_{FORST}. The $I_{NF}Vk$ averaged +4.6 cm d^{-1} in SEC_{FORST} condition, denoting possible differences in internal soil architecture by evolution over the time period of ~45 y since the last disruptive event.

Table 2. Results of physical characterization.

Property	FAD_{IST} 0–5	FAD_{IST} 5–20	SEC_{FORST} 0–5	SEC_{FORST} 5–20
^{A} $I_{NF}Vk$ *	13.05 ± 0.3 A	13.05 ± 0.3 A	18.1 ± 0.66 B	18.1 ± 0.66 B
WHC **	40.42 ± 4.35 Aa	38.05 ± 4.42 Aa	39.30 ± 1.68 Aa	34.86 ± 6.07 Aa
BD ***	0.63 ± 0.02 Aa	0.59 ± 0.02 Aa	0.57 ± 0.02 Aa	0.56 ± 0.01 Aa
PD ***	1.94 ± 0.01 Aa	1.92 ± 0.02 Aa	1.95 ± 0.03 Aa	1.93 ± 0.03 Aa
^{B} P_{OR} (%) **	67.19 ± 0.69 Aa	70.77 ± 1.18 Aa	69.59 ± 1.61 Aa	70.98 ± 0.89 Aa
WSA **	49.77 ± 0.36 Aa	49.80 ± 0.14 Aa	52.47 ± 4.77 Aa	50.83 ± 1.64 Aa
PEN_{RES} ****	350.0 ± 0.0 Aa	316.67 ± 57.7 Ab	250.0 ± 0.0 Bb	250.0 ± 0.0 Bb

^{A} Relative to total depth (0–20), ^{B} estimated through the Formula (1), *: (cm d^{-1}); **: (%); ***: (g cm^{-3}); ****: (PSI). Disparity in capital letters indicates significant differences ($p < 0.05$) between PEN_{RES} and FAD_{IST}, while in the case of lowercase letters, it refers to differences between depths.

With respect to PEN_{RES}, the high mechanical and animal loads in the recent past within FAD_{IST} have caused compaction exceeding critical levels for plant development (≥300 psi) [50], which can be evidenced in the area by bare soil patches and the presence of gullies. However, a decrease of 33.4 psi was observed in FAD_{IST} (0–5 cm) in relation to the 5–20 cm depth, suggesting a significant effect ($p < 0.05$) of bioturbation root developed to reverse compaction itself after 5 y, whilst the −100 psi difference between SEC_{FORST} and FAD_{IST} (0–5 cm) mirrors these bioactive effects.

However, a numeric positive trend was observed in the upper depth for the properties WHC, BD, and PD in both FAD_{IST} and SEC_{FORST} and P_{OR} and WSA in SEC_{FORST} and PEN_{RES} in FAD_{IST}, which is directly related to water retention–supply capacity. In relation to total depth (0–20 cm), we observed a net decrease in $I_{NF}Vk$ (26.52%), WSA (2.8%), PD (0.5%), and P_{OR} (2.4 %) (FAD_{IST} < SEC_{FORST}), contrary to BD (7.0), WHC (3.08), and PEN_{RES} (23.08 %) (FAD_{IST} > SEC_{FORST}).

The WSA decline in FAD_{IST} may also be influenced by the progressive depletion of Ca^{2+}/native lack in Andisols, which is considered the most relevant inorganic binding agent (e.g., clay–polyvalent cation–SOC aggregation mechanisms) [22,51].

Both BD and PD showed typical values for volcanic soils (BD ≤ 1 g cm^{-3}) [52], which is intrinsically linked to the physical high internal porosity of parent material (e.g., allophane) [31,53]. Consequently, the estimated P_{OR} values (Equation (1)) clearly exceeded the typical values (>40) [52], which is in compliance with the high WHC observed, globally promoting water cycling–bioavailability [54].

3.2. Chemical Properties

Chemical properties resume in Table 3, showing significant variations ($p < 0.05$) only between SOC and Al_{SAT} among FAD_{IST}-SEC_{FORST} and depths comparisons. However, other significant differences were detected between the comparisons as follows: (i) FAD_{IST}-SEC_{FORST}:—pH, N, and Al_{EXCH} varied; (ii) similar depths—Mg^{2+} and ECEC; (iii) FAD_{IST} (0–5) and SEC_{FORST} (5–20) depths—NO_3^- and Ca^{2++}, (iv) depths in FAD_{IST} with respect to the SEC_{FORST} condition (which did not vary between depths K and S, (v) 0–5 depths of both FAD_{IST} and SEC_{FORST}—P^+ and (vi), FAD_{IST}-SEC_{FORST}, or depths: C:N and NH_4^+. In addition, as occurred with the physical properties, better chemical-fertility conditions were generally observed for the 0–5 cm depth, except for Al_{EXCH}, Al_{SAT}, N (only for FAD_{IST}), and S (only for SEC_{FORST}).

Table 3. Soil chemical properties.

Property	Condition			
	FAD_{IST} 0–5	FAD_{IST} 5–20	SEC_{FORST} 0–5	SEC_{FORST} 5–20
pH	6.19 ± 0.05 Aa	6.15 ± 0.05 Aa	5.9 ± 0.06 Bb	5.86 ± 0.07 Bb
SOC *	10.10 ± 0.25 Aa	8.47 ± 0.66 Bb	11.03 ± 1.01 Cc	10.46 ± 0.96 Dd
N *	0.73 ± 0.06 Aa	0.63 ± 0.06 Aa	0.77 ± 0.06 Bb	0.77 ± 0.06 Bb
C:N	17.61 ± 1.34 Aa	16.71 ± 1.64 Aa	17.01 ± 1.67 Aa	16.39 ± 2.55 Aa
NH_4^+ **	5.94 ± 0.32 Aa	5.37 ± 0.66 Aa	6.11 ± 0.27 Aa	5.36 ± 0.41 Aa
NO_3^- **	5.4 ± 0.18 Aa	4.43 ± 0.39 Bb	3.72 ± 0.43 Bb	3.02 ± 0.28 Cc
P^+ **	2.25 ± 0.29 Ab	1.91 ± 0.31 Aa	2.62 ± 0.11 Bb	1.91 ± 0.24 Aa
K **	0.34 ± 0.06 Aa	0.19 ± 0.03 Bb	0.71 ± 0.05 Cc	0.59 ± 0.02 Cc
Ca^{2+} **	8.06 ± 2.11 Aa	3.57 ± 0.91 Bb	3.78 ± 0.23 Bb	2.05 ± 0.61 Cc
Mg^{2+} **	0.84 ± 0.07 Aa	0.30 ± 0.07 Bb	0.68 ± 0.09 Aa	0.33 ± 0.02 Bb
S **	2.83 ± 0.71 Aa	5.86 ± 1.14 Bb	12.70 ± 0.44 Cc	11.97 ± 0.60 Cc
ECEC ***	5.92 ± 0.73 Aa	4.44 ± 0.68 Bb	6.40 ± 0.80 Aa	2.81 ± 0.29 Bb
Al_{EXCH} ***	0.09 ± 0.05 Aa	0.08 ± 0.02 Aa	0.18 ± 0.03 Bb	0.22 ± 0.02 Bb
^{A} Al_{SAT} *	1.52 ± 0.07 Aa	1.80 ± 0.15 Bb	0.43 ± 0.60 Cc	7.83 ± 0.26 Dd

*: %; **: mg kg^{-1}; ***: cmol (+) kg^{-1}). Disparity in capital letters indicates significant differences ($p < 0.05$) between PEN_{RES} and FAD_{IST}, while in the case of lowercase letters, it refers to differences between depths. ^{A} Estimated through the Formula (2).

When comparing the main percentage variations from significantly different properties ($p < 0.05$) between FAD_{IST} and SEC_{FORST} (0–20 depth), the following was estimated: NO_3^- (+31.17% in FAD_{IST}), K^+ (+62.91% in SEC_{FORST}), Ca^{2+} (+46.33% in FAD_{IST}), S (+58.03% in SEC_{FORST}), ECEC (+22.87% in FAD_{IST}), and potential toxicity with Al_{EXCH} (+61.91% in SEC_{FORST}) and Al_{SAT} (+83.65% in SEC_{FORST}).

The implications and effects of the concentrations observed are described as follows: (i) pH values are adequate for plant development in all cases [55] and the ranges observed (5.86 to 6.1) could be attributable to the high precipitation local regime ($\geq$3000 mm y^{-1}), native basic poor volcanic materials, and desilication processes [31], and the more acidic values in SEC_{FORST} are due to slight shifts in precipitation patterns (north > south) [54,56]; (ii) the high SOC concentrations may be due to the particular mechanisms of SOC preservation in volcanic soils described in Section 1 (e.g., Al-SOC complexation, extracellular enzymes sorption, occlusion of organics) [7,17,31,32,57,58]. However, specific differences in SOC contents cycling between FAD_{IST} and SEC_{FORST} can be related to the possible influence of *Chusquea* sp., contributing to more labile SOC and rich lignin substrates exerted by the high tree density for SEC_{FORST} [59,60], and the presence of pyrogenic C in FAD_{IST} as a remnant of site historical use, which was evidenced by charcoal fragments, which has been accounted up to 5% of total SOC in national forests [61], and these materials are also able to neutralize pH values due to their diverse functional (e.g., COO–, $CaCO_3$, PO_4^{3-}) [53]; (iii) the N differences may be explained by the massive presence of *Chusquea* sp., leading to an increase in total-available N coupled with N-NO_3^- ratios, our results are in concordance with Pérez et al. (2019) [59], who studied a *Laureliopsis philippiana* (*Monimiaceae*) forest in Chiloé, Chile, which was logged for 10 y of having understory strata dominated by *Chusquea* sp. However, N uptake by *Nothofagus* individuals readily counteracted this effect, which could also explain the lower NO_3^- compared to FAD_{IST}, which in turn had a less dense understory. The lower numerical NH_4^+ levels in FAD_{IST} indicate a greater demand by vigorous understory herbaceous strata, which uptake this bioavailable N species due to the minor uptake energetic cost compared to NO_3^- [62]. Regarding C:N ratios, numerical differences may be related to limiting N bioavailability and the presence of recalcitrant C and low substrate quality; (iv) the low critical P levels observed under all criteria correspond to typical volcanic soils due to the high PO_4^3 fixation of at least 70% [63,64]; however, the significant differences ($p < 0.05$) between 0 and 5 cm depths may be related to greater nutrient cycling in SEC_{FORST}; and (v) the S values were under adequate levels under all criteria; however, the $FAD_{IST} < SEC_{FORST}$ pattern observed, near to the lower fertile limit (<1 mg kg^{-1}) [65], suggests a deeper and well-established root system by both *Nothofagus obliqua* and *Chusquea* sp. in SEC_{FORST} and a greater uptake-mining by root systems in FAD_{IST}, along with a possible SO_4^2-retention/leaching processes exerted by charcoal fragments having anion exchange capacity [66], and percolation by competition with NO_3^- for interchange sites [67].

(vi) Regarding basic cations, the critical deficiencies of K, Ca^{2++}, and Mg^{2+} could be related to three main mechanisms: (a) soil solution percolating due to high local precipitation regime and suitable native hydraulic–structural properties [31,54], (b) uptaking for overall metabolic processes, (c) desilication, leading to a progressive lowering on cation concentrations [17], and/or (d) higher soil organic matter content and nutrient cycling [55], as in the case of K ($SEC_{FORST} > FAD_{IST}$), where, however, the higher Ca^{2+} values in FAD_{IST} may be related to the high specific area of the present pyrogenic materials (50–≥500 m^2 g^{-1}) [68]; (vii) the ECEC values ranged from moderate in all criteria, except for SEC_{FORST} (0–5 cm) corresponding to an adequate level (>6.27 cmol kg^{-1}) [65], which could be related to the more developed deep root systems in SEC_{FORST} of both *Chusquea* sp. and *Nothofagus obliqua* mining nutrients to depths greater than 5 cm, whether they can compete for nutrients or not; and (viii) the variables Al_{EXCH} and Al_{SAT}, expressing potential toxic conditions for plants, showed adequate values in FAD_{IST}, indicating the effect of liming practices along the total depth studied (0–20 cm), whereas moderate Al_{EXCH} and moderate Al_{SAT} levels were determined in all conditions (except for SEC_{FORST} 5–20 cm depth, critical). However, in all cases, Al_{EXCH} varied significantly between the depths of both FAD_{IST} and SEC_{FORST}, expressing a lesser capacity at lower depths to form Al–organic complexes due to the decreasing SOC content [69]. Finally, the higher absolute values observed at depth 0–5 (except for S in FAD_{IST} and desirably in the Al-related variables), suggest a propensity for the development/thickening of A horizon–SOC stock enlargement and the associated direct effects in terms of fertility status and ecosystem services provision.

3.3. *Soil Organic C Sequestration* ($CO_q \rightarrow SOC$)

According to earlier research in areas near our study site (Dube, pers. Commun.) [34], temporal variations in C content and stocks could be estimated at total depth (0–20 cm) and the relative temporal increment or decrease rates (Table 4). The carbon concentration and SOC stock were greater at SEC_{FORST}; however, larger CO_2→SOC was estimated at FAD_{IST}. Total variations on SOC stocks from the 6 y time lapse were 26.1 and 15.96 Mg SOC ha^{-1} for FAD_{IST} and SEC_{FORST}, respectively.

Table 4. Temporal variation in SOC concentration, SOC stocks, and CO_2→SOC.

	FAD_{IST}	SEC_{FORST}
Previous SOC (%)	8.2 *	10.1 *
SOC_{2020}	8.9	10.6
^{A} Previous SOC stock (Mg ha^{-1})	81.18 *	103.53 *
ASOC stock 2020 (Mg ha^{-1})	106.53	120.07
Theoretical annual CO_2→SOC	4.23	2.75

The data listed correspond to the weighted values for the depths 0–5 and 5–20 as follows: * for the year 2014 (source: Dube, pers. Commun.) [34], as the nearest referential records to our study conditions. ^{A} Estimated through the Formula (3).

These positive differences on SOC stocks reaching +21.3 and +16.5 Mg ha^{-1} y^{-1}, (2014–2020 period) for FAD_{IST} and SEC_{FORST}, respectively, resulted in the CO_2→SOC potentials expressed in Table 4. In consequence, a +1.48 Mg SOC ha y^{-1} was estimated for FAD_{IST}, compared to SEC_{FORST}, despite similar SOC % concentrations. This can be explained by the following: In SEC_{FORST}, the particular role of *Chusquea* sp. on soil fertility is of major relevance, since its chemical characteristics result in up to 67.2% of alpha cellulose + hemicellulose and only 13.7% of lignin content, reaching a height of 2.0 m and a density of 290 kg m^3 [70], implying a potential massive input of labile SOC (glucose-based), facilitating its rapid mineralization, not only by their chemical nature per se, but also by their effect in reactivating nutrient mining processes by decomposers of the anterior organic matter (priming effect), resulting in limited CO_2→SOC rates [20].

In the particular scenario of FAD_{IST}, and despite the significantly lower tree density, this condition showed higher CO_2→SOC capacity, which can be related to the following factors: (i) C inputs derived from the debris of emergent vegetation (understory and herbaceous strata), (ii) presumably higher SOC–metal (Fe^{3+}, Al^{3+}) complexation potentials than SEC_{FORST}, (iii) an appreciable enrichment of energy-rich reduced SOC residues (aromatics) derived from previous ignition events and practices (wildfires and agricultural slash-and-burn) resulting in the most stable recalcitrant SOC pool (centennial–millennial mean residence time), and (iv) because it is located in an area of accumulation (Section 2.1).

These findings are comparable with data reported by Ortiz et al. [71], who estimated annual SOC increases of 7.5, 4.8, and 1.6 Mg ha^{-1} y^{-1} in nearby silvopastoral systems based on *Nothofagus obliqua* to our study sites at 3 different tree densities of 134, 60, and 258 stems ha^{-1}, respectively. Similarly, Ortiz et al. [72] reported CO_2→SOC of 5.50 and 5.48 Mg C y^{-1} in two conterminous agroforestry systems corresponding to silvopasture under *N. obliqua* and *N. dombeyi* of a tree density of 173.3 stems ha^{-1} (133.3:40, respectively) and an agroforest growing oats (*Avena sativa*)-vetch (*Vicia atropurpurea*) associations below *N. obliqua* at a tree density of 446.6 stems ha^{-1}. The authors concluded that these systems promote nutrient cycling, SOC accumulation, and a progressive desirable increase/decrease in physical and chemical properties, except those that are closely related to the intrinsic characteristics of volcanic soils (for example, P) and at the ecosystem level (e.g., total and bioavailable N). In addition, Muñoz et al. [73] estimated SOC increases of 33.1 to 35.5 Mg ha^{-1} in a monoculture agro-system under no-tillage after 16 y in the same soil series as our study sites. In different managed forests (e.g., plantations) developing over Ultisols from Sothern China, Li et al. [74] did not find variations in SOC associated with minerals, but critical changes in SOC related to stable aggregates. The above suggests that well-planned and operational productive management may be arguably better alternatives for land reclamation than the mere cessation of human activity in degraded native forests of south-central Chile, not only as a better $CO_2 \rightarrow SOC$ option, but in order to control the occurrence and dominance of opportunistic species that represent a potential threat, such as (i) underbrush for farming proposes, (ii) a habitat of *Oligoryzomys longicaudatus*, a vector of the hantavirus *Bunyaviridae* of major medical importance, (iii) a causal agent of wildfires during summer seasons as considerable dry biomass aboveground stores, and (iv) impeding the re-consolidation of the previous understory and therefore of the original ecosystem biocomplexity and/or long-term sustainable management of forests such as agroforestry (for example, [35,75–78]) and planted forests [79].

3.4. Interactions Among the Analyzed Soil Properties

A total of 93 relations ($r \geq 0.80$) were observed among multiple variables (including inverse), of which 37, 11, and 45 occurred between chemical (4 inverse), physical (5 inverse), and 45 involving physicochemical properties (12 inverse), respectively (Figure 3). pH was the most influential property correlated with the other nine, explaining, for instance, their inverse relationship with Al_{EXCH} and Al_{SAT} (r = 98 and r = 89, respectively). SOC was correlated with N availability (r = 98) and exerted an influence on other nutrients (for example, P^+, K, Mg^{2+}); however, it did not meet the $r \geq 0.80$ criteria.

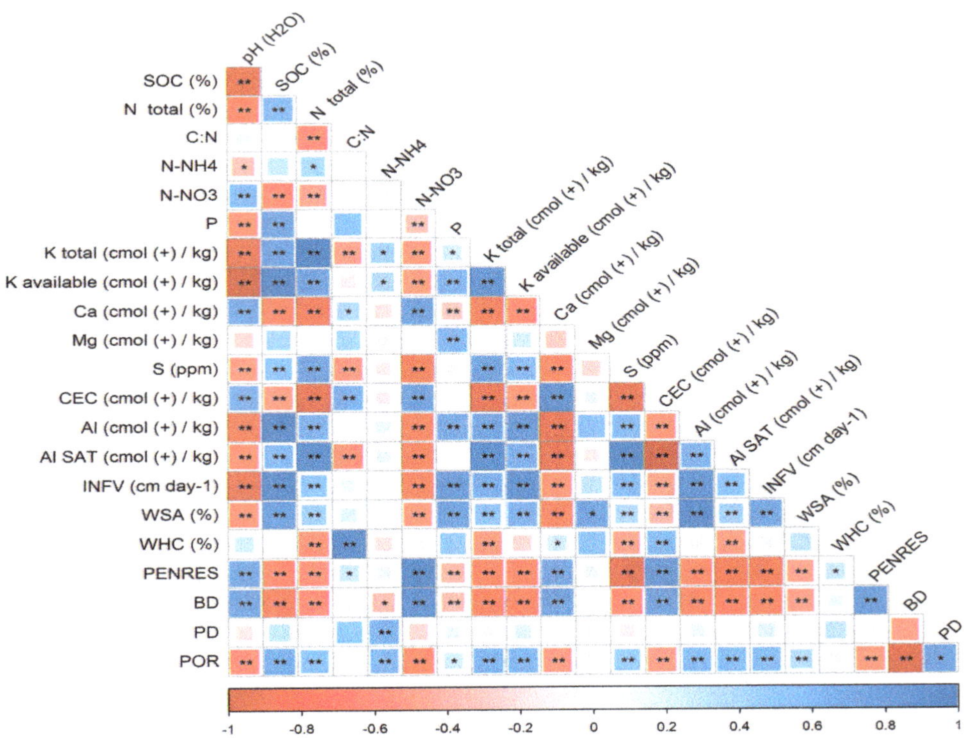

Figure 3. Heat map illustrating Spearman's correlation coefficients among the evaluated physical and chemical properties. The symbols * and ** represent *p*-values below 0.05 and 0.01, respectively. Reddish tones correspond to negative correlations, blue tones refer to positive correlations, and color intensity represents levels of correlation.

The C:N ratio clearly influenced both N occurrence and availability (r = 97, r = 75, and r = 86) for total N, NH_4^+, and NO_3^-, respectively, as well as the nutrient parameters Ca^{2+}, Mg^{2+}, and ECEC. Interactions between physical properties, including $I_{NF}Vk$, were well correlated with P_{OR} and WSA (r = 98 and 85), respectively, and inversely correlated with BD (r = −98) and PEN_{RES} (r = −96); P_{OR} was correlated with WSA and PEN_{RES} (r = 98 and r = 85), and WSA had an inverse correlation with PEN_{RES} (r = −82). With respect to physicochemical interactions, WHC was well correlated with the C:N ratio (r = 92) and nutrient levels (r = 83, r = 81, r = 95) for Ca^{2+}, Mg^{2+}, and ECEC, respectively.

Our physical results are consistent with those reported by Ortiz et al. [71], who reported in conterminous plots under silvopastoral management correlations of ($r \geq 0.8$) for %WSA, POR, and INFV, indicating suitable hydraulic conditions in these soils. In relation to chemical properties, our results are contrary to those reported by Ortiz et al. [71] and Ortiz et al. [72] (both under agroforestry), where SOC was the core property, which correlated to nutrient availability; in our case, the presumably higher pH values tend to drive such a role.

3.5. Soil Microbial Communities

The metabarcoding results showed that changes in soil microbial communities in degraded (FAD) and non-degraded (SEC) soils at two depths (20 and 5 cm) revealed distinct patterns in the relative abundance and diversity of microbial classes (Figure 4), affecting both prokaryotic and eukaryotic microorganisms. However, the eukaryotic 18S rRNA sequences obtained from the study fields were predominantly composed of fungi. Therefore, we focused on the analysis of bacterial communities based on 16S rRNA sequences, and fungal communities based on ITS-specific sequences.

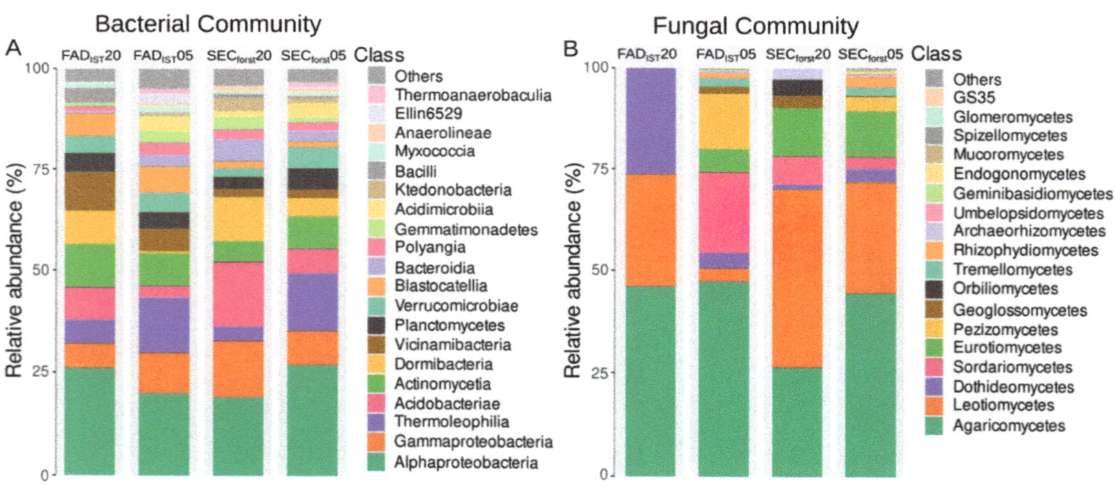

Figure 4. Composition of microbial communities in degraded and non-degraded soils at different depths. (**A**) Bacterial community and (**B**) fungal community. Bars represent the relative abundance (%) of different microbial classes in degraded soils at 20 cm ($FAD_{ist}20$) and 5 cm ($FAD_{ist}5$) depths, and in non-degraded soils at 20 cm ($SEC_{forst}20$) and 5 cm ($SEC_{forst}5$) depths. Different microbial classes are indicated by specific colors, as shown in the legend. Differences in the abundance and diversity of microbial classes reflect the influence of both soil degradation and depth.

For bacterial communities, the data showed that Alphaproteobacteria and Gammaproteobacteria were dominant in bacterial communities across all samples, with notable variations between depths and soil types. In deeper soils (20 cm), there was a higher relative abundance of Acidobacteria, especially in non-degraded soils (SEC). Conversely, degraded soils (FAD) exhibited a higher presence of Bacilli and Thermoleophilia, indicating a potential adaptation to altered soil conditions due to degradation. Several authors have reported that the presence of Bacilli, known for their ability to form endospores, can help them survive harsh conditions, including soil degradation [80,81]. Thermoleophilia, a class of bacteria within the phylum Actinobacteria, is typically found in hot and nutrient-poor environments, indicating its adaptability to stress conditions [82]. The bar plots (Figure 4A) showed that the dominant class in both SEC_{FORST} and FAD_{IST} was Alphaproteobacteria, followed by Gammaproteobacteria, which was more represented in SEC_{Frost} 5–20. Thermoleophilia was particularly abundant at 0–5 m in both FAD_{IST} and SEC_{FORST} (13.6 and 14.2%, respectively), while Acidobacterieae was more abundant at depth, representing 7.7% of the community at FAD_{ist} and 15.7% at SEC_{Frost}, a similar distribution pattern was observed for Dormibacteria. Viciniamibacteria was more abundant at the FAD_{ist} site at both depths.

When analyzing the bacterial community structure, we observed that the representation of *Dormibacteriaceae* and *Steroidobacteriaceae* negatively correlates with the Mg^{2+} concentration, while other families as *Solirubrobacteriaceae* strongly correlated with PD, P, and C:N. *Pyrimonadaceae* was positively correlated with pH but negatively correlated with total S, Al, INFV, and K. Finally, UBA2999 negatively correlates with total N (Figure A1).

The bacterial community composition displayed a similar representation as those reported by Navarrete et al. [21], who observed within the bacterial community a prevailing occurrence of Proteobacteria (45.35 ± 0.89%), Acidobacteria (20.73 ± 1.48%), Actinobacteria (12.59 ± 0.34%), and Bacteroidetes (7.32 ± 0.36%). Among Proteobacteria, the most abundant Alphaproteobacteria corresponded to Bradyrhizobium. Bacteria of this genus penetrate the roots of many legume species, producing root nodules in which N fixation occurs. Symbiotic N fixation is most important in forestry, where trees are closely associated with many wild, herbaceous legumes [83]. It has also been demonstrated that soil pH and soil C:N ratios significantly vary with soil microbial composition [84]. In our work, *Pyrinomonadaceae* was strongly correlated with pH, while *Solirubrobacteraceae* was strongly correlated with C:N, as well as with total P. Both pH and C:N were heavily influenced by symbiotic N_2 fixation. Recent research has revealed a link between the phosphate solubilization capacity and increased potential for polysaccharide hydrolysis and carbohydrate metabolism. This unique link strongly corresponds to the positive relationship between the population density of PSB and available dissolved organic carbon in the soil [85]. In our work, Pyrinomonadaceae strongly correlated with pH, while Solirubrobacteraceae strongly correlated with C:N, but also with total P. Both the pH and the C:N are heavily influenced by symbiotic N2 fixation. Recent research has revealed a link between the phosphate solubilization capacity and increased potential for polysaccharide hydrolysis and carbohydrate metabolism (Figure A1). This unique link corresponds strongly to the positive relationship between the population density of PSB and available dissolved SOC [85].

For fungal communities, Figure 4B displays a significant shift in composition between depths and soil types. The phyla Ascomycota and Basidiomycota were dominant in both FAD_{IST} and SEC_{FORST}. The topsoil layer (0–5 cm) was dominated by Ascomycota (49% and 50% of the fungal community) and Basidiomycota (50–49% of the fungal community). In the second soil layer (5–20 cm), both FAD_{IST} and SEC_{FORST} were also dominated by Ascomycota, accounting for 54% and 74% of the fungal community, respectively, along with Basidiomycota (45% and 25% of the fungal community, respectively). In non-degraded soils, classes such as Agaricomycetes and Leoiomycetes were more prevalent, suggesting a rich saprotrophic community. In degraded soils, there was a marked increase in Dothideomycetes in the second soil layer (5–20 cm), which could be associated with altered organic matter composition and nutrient availability. Interestingly, the soils of the second layer exhibited a lower diversity of fungi than the non-degraded soils. This could indicate the influence of agronomic management practices and the negative effects of degradation on soil nutrient cycling [86–88].

The functional analysis using FAPROTAX (Figure 5) highlights key in silico processes occurring in these soils. Soil depth and degradation status significantly influence the structure and functional potential of microbial communities. Non-degraded soils, particularly at greater depths, support more diverse and functionally rich microbial communities, which are crucial for maintaining soil health and ecosystem stability. Degraded soils, on the other hand, exhibit a shift towards less diverse and more pathogenic microbial communities, along with a potential reduction in key nutrient cycling processes. This study underscores the importance of soil management practices in preserving soil microbial diversity and functionality, which are essential for sustainable ecosystem services. On the other hand, non-degraded soils exhibit a higher diversity of saprotrophic fungi, which play a crucial role in decomposing complex organic matter. This is consistent with the higher organic matter content typically found in non-degraded soils. However, degraded soils have a higher representation of fungal pathogens and mycorrhizal fungi, possibly reflecting a stressed environment in which plants are more susceptible to diseases and rely heavily on symbiotic relationships for nutrient acquisition.

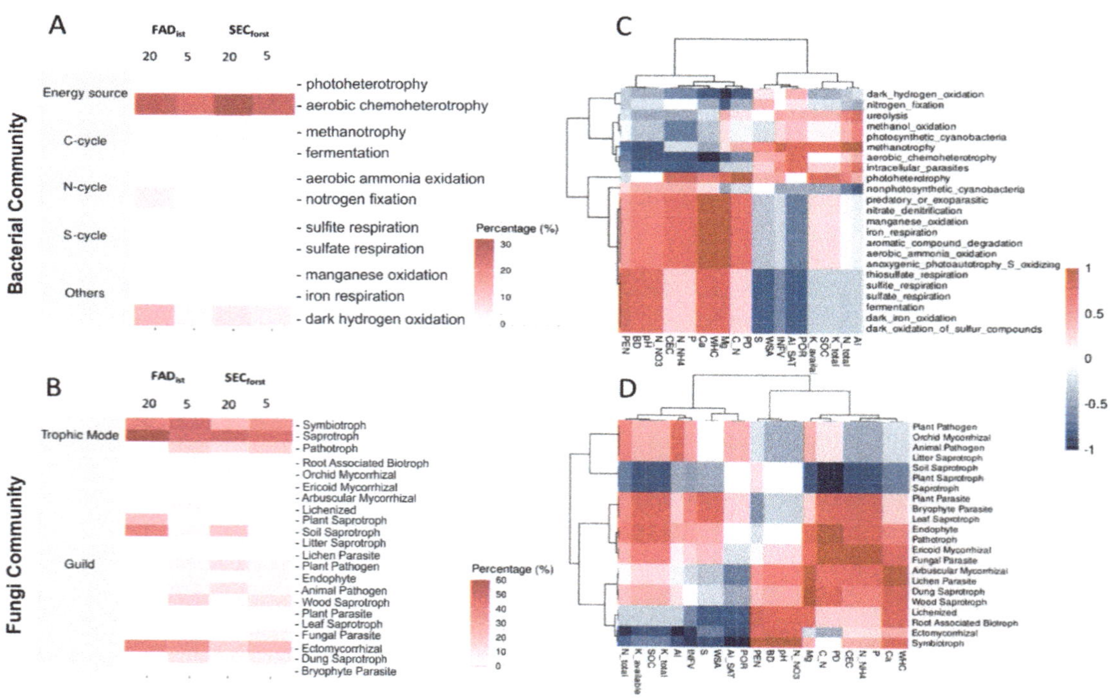

Figure 5. The figure shows a comparison of bacterial and fungal communities across soil samples with different levels of organic management (FAD_{IST} and SEC_{FORST}). Panels (**A**,**B**) display the distribution of bacterial and fungal communities, respectively, categorized by their energy sources, biogeochemical cycles, trophic modes, and guilds, with color intensity reflecting the percentage of each functional group. Panels (**C**,**D**) present heatmaps illustrating the correlations between microbial community functions (bacterial and fungal) and soil characteristics, with color gradients indicating the strength and direction of these correlations (red for positive and blue for negative). These analyses highlight how varying soil management practices influence the composition and functional dynamics of microbial communities in agricultural soils.

We also observed a strong partitioning of the functions related to the soil properties (Figure 5C,D, Appendix A). The sulfur cycle was strongly represented when PEN_{RES}, BD, pH, and NO_3^- were higher. In addition, N-related processes, such as denitrification and aerobic ammonia oxidation, were more represented when the concentrations of Ca, Mg, and WHC were higher. On the other hand, methanol oxidation, metanothrophy, ureolysis, and photoheterotrophy were more represented when available K, K total, and SOC were higher.

In bacterial communities, deeper, non-degraded soils showed enhanced capacities for nitrogen fixation and sulfur cycle processes, such as sulfite respiration and sulfate reduction. This indicates a robust biogeochemical cycling capacity, possibly supporting a more stable ecosystem [89]. Degraded soils, particularly at 5 cm depth, show increased potential for fermentation and aerobic chemoheterotrophy, suggesting a shift towards more anaerobic and less efficient energy-yielding processes, likely due to soil compaction and reduced oxygen availability. Fungal functional profiles (Figure 5D) also differ significantly between soil types and depths. The aerobic chemoheterothrophy was the most represented energy source. The N fixation appears to be more represented at the FAD_{IST}, as well as the dark hydrogen oxidation. Other metabolisms represented corresponded to metanothrophy sulfite respiration and sulfate oxidation, iron respiration, and aerobic ammonia oxidation.

Microbial communities play a fundamental role in soil health, influencing nutrient cycling, organic matter decomposition, and overall ecosystem resilience [2,90]. The observed differences in bacterial and fungal composition between FAD_{IST} and SEC_{FORST} soils suggest that post-disturbance microbial dynamics may drive long-term soil recovery [91]. The dominance of Alphaproteobacteria and Gammaproteobacteria, known for their roles in organic matter decomposition and nitrogen cycling, indicates potential microbial contributions to nutrient turnover [92,93]. Conversely, the higher presence of Acidobacteria in deeper soils suggests microbial adaptation to resource-limited conditions, which may affect carbon stabilization processes over time [94].

Additionally, the shifts in fungal communities, particularly the balance between Ascomycota and Basidiomycota, highlight potential changes in decomposition pathways that could influence soil organic matter persistence [95]. These microbial trends align with broader ecological resilience frameworks, where community stability and functional redundancy are key indicators of a system's ability to recover from disturbance [96]. The functional analysis further supports this perspective, as degraded soils showed increased microbial metabolic activity related to stress adaptation, while non-degraded soils maintained a more diverse functional profile, reinforcing their potential for long-term stability [97]. Understanding these microbial shifts is critical for assessing the trajectory of soil health in post-disturbance ecosystems. Future studies should further explore the long-term implications of these microbial dynamics in soil carbon storage and nutrient cycling, providing a more comprehensive view of their ecological impact [98].

Further research is necessary in terms of similar tests on different biomes-passive management, temporal variations in tested properties in other secondary *Nothofagus* sp. forests, as well as greater depths of present studies sites in order to elucidate the effects of both systems on the soil profile, as well as the variations in their physical, chemical, and biological properties, taking into account that this study seeks to be an approximation to these conditions widely distributed at national level.

4. Conclusions

Our study highlights that post-disturbance passive management of native *Nothofagus* forests in south-central Chile had similar effects on nutrient status, physical aptitudes for water cycling–plant utilization, a decline of soil carbon sequestration rates (1.48 Mg SOC ha^{-1} y^{-1}), and shifts only on abundance of soil microbiome for a ~45 y period. Such edaphic proximity could be attributed to three concurrent factors, the dominance of *Chusquea* sp., which is a widespread regional opportunistic genre, able to inhibit understory biocomplexity, transitioning SEC_{FORST} into to a more dynamic SOC system, the SOC and nutritional residual effects of pyrogenic materias and recent agricultural activities in FAD_{IST} and the overall influence of disctintive mineral fraction of volcanic soils on nutrient availability-limitations and hydraulic properties. In addition, these soils exhibit remarkable resilience traits, since a generalized upper desirable level (at 0–5 cm depth) of most variables was analyzed, despite being statistically significant ($p < 0.05$), a trend also observed in the SEC_{FORST} > FAD_{IST} comparison. Consequently, an a priori scarification processes that promote ecosystem recovery (>25% of surface area) is highly recommended, because, although *Chusquea* sp. promotes soil protection (e.g., soil cover, pedogenesis), it impedes ecosystem recovery by setting severe limitations to original regeneration patterns. In the case of productive management, agroforestry techniques are very suitable, since in a direct comparison between natural regeneration vs. agroforestry systems established under similar conditions, the latter also has great social and ecological advantages, potentially serving as ecotones and/or major hillside production systems. Knowledge gaps involve seasonal and forest evolution in relation to robust metagenomics and belowground carbon dynamics studies (including pyrogenic) coupled with *Chusquea* sp. occurrence in order to discriminate the effective C stabilization ($CO_2 \rightarrow SOC$).

Author Contributions: J.O.: conceptualization, investigation, methodology, data curation, software, validation, visualization, writing—original draft, and writing—review and editing; M.P.: conceptualization, methodology, validation, data curation, and funding acquisition; P.N.: methodology, data curation, formal analysis, software, validation, visualization, and writing; C.H.-C.: methodology, data curation, formal analysis, software, validation, visualization, writing—review and editing, and funding acquisition; R.E.G.J.: investigation, validation, visualization, and writing—review and editing; R.R.: data curation, formal analysis, methodology, and writing—review and editing; A.M.: data curation, methodology, software, validation, and visualization; C.R.: investigation, data curation, methodology, and validation; W.E.: methodology, formal analysis, visualization, writing—review and editing; R.P.-K.: visualization, data curation, and formal analysis; E.Z.: validation, writing—review and editing, project administration, and funding acquisition; N.S.: conceptualization, supervision, validation, writing—review and editing, and project administration; M.S. (Mauricio Schoebitz): formal analysis, methodology, and writing—review and editing; M.S. (Marco Sandoval): data curation, formal analysis, methodology, validation, and visualization; F.D. (Francis Dube): conceptualization, writing—review and editing, project administration, and funding acquisition. All authors have read and agreed to the published version of the manuscript.

Funding: This research was funded by Concurso VRID Multidiscliplinaria (UdeC), no. 219.142.040-M and CONAF (Fondo de Investigación del Bosque Nativo), no. 001/2014.

Data Availability Statement: Data are contained within the article.

Acknowledgments: We would like to express our sincere thanks to the staff of the Ranchillo Alto Research Forest, administered by the Faculty of Forest Sciences at the University of Concepción. We thank the Agroecology laboratory of INIA (Quilamapu) and, in particular, Cecilia Céspedes, for the support provided for this research, as well as the Soil and Environment Research Laboratory of the Faculty of Agronomy at the University of Concepción and, in particular, Katherine Rebolledo, for her valuable collaboration.

Conflicts of Interest: The authors declare that they have no known competing financial interests or personal relationships that could have appeared to influence the work reported in this paper.

Appendix A

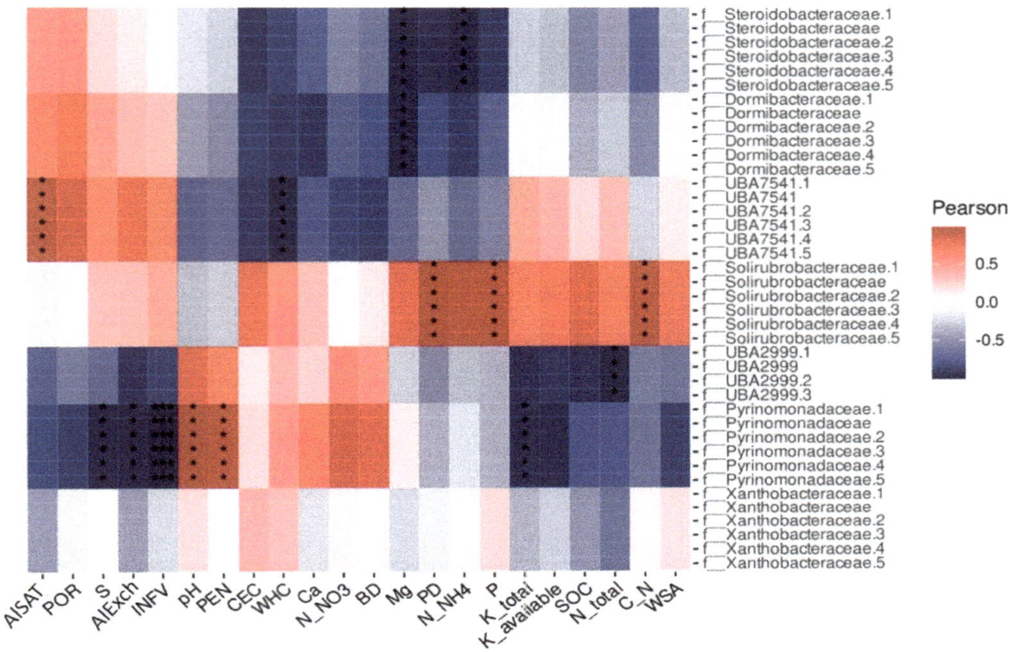

Figure A1. Relationship between bacterial orders and soil properties. The symbols * and *** represent p-values below 0.05 and 0.01, respectively.

References

1. Chaplin-Kramer, R.; Neugarten, R.A.; Sharp, R.P.; Collins, P.M.; Polasky, S.; Hole, D.; Schuster, R.; Strimas-Mackey, M.; Mulligan, M.; Brandon, C.; et al. Mapping the planet's critical natural assets. *Nat. Ecol. Evol.* **2022**, *7*, 51–61. [CrossRef] [PubMed]
2. van Bruggen, A.H.C.; Semenov, A.M. In search of biological indicators for soil health and disease suppression. *Appl. Soil Ecol.* **2000**, *15*, 13–24. [CrossRef]
3. IPCC. *Climate Change 2014: Mitigation of Climate Change. Working Group III Contribution to the Fifth Assessment Report of the Intergovernmental Panel on Climate Change*; Edenhofer, O., Pichs-Madruga, R., Sokona, Y., Farahani, E., Kadner, S., Seyboth, K., Adler, A., Baum, I., Brunner, S., Eickemeier, P., et al., Eds.; Cambridge University Press: Cambridge, UK; New York, NY, USA, 2014.
4. Sünnemann, M.; Beugnon, R.; Breitkreuz, C.; Buscot, F.; Cesarz, S.; Jones, A.; Lehmann, A.; Lochner, A.; Orgiazzi, A.; Reitz, T.; et al. Climate change and cropland management compromise soil integrity and multifunctionality. *Commun. Earth Environ.* **2023**, *4*, 394. [CrossRef]
5. Don, A.; Seidel, F.; Leifeld, J.; Kätterer, T.; Martin, M.; Pellerin, S.; Emde, D.; Seitz, D.; Chenu, C. Carbon sequestration in soils and climate change mitigation—Definitions and pitfalls. *Glob. Change Biol.* **2024**, *30*, e16983. [CrossRef] [PubMed]
6. Calvin, K.; Dasgupta, D.; Krinner, G.; Mukherji, A.; Thorne, P.W.; Trisos, C.; Romero, J.; Aldunce, P.; Barrett, K.; Blanco, G.; et al. *IPCC, 2023: Climate Change 2023: Synthesis Report. Contribution of Working Groups I, II and III to the Sixth Assessment Report of the Intergovernmental Panel on Climate Change*, 1st ed.; Core Writing Team, Lee, H., Romero, J., Eds.; IPCC: Geneva, Switzerland, 2023. Available online: https://www.ipcc.ch/report/ar6/syr/ (accessed on 12 March 2024).
7. Ren, S.; Wang, T.; Guenet, B.; Liu, D.; Cao, Y.; Ding, J.; Smith, P.; Piao, S. Projected soil carbon loss with warming in constrained Earth system models. *Nat. Commun.* **2024**, *15*, 102. [CrossRef] [PubMed]
8. Georgiou, K.; Jackson, R.B.; Vindušková, O.; Abramoff, R.Z.; Ahlström, A.; Feng, W.; Harden, J.W.; Pellegrini, A.F.A.; Polley, H.W.; Soong, J.L.; et al. Global stocks and capacity of mineral-associated soil organic carbon. *Nat. Commun.* **2022**, *13*, 3797. [CrossRef]
9. Bramble, D.S.E.; Ulrich, S.; Schöning, I.; Mikutta, R.; Brandt, L.; Poll, C.; Kandeler, E.; Mikutta, C.; Konrad, A.; Siemens, J.; et al. Formation of mineral-associated organic matter in temperate soils is primarily controlled by mineral type and modified by land use and management intensity. *Glob. Change Biol.* **2024**, *30*, e17024. [CrossRef] [PubMed]

10. Bossio, D.A.; Cook-Patton, S.C.; Ellis, P.W.; Fargione, J.; Sanderman, J.; Smith, P.; Wood, S.; Zomer, R.J.; von Unger, M.; Emmer, I.M.; et al. The role of soil carbon in natural climate solutions. *Nat. Sustain.* **2020**, *3*, 391–398. [CrossRef]
11. Pan, Y.; Birdsey, R.A.; Fang, J.; Houghton, R.; Kauppi, P.E.; Kurz, W.A.; Phillips, O.L.; Shvidenko, A.; Lewis, S.L.; Canadell, J.G.; et al. A Large and Persistent Carbon Sink in the World's Forests. *Science* **2011**, *333*, 988–993. [CrossRef]
12. de Gouvenain, R.C.; Silander, J.A. Temperate Forests. In *Reference Module in Life Sciences*; Elsevier: Amsterdam, The Netherlands, 2017; p. 9780128096338022000. Available online: https://linkinghub.elsevier.com/retrieve/pii/B9780128096338023104 (accessed on 15 January 2024).
13. Mayer, M.; Prescott, C.E.; Abaker, W.E.A.; Augusto, L.; Cécillon, L.; Ferreira, G.W.; James, J.; Jandl, R.; Katzensteiner, K.; Laclau, J.-P.; et al. Tamm Review: Influence of forest management activities on soil organic carbon stocks: A knowledge synthesis. *For. Ecol. Manag.* **2020**, *466*, 118127. [CrossRef]
14. Currie, W.S.; Bergen, K.M. Temperate forest. In *Ecosystem Ecology*; Jorgenson, S.E., Ed.; Elsevier: Amsterdam, The Netherlands, 2008; pp. 607–626.
15. McCarragher, S.; Rigg, L.S. Temperate Forest Ecosystems. In *The International Encyclopedia of Geography*, 1st ed.; Richardson, D., Castree, N., Goodchild, M.F., Kobayashi, A., Liu, W., Marston, R.A., Eds.; Wiley: Hoboken, NJ, USA, 2017; pp. 1–14. Available online: https://onlinelibrary.wiley.com/doi/10.1002/9781118786352.wbieg0508 (accessed on 19 January 2024).
16. Amerongen-Madison, J.; Bulmer, C.; Trofymow, T.; Prescott, C.; Wallace, B.; Philpott, T.; Dymond, C.; Fredeen, A. Soil Carbon in Forest Ecosystems: Pools and Processes. Forest Carbon Initiative. 2021. Available online: https://a100.gov.bc.ca/pub/eirs/finishDownloadDocument.do;jsessionid=8F62A9315428CC8F653587E211134F7F?subdocumentId=21641 (accessed on 23 January 2024).
17. Sollins, P.; Robertson, G.P.; Uehara, G. Nutrient mobility in variable- and permanent-charge soils. *Biogeochemistry* **1988**, *6*, 181–199. [CrossRef]
18. Blanco-Canqui, H.; Lal, R. Soil and Water Conservation. In *Principles of Soil Conservation and Management*; Springer: Dordrecht, The Netherlands, 2010; pp. 1–19. Available online: http://link.springer.com/10.1007/978-1-4020-8709-7_1 (accessed on 13 January 2025).
19. Juhos, K.; Madarász, B.; Kotroczó, Z.; Béni, Á.; Makádi, M.; Fekete, I. Carbon sequestration of forest soils is reflected by changes in physicochemical soil indicators—A comprehensive discussion of a long-term experiment on a detritus manipulation. *Geoderma* **2021**, *385*, 114918. [CrossRef]
20. Gunina, A.; Kuzyakov, Y. From energy to (soil organic) matter. *Glob. Change Biol.* **2022**, *28*, 2169–2182. [CrossRef]
21. Navarrete, A.A.; Aburto, F.; González-Rocha, G.; Guzmán, C.M.; Schmidt, R.; Scow, K. Anthropogenic degradation alter surface soil biogeochemical pools and microbial communities in an Andean temperate forest. *Sci. Total Environ.* **2023**, *854*, 158508. [CrossRef] [PubMed]
22. Prăvălie, R.; Nita, I.A.; Patriche, C.; Niculiță, M.; Birsan, M.V.; Roșca, B.; Bandoc, G. Global changes in soil organic carbon and implications for land degradation neutrality and climate stability. *Envr. Res.* **2021**, *201*, 111580. [CrossRef]
23. Seydewitz, E.; Mulrennan, M.; García, M. A tale of two trails: Lessons from a comparative account of the Trans Canada Trail and the Sendero de Chile. *Can. Geogr./Géogr. Can.* **2023**, *67*, 288–303. [CrossRef]
24. Keenan, R.J.; Reams, G.A.; Achard, F.; de Freitas, J.V.; Grainger, A.; Lindquist, E. Dynamics of global forest area: Results from the FAO Global Forest Resources Assessment 2015. *For. Ecol. Manag.* **2015**, *352*, 9–20. [CrossRef]
25. Miranda, A.; Altamirano, A.; Cayuela, L.; Lara, A.; González, M. Native forest loss in the Chilean biodiversity hotspot: Revealing the evidence. *Reg. Environ. Change* **2017**, *17*, 285–297. [CrossRef]
26. Mohr Bell, D.; Díaz, G.; Príncipe, R.; González, C.; Bono, J.; Ciuffoli, L.; Strada, M.; Parmuchi, G.; Chomnalez, F.; Montenegro, C.; et al. *Monitoreo de la Superficie de Bosque Nativo de la República Argentina, Región Forestal Bosque Andino Patagónico*; Tomo I. Esquel; Secretaría de Ambiente y Desarrollo Sustentable de la Nación: Buenos Aires, Argentina, 2019.
27. CONAF. *Superficie de uso de Suelo Regional (Actualizado a Agosto 2017)*; CONAF: Santiago, Chile, 2021. Available online: https://www.conaf.cl/?s=Superficie+de+uso+de+Suelo+Regional+(Actualizado+a+Agosto+2017) (accessed on 20 February 2024).
28. Donoso, P.J.; Promis, A.; Loguercio, G.A.; Beltrán, H.A.; Casseli, M.; Chauchard, L.M.; Cruz, G.; Peñalba, M.G.; Pastur, G.M.; Navarro, C.; et al. Silviculture of South American temperate native forests. *N. Z. J. For. Sci.* **2022**, *52*. [CrossRef]
29. Dube, F.; Sotomayor, A.; Loewe, V.; Müller-Using, B.; Stolpe, N.; Zagal, E.; Doussoulin, M. Silvopastoral Systems in Temperate Zones of Chile. In *Silvopastoral Systems in Southern South America*; Peri, P.L., Dube, F., Varella, A., Eds.; Advances in Agroforestry; Springer International Publishing: Cham, Switzerland, 2016; Volume 11, pp. 183–211. Available online: https://link.springer.com/chapter/10.1007/978-3-319-24109-8_8 (accessed on 28 January 2024).
30. Valle, S.R.; Carrasco, J. Soil quality indicator selection in Chilean volcanic soils formed under temperate and humid conditions. *CATENA* **2018**, *162*, 386–395. [CrossRef]
31. Nanzyo, M. Unique properties of ash volcanic soils. *Glob. Environ. Res.* **2002**, *6*, 99–112.
32. Matus, F.; Rumpel, C.; Neculman, R.; Panichini, M.; Mora, M.L. Soil carbon storage and stabilisation in andic soils: A review. *CATENA* **2014**, *120*, 102–110. [CrossRef]

33. San Martín, P. Reserva de la Biosfera Corredor Biológico Nevados de Chillán—Laguna del Laja: De la amenaza de la extinción al desarrollo sustentable. In *Reservas de la Biosfera de Chile: Laboratorios para la Sustentabilidad*; Moreira-Muñoz, A., Borsdorf, A., Eds.; Serie Geolibros; Academia de Ciencias Austriaca, Pontificia Universidad Católica de Chile, Instituto de Geografía: Santiago, Chile, 2014; Volume 17, pp. 146–160.
34. Dube, F. Universidad de Concepción, Concepción, Chile. 2020, *manuscript in preparation*.
35. Alfaro, M.; Dube, F.; Zagal, E. Soil quality indicators in an Andisol under different tree covers in disturbed Nothofagus forests. *Chil. J. Agric. Res.* **2018**, *78*, 106–116. [CrossRef]
36. Muñoz, A.A.; González, M.E. Patrones de Regeneración Arbórea en Claros a una Década de la Floración y Muerte Masiva de *Chusquea quila* (Poaceae) en un Remanente de Bosque Antiguo del valle Central en el Centro-Sur de Chile. *Rev. Chil. Hist. Nat.* **2009**, *82*, 185–198. [CrossRef]
37. United States Department of Agriculture (USDA). *Soil Survey Staff: Keys to Soil Taxonomy*; USDA: Washington, DC, USA, 2022.
38. Stolpe, N. *Descripciones de los Principales Suelos de la VIII Región de Chile*; Departamento de Suelos y Recursos Naturales, Facultad de Agronomia, Universidad de Concepción: Chillán, Chile, 2006; pp. 17–18.
39. Blake, G.R.; Hartge, K.H. Particle Density 1. In *Methods of Soil Analysis: Part 1—Physical and Mineralogical Methods*, 2nd ed.; Klute, A., Ed.; American Society of Agronomy; Soil Science Society of America Inc.: Madison, WI, USA, 1986; pp. 377–382.
40. Zhang, R. Determination of soil sorptivity and hydraulic conductivity from the disk infiltrometer. *Soil Sci. Soc. Am. J.* **1997**, *61*, 1024–1030. [CrossRef]
41. Zagal, E.; Longeri, L.; Vidal, I.; Hoffman, G.; González, R. Influencia de la Adición de Nitrógeno y Fósforo sobre la Descomposición de Paja de Trigo en un Suelo Derivado de Cenizas Volcánicas. *Agric. Téc.* **2003**, *63*, 403–415. [CrossRef]
42. Kemper, W.D.; Rosenau, R.C. *Aggregate Stability and Size Distribution*; Klute, A., Ed.; SSSA Book Series; Soil Science Society of America, American Society of Agronomy: Madison, WI, USA, 2018; pp. 425–442.
43. Sadzawka, A.; Carrasco, M.A.; Grez, R.; Mora, M.L.; Flores, H.; Neaman, A. *Métodos de Análisis Recomendados para los Suelos Chilenos*; Comisión de Normalización y Acreditación, Sociedad Chilena de la Ciencia del Suelo: Santiago, Chile, 2006; p. 113.
44. Wright, A.F.; Bailey, J.S. Organic carbon, total carbon, and total nitrogen determinations in soils of variable calcium carbonate contents using a Leco CN-2000 dry combustion analyzer. *Commun. Soil Sci. Plant Anal.* **2001**, *32*, 3243–3258. [CrossRef]
45. Brody, J.R.; Kern, S.E. Sodium boric acid: A Tris-free, cooler conductive medium for DNA electrophoresis. *BioTechniques* **2004**, *36*, 214–216. [CrossRef]
46. Mallott, E.K.; Malhi, R.S.; Amato, K.R. Assessing the Comparability of Different DNA Extraction and Amplification Methods in Gut Microbial Community Profiling. *Access Microbiol.* **2019**, *1*, e000060. [CrossRef]
47. Louca, S.; Parfrey, L.W.; Doebeli, M. Decoupling function and taxonomy in the global ocean microbiome. *Science* **2016**, *353*, 1272–1277. [CrossRef]
48. Põlme, S.; Bahram, M.; Jacquemyn, H.; Kennedy, P.; Kohout, P.; Moora, M.; Oja, J.; Öpik, M.; Pecoraro, L.; Tedersoo, L. Host preference and network properties in biotrophic plant–fungal associations. *New Phytol.* **2018**, *217*, 1230–1239. [CrossRef] [PubMed]
49. Kay, M.; Elkin, L.A.; Higgins, J.J.; Wobbrock, J.O. *ARTool: Aligned Rank Transform. [R] Package with Documentation and Two Vignettes*; CRAN: Wien, Austria, 2021.
50. Reynolds, W.D.; Yang, X.M.; Drury, C.F.; Zhang, T.Q.; Tan, C.S. Effects of selected conditioners and tillage on the physical quality of a clay loam soil. *Can. J. Soil Sci.* **2003**, *83*, 381–393. [CrossRef]
51. Clough, A.; Skjemstad, J.O. Physical and chemical protection of soil organic carbon in three agricultural soils with different contents of calcium carbonate. *Soil Res.* **2000**, *38*, 1005–1016. [CrossRef]
52. Arshad, M.A.C.; Lowery, B.; Grossman, B. *Physical Tests for Monitoring Soil Quality*; Doran, J.W., Jones, A.J., Eds.; SSSA Special Publications; Soil Science Society of America: Madison, WI, USA. 2015; pp. 123–141. Available online: https://acsess.onlinelibrary.wiley.com/doi/abs/10.2136/sssaspecpub49.c7 (accessed on 26 February 2024).
53. Geng, N.; Kang, X.; Yan, X.; Yin, N.; Wang, H.; Pan, H.; Yang, Q.; Lou, Y.; Zhuge, Y. Biochar mitigation of soil acidification and carbon sequestration is influenced by materials and temperature. *Ecotoxicol. Environ. Saf.* **2022**, *232*, 113241. [CrossRef] [PubMed]
54. Nissen, J.; Quiroz, C.; Seguel, O.; Roberto Mac Donald, H.; Sch. A.E. Variación del potencial mátrico durante el movimiento de agua en andisoles. *Agro Sur* **2005**, *33*, 36–47. [CrossRef]
55. Amacher, M.C.; O'Neil, K.P.; Perry, C.H. *Soil Vital Signs: A New Soil Quality Index (SQI) for Assessing Forest Soil Health*; RMRS-RP-65; U.S. Department of Agriculture, Forest Service, Rocky Mountain Research Station: Washington, DC, USA, 2007.
56. Hu, W.; Yao, J.; He, Q.; Chen, J. Elevation-Dependent Trends in Precipitation Observed Over and Around the Tibetan Plateau from 1971 to 2017. *Water* **2021**, *13*, 2848. [CrossRef]
57. Dahlgren, R.A.; Saigusa, M.; Ugolini, F.C. The Nature, Properties and Management of Volcanic Soils. In *Advances in Agronomy*; Elsevier: Amsterdam, The Netherlands, 2004; pp. 113–182. Available online: https://linkinghub.elsevier.com/retrieve/pii/S0065211303820035 (accessed on 25 February 2024).
58. Takahashi, T.; Dahlgren, R.A. Nature, properties and function of aluminum–humus complexes in volcanic soils. *Geoderma* **2016**, *263*, 110–121. [CrossRef]

59. Pérez, C.A.; Carmona, M.R.; Fariña, J.M.; Armesto, J.J. Selective logging of lowland evergreen rainforests in Chiloé Island, Chile: Effects of changing tree species composition on soil nitrogen transformations. *For. Ecol. Manag.* **2019**, *258*, 1660–1668. [CrossRef]
60. Binkley, D.; Fisher, R.F. Soil Organic Matter. In *Ecology and Management of Forest Soils*, 4th ed.; Wiley-Blackwell: Hoboken, NJ, USA, 2013; p. 43.
61. Rivas, Y.; Matus, F.; Rumpel, C.; Knicker, H.; Garrido, E. Black carbon contribution in volcanic soils affected by wildfire or stubble burning. *Org. Geochem.* **2012**, *47*, 41–50. [CrossRef]
62. Liu, G.; Du, Q.; Li, J. Interactive effects of nitrate-ammonium ratios and temperatures on growth, photosynthesis, and nitrogen metabolism of tomato seedlings. *Sci. Hortic.* **2017**, *214*, 41–50. [CrossRef]
63. Parfitt, R.L. Phosphate reactions with natural allophane, ferrihydrite and goethite. *J. Soil Sci.* **1989**, *40*, 359–369. [CrossRef]
64. Borie, F.; Aguilera, P.; Castillo, C.; Valentine, A.; Seguel, A.; Barea, J.M.; Cornejo, P. Revisiting the Nature of Phosphorus Pools in Chilean Volcanic Soils as a Basis for Arbuscular Mycorrhizal Management in Plant P Acquisition. *J. Soil Sci. Plant Nutr.* **2019**, *19*, 390–401. [CrossRef]
65. Villarroel, R.B. *Diagnóstico de la Fertilidad del Suelo*; Instituto de Investigaciones Agropecuarias–Centro Regional de Investigación Remehue Serie Acta: Osorno, Chile, 2000; p. 71.
66. Lehmann, J.; Rillig, M.C.; Thies, J.; Masiello, C.A.; Hockaday, W.C.; Crowley, D. Biochar effects on soil biota—A review. *Soil Biol. Biochem.* **2011**, *43*, 1812–1836. [CrossRef]
67. Qafoku, N.P.; Ranst, E.V.; Noble, A.; Baert, G. Variable Charge Soils: Their Mineralogy, Chemistry and Management. In *Advances in Agronomy*; Elsevier: Amsterdam, The Netherlands, 2004; pp. 159–215. Available online: https://linkinghub.elsevier.com/retrieve/pii/S0065211304840045 (accessed on 29 February 2024).
68. Jia, Y.; Shi, S.; Liu, J.; Su, S.; Liang, Q.; Zeng, X.; Li, T. Study of the Effect of Pyrolysis Temperature on the Cd2+ Adsorption Characteristics of Biochar. *Appl. Sci.* **2018**, *8*, 1019. [CrossRef]
69. Gillespie, C.J.; Antonangelo, J.A.; Zhang, H. The Response of Soil pH and Exchangeable Al to Alum and Lime Amendments. *Agriculture* **2021**, *11*, 547. [CrossRef]
70. Oliveira, P.E.; Cunha, A.G.; Reyes, G.; Gacitúa, W.; Petit-Breuilh, X. *Chusquea quila*, a Natural Resource from Chile: Its Chemical, Physical, and Nanomechanical Properties. *BioResources* **2016**, *11*, 10057–10069. [CrossRef]
71. Ortiz, J.; Dube, F.; Neira, P.; Panichini, M.; Stolpe, N.B.; Zagal, E.; Martínez-Hernández, P.A. Soil Quality Changes within a (*Nothofagus obliqua*) Forest Under Silvopastoral Management in the Andes Mountain Range, South Central Chile. *Sustainability* **2020**, *12*, 6815. [CrossRef]
72. Ortiz, J.; Neira, P.; Panichini, M.; Curaqueo, G.; Stolpe, N.B.; Zagal, E.; Dube, F.; Gupta, S.R. Silvopastoral Systems on Degraded Lands for Soil Carbon Sequestration and Climate Change Mitigation. In *Agroforestry for Sustainable Intensification of Agriculture in Asia and Africa*; Dagar, J.C., Gupta, S.R., Sileshi, G.W., Eds.; Sustainability Sciences in Asia and Africa; Springer Nature: Singapore, 2023; pp. 207–242. Available online: https://link.springer.com/10.1007/978-981-19-4602-8_7 (accessed on 12 January 2024).
73. Muñoz, C.; Torres, P.; Alvear, M.; Zagal, E. Physical protection of C and greenhouse gas emissions provided by soil macroaggregates from a Chilean cultivated volcanic soil. *Acta Agric. Scand. Sect. B—Soil Plant Sci.* **2012**, *62*, 739–748. [CrossRef]
74. Li, X.; Zhang, Q.; Feng, J.; Jiang, D.; Zhu, B. Forest management causes soil carbon loss by reducing particulate organic carbon in Guangxi, Southern China. *For. Ecosyst.* **2023**, *10*, 100092. [CrossRef]
75. Dube, F.; Thevathasan, N.V.; Zagal, E.; Gordon, A.M.; Stolpe, N.B.; Espinosa, M. Carbon Sequestration Potential of Silvopastoral and Other Land Use Systems in the Chilean Patagonia. In *Carbon Sequestration Potential of Agroforestry Systems*; Kumar, B.M., Nair, P.K.R., Eds.; Advances in Agroforestry; Springer: Dordrecht, The Netherlands, 2011; Volume 8, pp. 101–127. Available online: https://link.springer.com/10.1007/978-94-007-1630-8_6 (accessed on 25 February 2024).
76. Alfaro, M.; Dube, F.; Zagal, E. The Influence of Overmature, Degraded Nothofagus Forests with Strong Anthropic Disturbance on the Quality of an Andisol and its Gradual Recovery with Silvopasture in Southwestern South America. In *Agroforestry for Degraded Landscapes*; Dagar, J.C., Gupta, S.R., Teketay, D., Eds.; Springer: Singapore, 2020; pp. 67–85. Available online: http://link.springer.com/10.1007/978-981-15-6807-7_3 (accessed on 15 January 2025).
77. Neira, P.; Henríquez-Castillo, C.; Ortiz, J.; Stolpe, N.; Dube, F. Do different densities of tree cover affect pasture biomass and soil microbial communities? *Agrofor. Syst.* **2021**, *95*, 1465–1478. [CrossRef]
78. Zerbe, S.; Storz, S.T.; Leitinger, G.; Joelson, N.Z.; Bava, J.; Heinrichs, S.; Leuschner, C.; Loguercio, G.; Simon, A.; Urretavizcaya, M.F.; et al. Regeneration of *Nothofagus dombeyi* (Mirb.) Ørst. in little to moderately disturbed southern beech forests in the Andes of Patagonia (Argentina). *For. Ecosyst.* **2023**, *10*, 100126. [CrossRef]
79. Diao, J.; Liu, J.; Zhu, Z.; Wei, X.; Li, M. Active forest management accelerates carbon storage in plantation forests in Lishui, southern China. *For. Ecosyst.* **2022**, *9*, 100004. [CrossRef]
80. Nicholson, W.L.; Fajardo-Cavazos, P.; Rebeil, R.; Slieman, T.A.; Riesenman, P.J.; Law, J.F.; Xue, Y. Bacterial endospores and their significance in stress resistance. *Antonie van Leeuwenhoek* **2002**, *81*, 27–32. [CrossRef] [PubMed]
81. Checinska, A.; Paszczynski, A.; Burbank, M. Bacillus and Other Spore-Forming Genera: Variations in Responses and Mechanisms for Survival. *Annu. Rev. Food Sci. Technol.* **2015**, *6*, 351–369. [CrossRef]

82. Ranawat, P.; Rawat, S. Radiation resistance in thermophiles: Mechanisms and applications. *World J. Microbiol. Biotechnol.* **2017**, *33*, 112. [CrossRef] [PubMed]
83. Kozlowski, T.T.; Pallardy, S.G. *Physiology of Woody Plants*, 2nd ed.; Elsevier: Amsterdam, The Netherlands, 1996; 42332p.
84. Epihov, D.Z.; Saltonstall, K.; Batterman, S.A.; Hedin, L.O.; Hall, J.S.; van Breugel, M.; Leake, J.R.; Beerling, D.J. Legume–microbiome interactions unlock mineral nutrients in regrowing tropical forests. *Proc. Natl. Acad. Sci. USA* **2021**, *118*, e2022241118. [CrossRef]
85. Wu, D.; Wei, Z.; Mohamed, T.A.; Zheng, G.; Qu, F.; Wang, F.; Zhao, Y.; Song, C. Lignocellulose biomass bioconversion during composting: Mechanism of action of lignocellulase, pretreatment methods and future perspectives. *Chemosphere* **2022**, *286*, 131635. [CrossRef] [PubMed]
86. McDaniel, M.D.; Tiemann, L.K.; Grandy, A.S. Does agricultural crop diversity enhance soil microbial biomass and organic matter dynamics? A meta-analysis. *Ecol. Appl.* **2014**, *24*, 560–570. [CrossRef]
87. Zhou, Z.; Gao, T.; Van Zwieten, L.; Zhu, Q.; Yan, T.; Xue, J.; Wu, Y. Soil Microbial Community Structure Shifts Induced by Biochar and Biochar-Based Fertilizer Amendment to Karst Calcareous Soil. *Soil Sci. Soc. Am. J.* **2019**, *83*, 398–408. [CrossRef]
88. Xue, Y.; Jonassen, I.; Øvreås, L.; Taş, N. Metagenome-assembled genome distribution and key functionality highlight importance of aerobic metabolism in Svalbard permafrost. *FEMS Microbiol. Ecol.* **2020**, *96*, fiaa057. [CrossRef]
89. Scarlett, K.; Denman, S.; Clark, D.R.; Forster, J.; Vanguelova, E.; Brown, N.; Whitby, C. Relationships between nitrogen cycling microbial community abundance and composition reveal the indirect effect of soil pH on oak decline. *ISME J.* **2021**, *15*, 623–635. [CrossRef]
90. Fierer, N. Embracing the unknown: Disentangling the complexities of the soil microbiome. *Nat. Rev. Microbiol.* **2017**, *15*, 579–590. [CrossRef]
91. Allison, S.D.; Martiny, J.B.H. Resistance, resilience, and redundancy in microbial communities. *Proc. Natl. Acad. Sci. USA* **2008**, *105* (Suppl. S1), 11512–11519. [CrossRef] [PubMed]
92. Goldfarb, K.C.; Karaoz, U.; Hanson, C.A.; Santee, C.A.; Bradford, M.A.; Treseder, K.K.; Wallenstein, M.D.; Brodie, E.L. Differential Growth Responses of Soil Bacterial Taxa to Carbon Substrates of Varying Chemical Recalcitrance. *Front. Microbio.* **2011**, *2*, 94. [CrossRef] [PubMed]
93. Lladó, S.; López-Mondéjar, R.; Baldrian, P. Forest Soil Bacteria: Diversity, Involvement in Ecosystem Processes, and Response to Global Change. *Microbiol. Mol. Biol. Rev.* **2017**, *81*, e00063-16. [CrossRef] [PubMed]
94. Fierer, N.; Bradford, M.A.; Jackson, R.B. Toward an ecological classification of soil bacteria. *Ecology* **2007**, *88*, 1354–1364. [CrossRef] [PubMed]
95. Treseder, K.K.; Lennon, J.T. Fungal Traits That Drive Ecosystem Dynamics on Land. *Microbiol. Mol. Biol. Rev.* **2015**, *79*, 243–262. [CrossRef] [PubMed]
96. Shade, A.; Peter, H.; Allison, S.D.; Baho, D.L.; Berga, M.; Bürgmann, H.; Huber, D.H.; Langenheder, S.; Lennon, J.T.; Martiny, J.B.H. Fundamentals of Microbial Community Resistance and Resilience. *Front. Microbio.* **2012**, *3*, 417. [CrossRef] [PubMed]
97. Evans, S.E.; Wallenstein, M.D. Climate change alters ecological strategies of soil bacteria. *Ecol. Lett.* **2014**, *17*, 155–164. [CrossRef]
98. Schimel, J.P.; Schaeffer, S.M. Microbial control over carbon cycling in soil. *Front. Microbio.* **2012**, *3*, 348. [CrossRef] [PubMed]

forests

MDPI

Article

Response of Soil Microbial Communities in Extreme Arid Deserts to Different Long-Term Management Methods

Yulin Zhang [1,2,3,4,†], Yi Du [1,2,3,5,†], Zhaobin Mu [1,2,3], Waqar Islam [1,2,3], Fanjiang Zeng [1,2,3,4,5,*], Zhihao Zhang [1,2,3] and Norela C. T. Gonzalez [6]

1 Xinjiang Key Laboratory of Desert Plant Roots Ecology and Vegetation Restoration, Xinjiang Institute of Ecology and Geography, Chinese Academy of Sciences, Urumqi 830011, China; zhangyl20201051213@163.com (Y.Z.); muzhaobin@ms.xjb.ac.cn (Z.M.); waqarislam@ms.xjb.ac.cn (W.I.); zhangzh@ms.xjb.ac.cn (Z.Z.)
2 State Key Laboratory of Desert and Oasis Ecology, Key Laboratory of Ecological Safety and Sustainable Development in Arid Lands, Xinjiang Institute of Ecology and Geography, Chinese Academy of Sciences, Urumqi 830011, China
3 Cele National Station of Observation and Research for Desert-Grassland Ecosystems, Cele 848300, China
4 College of Ecology and Environmental, Xinjiang University, Urumqi 830046, China
5 University of Chinese Academy of Sciences, Beijing 100049, China
6 College of Forestry, Central South University of Forestry and Technology, Changsha 410004, China; norelaclarizel@gmail.com
* Correspondence: fjzeng@ms.xjb.ac.cn
† These authors contributed equally to this work.

Abstract: (1) Background: As population growth accelerates, unsustainable practices such as excessive cutting and burning of desert plants in the transition zones between deserts and oases have led to widespread vegetation loss. (2) Methods: The experiment was conducted in the oasis transition zone on the southern edge of the Taklamakan Desert from 2010 to 2023 year. Among the treatments included a control group (CK), cutting in spring (CS), cutting in fall (CF), burning in spring (BS), and flood water irrigation (FI). We used high-throughput sequencing to determine soil microbial composition and diversity and routine laboratory methods to determine soil physical and chemical properties and enzyme activities. (3) Results: No significant differences in bacterial alpha diversity (Chao1, Dominance, Observed_features, Pielou_e, Shannon, and Simpson) across the different long-term disturbance patterns. In fungi, the CK treatment showed significantly higher Chao1, Shannon, and Observed_features indices compared to BS and FI. Principal component analysis revealed a substantial reduction in bacterial community diversity in BS compared to FI, while fungal communities were lower in CK and CS compared to BS, CF, and FI; (4) Conclusions: Soil moisture content, electrical conductivity, organic carbon, and the activity of the enzyme cellobiohydrolase as key factors shaping the bacterial community. For fungi, organic carbon and the β-1,4-glucosidase enzyme were the main drivers.

Keywords: desert ecosystems; arid regions; land management; soil nutrients; microbial diversity

1. Introduction

Soil microbes play a crucial role in terrestrial ecosystems, acting as key regulators and contributors to nutrient cycling within the soil and across the entire ecosystem [1,2]. These microorganisms, such as bacteria, fungi, and actinomycetes, aid in the conversion of nutrients and their release by breaking down organic material, mineralizing nutrients, and establishing symbiotic interactions with plant roots [2–4]. Their responsiveness to changes in the soil environment renders them crucial biological indicators for evaluating

https://doi.org/10.3390/f16020306

soil quality [3,5]. Moreover, the complex ecological interactions among diverse microbial groups within the plant-soil system enhance nutrient uptake, stimulate plant growth, and boost plant resilience to environmental stressors, thereby supporting plant health and ensuring ecosystem stability [2,4,6,7].

Desert ecosystems, common in dry and semi-dry areas, confront significant environmental difficulties, including extreme lack of water, elevated temperatures, increased soil salinity, and shortages of nutrients [4,8]. Desert plants have evolved unique adaptations, such as deep root systems, water-storing tissues, and reduced leaf surfaces, to survive severe water scarcity and high temperatures [9,10]. To withstand extreme conditions, desert plants often establish symbiotic relationships with drought-tolerant microorganisms, such as *Streptomyces*, which significantly enhance their survival capabilities [11,12]. *Alhagi sparsifolia* Shap, a deep-rooted desert phreatophyte, is widely distributed across the arid and saline regions of Central Asia. It is essential for maintaining the structure and function of ecosystems, in addition to aiding local livestock farming [4,11,13]. Grasping the diversity and community structure of soil microorganisms is crucial for illuminating the intricate relationships that exist within the nutrient cycle of the soil and its microorganisms. At the boundary where the desert meets the oasis on the outskirts of the Taklamakan Desert, there exists a dry ecosystem that is low in nutrients and primarily characterized by longstanding desert vegetation [9,13]. Changes in land use and plant litter inputs can significantly alter soil hydrothermal conditions and nutrient content [8]. These shifts, in turn, influence the composition and diversity of soil microbial communities [14,15]. In addition to influencing the abundance of microorganisms that utilize carbohydrates and aromatic compounds as carbon sources, different land use patterns also impact microbial communities that metabolize other compounds, such as nitrogen compounds, lipids and fatty acids, sulfur compounds, phenols, and lignin derivatives [16–18]. Although shifts in land use can change the composition and abundance of soil microbes, microbial diversity itself often remains relatively stable [13,19]. Fungi demonstrate a more significant resilience to water stress compared to bacteria and are essential in aiding bacterial populations in arid soils [20,21]. Thus, it is important to examine the diversity and community structure of both fungi and bacteria in these dry, nutrient-deficient ecosystems.

In the southern Tarim Basin, increasing human activities such as grazing, cutting, and burning have rapidly transformed the structure, function, and nutrient dynamics of ecosystems, profoundly impacting plant communities and broader global ecosystems [12,22]. Cutting and burning practices can reduce root respiration and exudates, limiting nutrient availability for soil microbes [23,24]. Selective cutting, however, has been shown to significantly increase soil enzyme activities (e.g., β-glucosidase, urease, phosphatase), as well as soil moisture, total nitrogen, phosphorus, and hydrolyzable nitrogen content [25]. Burning can alter soil pH and nutrient availability, leading to losses in carbon and nitrogen and influencing microbial metabolic activity [26,27]. While burning decreases hydrolase activities for carbon, nitrogen, and phosphorus acquisition significantly, it has been found to enhance oxidative enzyme activities [28].

In the desert-oasis transition zone of the southern Tarim Basin, *A. sparsifolia* is frequently cut for livestock fodder in spring and autumn, and burning is mainly used in agricultural areas [12,29]. Floodwater irrigation, using water stored during upstream summer floods, helps protect desert vegetation by providing ecological support [30]. This study applied long-term artificial disturbance treatments—cutting in spring (CS), cutting in fall (CF), burning in spring (BS), and flood water irrigation (FI)—at the Cele Desert Research Station from 2009 to 2023. The aim was to evaluate the effects of these treatments on the soil's physical and chemical properties and the microbial communities associated with *A. sparsifolia*. We hypothesized that (1) the various long-term disturbance treatments would alter the microbial community structure and composition, and (2) changes in soil organic carbon and available nutrients would not significantly affect soil extracellular enzyme activity. This study provides insights into how soil characteristics and microbial communities respond to long-term disturbances, enhancing our understanding of desert ecosystem management.

2. Materials and Methods

2.1. Site Description

The experiment was carried out at the Cele desert research station of the Chinese Academy of Sciences (80°43′45″ E, 37°00′57″ N). The average temperature in the region is 11.9 °C, with winter lows reaching −23.98 °C and summer highs reaching 41.98 °C. The average annual rainfall is 35.1 mm, while the average evaporation rate is significantly higher at 2600 mm [4]. On average, the region experiences 25.2 sandy days (sandstorms) annually, with a maximum of 59 days recorded. Additionally, there are between 3 to 9 instances of strong winds exceeding grade 8 each year [13]. *A. sparsifolia*, *Karelinia caspia*, *Calligonum caput-medusae*, and *Tamarix ramosissima* communities dominate the undisturbed dunes bordering the Cele oasis-desert transition area [9,13].

2.2. Experiment Design and Sampling

To protect and sustainably manage desert vegetation, particularly *A. sparsifolia*, which is both a dominant and constructive species, we established a long-term research site. The experimental treatment was conducted annually from 2009 to 2023. A random block design was implemented to establish an experimental area measuring 20 m × 20 m for each plot in 2009, which was enclosed by a wire fence. The research consisted of five treatments: a control group (CK), cutting in spring (CS), cutting in fall (CF), burning in spring (BS), and flood water irrigation (FI), with four replicates for each treatment, resulting in a total of 20 experimental plots. In 2009, the essential soil physicochemical properties examined included soil pH (7.72), soil organic matter (SOM) content (1.45 $g \cdot kg^{-1}$), total nitrogen (TN) (1.24 $g \cdot kg^{-1}$), total phosphorus (TP) (1.38 $g \cdot kg^{-1}$), total potassium (TK) (1.04 $g \cdot kg^{-1}$), available phosphorus (AP) (3.46 $mg \cdot kg^{-1}$), and available potassium (AK) (206.11 $mg \cdot kg^{-1}$)

The specific treatments included:

- CS: From 2010 to April 2023, all *A. sparsifolia* plants were cut at ground level, and the aboveground biomass was removed from the plot.
- CF: From 2010 to October 2023, all *A. sparsifolia* plants were cut at ground level, and the aboveground biomass was removed from the plot.
- BS: From 2010 to April 2023, all aboveground parts of *A. sparsifolia* were burned using artificial fire.
- FI: From 2010 to July 2023, heavy rainfall was simulated to cause flooding, with a water application rate of 1800 $m^3 \cdot hm^{-2}$.
- CK: From 2010 to 2023, this treatment allowed for the natural growth of *A. sparsifolia* without any interference.

In 2023, field research was carried out during the autumn (October) at long-term monitoring sites to gather soil samples. Before sampling, all tools and containers were strictly disinfected to prevent microbial cross-contamination. The five-point sampling method was used to drill 0~50 cm soil samples with a diameter of 7 cm [31]. Preferentially, the soil moisture content (SMC), pH, and EC (electrical conductivity) were measured. The samples were split into two sections, with one portion being left to dry in a shaded area for the analysis of soil organic carbon (SOC) and total nitrogen, phosphorus, potassium (TN, TP, and TK, respectively), and available nitrogen, phosphorus, potassium (AN, AP, AK, respectively), whereas the other segment was kept in a fridge at 4 °C to assess soil enzyme and microbial biomass. Simultaneously, soil microbial samples from depths of 0 to 50 cm were gathered, placed in dry ice for preservation, transported to the lab under cold conditions, and promptly stored in a freezer at −80 °C.

2.3. Soil Physicochemical Properties Analysis

To measure SOC, TN, AN, and AK, the following methods were used: the $K_2Cr_2O_7$-H_2SO_4 oxidation method, the Kjeldahl Nitrogen Analyzer (K1160, Jinan Hanon Instruments Co. Ltd., Jinan, China), the alkali hydrolyzable method, and the NH_4OAc extraction method. Before measuring TP and TK with the Inductively Coupled Plasma-Optical Emission Spectrometer (iCAP 6300, Thermo Elemental, Waltham, MA, USA), the samples were digested in concentrated HNO_3. AP extracted with HCl/NH_4F was determined colorimetrically using ascorbic acid molybdate on a continuous-flow autoanalyzer (Autoanalyzer 3, Bran and Luebbe, Norderstedt, Germany) [32–34]. The pH of the soil {soil/water ratio of 1:2.5 (w/v)} and its EC {soil/water ratio of 1:5 (w/v)} were measured using a pH and EC meter (PHSJ-6 L and DDSJ-319 L, INESA Scientific Instrument Co. Ltd., Shanghai, China) manufactured, respectively.

2.4. Soil Microbial Biomass and Enzyme Activity Analysis

Microplate fluorescence technique was used to analyze the levels of Cellobiohydrolase (CBH), β-1,4-glucosidase (BG), β-1,4-N-acetylglucosaminidase (NAG), Lleucine aminopeptidase (LAP), and Alkaline phosphatase (ALP) in the soil [35,36]. Information on the hydrolysis substrate can be found in Table 1. Specifically, weigh 1 g of fresh soil into a plastic bottle (250 m), add 125 mL distilled water, and shake for 2 h (25 °C; 180 r·min^{-1}). First, 1 mL of suspension was absorbed into the centrifuge tube (2 mL), 0.25 mL of fluorescent substrate was added, three parts were absorbed (adding different fluorescent substrates) and shaken well. Secondly, culture at 25 °C for 4 h away from light. Set blank sample and quenching standard sample at the same time. Then, 50 uL 0.5 mol·L^{-1} NaOH was added to terminate the reaction and then shaken. Transfer 250 uL to 96 well plate. Finally, the excitation wavelength of 4 MUB is 365 nm, and the detection wavelength is 450 nm. The chloroform fumigation extraction method was utilized to measure microbial biomass carbon, nitrogen, and phosphorus (MBC, MBN, and MBP, respectively) [37], with subse-quent adjustment using a conversion factor of 0.45, 0.45, and 0.40 [38].

Table 1. Name, function, and corresponding substrates of soil enzyme activities measured in this study.

Enzyme	Abbreviation	Function	Substrate
Cellobiohydrolase	CBH	By acting on the Β-1,4-glycosidic bond in the cellulose molecule, the cellulose is decomposed into smaller cellobiose units.	4-methylumbelliferyl-cellobioside
β-1,4-glucosidase	BG	Catalyzes the hydrolysis of terminal 1,4-linked β-D-glucose residues from β-D-glucosides, including short-chain cellulose oligomers.	4-methylumbelliferyl-glucoside
β-1,4-N-acetylglucosaminidase	NAG	Catalyzes the hydrolysis of terminal 1,4 linked *N*-acetyl-β-D-glucosaminide residues in chitooligosaccharides.	4-methylumbelliferyl-acetyl-glucosaminidase
Lleucine aminopeptidase	LAP	It is a hydrolase, which can decompose proteins into amino acids.	L-Leucine- 7-amino-4-methylcoumarin
Alkaline phosphatase	ALP	It mainly acts on the organic phosphorus in the soil and hydrolyzes it into phosphate.	4-methylumbelliferyl-phosphate

2.5. DNA Extraction, PCR Amplification, Illumina Sequencing, and Bioinformatics Analysis

To obtain the complete genomic DNA from bulk soil, the DNeasy Power Soil DNA Isolation Kit (Qiagen, Inc., Venlo, The Netherlands) was utilized. The amount of the isolated DNA was determined by employing the NanoDrop ND-1000 spectrophotometer (Thermo Fisher Scientific, Waltham, MA, USA), while the quality was evaluated through agarose gel electrophoresis. PCR (Polymerase Chain Reaction) amplification of bacterial 16S rRNA genes V3–V4 region and fungal ITS 1–5 F (Internal Transcribed Spacer) region was performed using universal primers 341 F (5-CCTAYGGGRBGCASCAG-3) and 806R (5-GGACTACNNGGGTATCTAAT-3); ITS 5-1737F (5-GGAAGTAAAAGTCGTAACAAGG-3) and ITS 2-2043R (5-GCTGC GTTCTTCATCGATGC-3), respectively.

According to the Barcode sequence and PCR amplified primer sequence, each sample data was separated from the dismounting data. After the Barcode sequence and primer sequence were cut off, the reads of each sample were spliced using FLASH (v1.2.11) software [39]. The spliced sequences were Raw Tags data. Then, FASTP (v0.23.1) software is used to process the Raw Tags obtained by splicing through strict filtering to obtain high-quality Clean Tags data [40]. The Tags obtained after the above processing need to be processed to remove the chimeric sequence. The Tag sequence is compared with the species annotation database through VSEARCH (v2.16.0) software to detect the chimeric sequence, and the chimeric sequence is finally removed to obtain the final Effective Tags [41].

Additional sequence filtering was performed using the QIIME-II software (v202202) [42]. Operational taxonomic units (OTUs) were generated using the Uparse algorithm (v7.0.1001) with a 97% similarity threshold [43]. Bacterial OTUs were classified with an RDP (Ribosomal Database Project) classifier (confidence threshold of 70%) based on the Mothur method and SSUrRNA database of Silva database (version 138.1) (Set threshold to 0.8~1) [44,45]. Fungal OTUs were classified based on the UNITE database [46] were utilized. The statistical results of data processing are shown in Table 2. The raw sequencing data can be accessed in the National Center for Biotechnology Information (NCBI) Short Read Archive under BioProject ID PRJNA1149307.

Table 2. Data preprocessing statistics and quality control (mean).

	Treatments	RawPE	Combined	Qualified	Nochime	Base (nt)	Avglen (nt)	GC (%)	Q20 (%)	Q30 (%)	Effective (%)
Bacteria	CK	97,307.25	87,601.75	87,292.25	84,882.50	18,956,465.75	222.95	49.84	98.34	95.67	86.79
	CS	104,739.00	99,001.00	98,758.25	96,509.25	21,395,038.00	221.82	48.48	98.90	96.93	92.08
	CF	161,639.75	153,178.50	153,038.25	151,282.75	32,896,311.75	215.24	48.41	99.74	98.73	93.82
	BS	140,181.50	134,873.25	134,720.25	133,191.50	29,453,824.00	220.52	48.12	99.67	98.55	95.27
	FI	105,605.75	102,683.25	102,634.50	101,557.50	21,697,796.00	213.67	48.50	99.81	99.08	96.17
Fungi	CK	116,928.75	115,880.75	114,370.00	77,550.50	31,901,248.25	411.51	57.96	98.79	95.88	66.36
	CS	116,551.00	115,295.75	113,758.00	71,316.50	29,449,188.50	412.88	57.54	98.76	95.87	61.19
	CF	117,475.25	116,139.25	114,738.75	88,316.75	36,450,969.50	412.76	57.37	98.85	96.07	75.09
	BS	124,760.75	123,477.50	121,828.00	82,082.00	33,944,259.50	413.50	56.96	98.71	95.70	65.99
	FI	116,183.00	114,639.50	113,148.25	91,473.75	37,500,255.50	410.08	57.38	98.76	95.82	78.88

Note: (1) RawPE indicates the original PE reads from the machine; (2) Combined is the sequence of Tags obtained by connection; (3) Qualified is the sequence of Raw Tags after filtering low quality and short length; (4) Nochime is the Tags sequence that is ultimately used for subsequent analysis after filtering the chimera, that is, Effective Tags; (5) The Base is the number of bases in the final Effective Tags; (6) AvgLen is the average length of Effective Tags; Q20 and Q30 are the percentage of bases with base mass values greater than 20 (sequencing error rate less than 1%) and 30 (sequencing error rate less than 0.1%) in Effective Tags. (7) GC(%) represents the content of GC bases in Effective Tags; (8) Effective(%) is the percentage of the number of Nochime to the number of rawPE.

2.6. Statistical Analysis

Statistical analysis utilized R v4.1.0. Most of the findings were displayed using packages such as 'ggplot2', 'vegan', 'corrplot', 'randomForest', and 'rfPermute' [47–50]. Before applying parametric tests, data was tested for normality and equality of variances. A one-way ANOVA was employed to analyze the impact of various long-term artificial disturbance treatments {control group (CK), cutting in spring (CS), cutting in fall (CF), burning in spring (BS), and flood water irrigation (FI)} on soil bacterial and fungal alpha diversity (Chao1, Observed_features, Dominance, Pielou_e, Shannon, and Simpson index) using Duncan's method ($p < 0.05$; mean $\pm$ standard error) [51]. The microeco package was used for LEfSe (Linear Discriminant Analysis Effect Size) analysis to identify unique biomarkers that highlight differences in soil microbial communities (based on phylum, class, order, family, genus, and species) [52]. Principal component analysis (PCA) and non-metric multidimensional scaling (NMDS) were employed to analyze variations in soil bacterial and fungal communities (based on OTU levels). The first and second axis data of principal components were extracted to compare the differences between different treatments, and the histogram was used to show (using PERMANOVA with 999 permutations). Using redundancy analysis (RDA) and Venn diagrams, the most suitable model was identified using both selections with 999 permutations and the significant influence of soil properties on bacterial and fungal communities (based on OTU levels). The Mantel test, employing 999 permutations, was conducted to assess Spearman's correlations between soil bacterial and fungal alpha diversity and various soil properties. Additionally, random forest models were utilized to identify the key soil properties that significantly influence bacterial and fungal alpha diversity under different artificial long-term disturbance conditions [49].

3. Results

3.1. Effects of Long-Term Disturbance on Soil Physicochemical Properties, Microbial Biomass, and Enzyme Activity

Significantly higher SMC was observed in the FI management measure in comparison to the CK, CS, CF, and BS management measures. No notable variations were observed in soil pH, EC, TN, TP, TK, and AP across the various management measures (CK, CS, CF, BS, and FI). The soil SOC and AK contents were notably higher under the CS than under the FI management measure. Furthermore, the soil AN in the CK was notably greater than that of the FI management measure. There were no significant changes in soil MBC, MBN, and MBP across various management strategies (CK, CS, CF, BS, and FI). The study revealed that there were no significant changes in soil enzyme activities (BG, CBH, NAG, LAP, and ALP) related to the cycling of C, N, and P under different management practices (CK, CS, CF, BS, and FI) (Table 3).

3.2. Alterations in the Relative Abundance and Diversity of Soil Microbial Phyla Under Various Artificial Long-Term Disturbance Treatments

Under different artificial long-term disturbance treatments (CK, CS, CF, BS, and FI), for bacteria, *Actinomycetota* and *Pseudomonadota* accounted for the largest proportion (Figure 1a,b), and for fungi, *Ascomycota* accounted for the largest proportion (Figure 1c,d). Bacterial alpha diversity (Chao1, Dominance, Observed_features, Pielou_e, Shannon, and Simpson) did not show significant variation across various artificial long-term disturbance treatments in the study (Figure 2a–f). In the CK treatment, the fungal Chao1 index demonstrated significantly higher values when contrasted with the CS, BS, and FI treatments (Figure 2g). Likewise, the fungal Observed_features index observed in the CK treatment was notably greater than in the CS, CF, BS, and FI treatments (Figure 2i). Moreover, the fungal Pielou_e index in CK treatment exhibited a significant increase relative to CF and FI treatments (Figure 2j). Furthermore, there was a significant rise in the fungal Shannon index in both the CK and CS groups compared to the CF, BS, and FI groups (Figure 2k). Conversely, the fungal Dominance index in FI treatment significantly exceeded that in CK and CS treatments (Figure 2h). Compared to the CK and CS treatments, the fungal Simpson index in the FI treatment was significantly the lowest (Figure 2l).

Table 3. Soil physical and chemical properties, microbial biomass, and enzyme activity.

Treatment	SMC (%)	pH	EC (μs·cm^{-1})	SOC (g·kg^{-1})	TN (g·kg^{-1})	TP (g·kg^{-1})	TK (g·kg^{-1})	AN (mg·kg^{-1})	AP (mg·kg^{-1})
CK	0.12 ± 0.03 b	8.20 ± 0.16 a	781.30 ± 302.76 a	1.82 ± 0.19 ab	0.18 ± 0.01 a	0.58 ± 0.02 a	18.76 ± 0.30 a	9.08 ± 1.56 a	4.48 ± 1.00 a
CS	0.13 ± 0.02 b	8.30 ± 0.16 a	265.98 ± 64.30 a	1.92 ± 0.24 a	0.19 ± 0.03 a	0.57 ± 0.01 a	18.81 ± 0.18 a	8.60 ± 2.17 ab	4.79 ± 1.03 a
CF	0.17 ± 0.04 b	8.52 ± 0.10 a	206.68 ± 29.26 a	1.61 ± 0.11 ab	0.16 ± 0.02 a	0.57 ± 0.02 a	18.44 ± 0.35 a	6.95 ± 0.79 ab	3.80 ± 0.37 a
BS	0.15 ± 0.05 b	8.50 ± 0.09 a	252.18 ± 32.72 a	1.39 ± 0.16 ab	0.14 ± 0.02 a	0.57 ± 0.01 a	18.32 ± 0.24 a	6.61 ± 0.98 ab	3.57 ± 0.70 a
FI	1.78 ± 0.92 a	8.46 ± 0.22 a	495.83 ± 281.19 a	1.28 ± 0.09 b	0.12 ± 0.01 a	0.59 ± 0.03 a	18.26 ± 0.16 a	4.47 ± 0.69 b	2.64 ± 0.33 a

Treatment	AK (mg·kg^{-1})	MBC (mg·kg^{-1})	MBN (mg·kg^{-1})	MBP (mg·kg^{-1})	BG (nmol·g^{-1}·h^{-1})	CBH (nmol·g^{-1}·h^{-1})	NAG (nmol·g^{-1}·h^{-1})	LAP (nmol·g^{-1}·h^{-1})	ALP (nmol·g^{-1}·h^{-1})
CK	191.25 ± 16.63 ab	60.50 ± 7.41 a	3.28 ± 0.32 a	2.65 ± 0.16 a	10.45 ± 1.01 a	1.15 ± 0.27 a	3.95 ± 0.14 a	59.75 ± 2.80 a	7.68 ± 0.52 a
CS	207.00 ± 25.25 a	58.32 ± 8.20 a	3.08 ± 0.41 a	2.46 ± 0.41 a	12.35 ± 2.17 a	1.64 ± 0.55 a	4.20 ± 0.12 a	64.71 ± 4.61 a	7.87 ± 0.59 a
CF	183.50 ± 20.67 ab	61.40 ± 3.53 a	3.31 ± 0.22 a	2.67 ± 0.47 a	10.30 ± 0.68 a	1.29 ± 0.32 a	4.03 ± 0.68 a	61.91 ± 2.87 a	6.66 ± 0.45 a
BS	188.25 ± 6.83 ab	52.88 ± 4.40 a	3.29 ± 0.40 a	2.27 ± 0.31 a	10.18 ± 1.18 a	1.26 ± 0.19 a	3.59 ± 0.11 a	63.34 ± 3.44 a	6.44 ± 0.71 a
FI	136.00 ± 28.67 b	52.24 ± 8.01 a	3.21 ± 0.48 a	2.06 ± 0.26 a	9.17 ± 0.77 a	0.72 ± 0.13 a	3.56 ± 0.06 a	58.19 ± 2.47 a	6.04 ± 0.63 a

Note: Different lowercase letters (a and b) indicate that the different artificial long-term disturbance treatments have significant differences (Duncan test, $p < 0.05$). CK, control group; CS, cutting in spring; CF, cutting in fall; BS, burning in spring; FI, flood water irrigation.

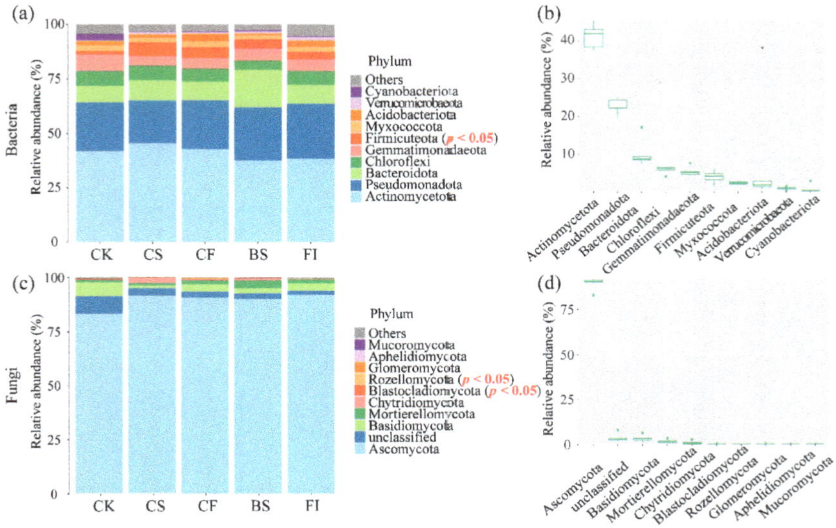

Figure 1. Relative abundance and proportion of soil microbial phyla (bacteria and fungi) under different artificial long-term disturbance treatments. (**a**,**c**) indicates the relative abundance proportion of soil bacteria and fungi under different artificial long-term disturbance treatments. (**b**,**d**) show that under different artificial long-term interference treatments, the relative abundance of phyla levels is ordered from large to small.

Notably, both burning and spring-cutting treatments significantly enhanced soil fungal alpha diversity.

3.3. LEfSe Analysis of Soil Microbial Communities Under Various Artificial Long-Term Disturbance Treatments

The *o_Frankiales*, *f_Geodermatophilaceae*, *f_Longimicrobiaceae*, *o_Longimicrobiales*, *c_Longimicrobia*, *g_Blastococcus*, *g_Geodermatophilus*, *o_Acetobacterales*, *f_Acetobacteraceae*, *f_Microtrichaceae*, *g_Roseomonas*, and *g_Modestobacter* were identified as biomarkers of soil bacterial communities in the CK. In the CS treatment, the *o_Bacillales*, *f_Planococcaceae*, *g_Domibacillus*, *f_Azospirillaceae*, *o_Azospirillales*, *g_Actinomycetospora*, *f_Streptosporangiaceae*, *s_planctomycete_WY108*, and *g_unidentified_WD2101_soil_group* emerged as the respective biomarkers. For the CF treatment, the *g_Saccharothrix*, *c_Clostridia*, *g_Haliangium*, *f _Haliangiaceae*, *o_Haliangiales*, *s_Massilia_albidiflava*, and *f_Methylopilaceae* served as the biomarkers for soil bacterial communities. In the BS treatment, *p_Armatimonadota*, *o_Armatimonadales*, *c_Armatimonadia*, and *o_Sphingobacterialesas* biomarkers for soil bacterial communities. In the FI treatment, *c_Gammaproteobacteria*, *o_Pseudomonadales*, *f_Sandaracinaceaes*, *f_Sandaracinaceae*, *s_Nocardioides_dilutes*, *o_Salinisphaerales*, and *g_Azoarcus* was the biomarkers for soil bacterial communities (Figure 3A).

The *s_Issatchenkia_orientalis*, *c_Saccharomycetes*, *f_Saccharomycetaceae*, *g_Issatchenkia*, *o_Saccharomycetales*, *f_Sporormiaceae*, *s_Sporormiella_megalospora*, *f_Plectosphaerellaceae*, *o_Glomerellales*, *s_Aspergillus_kalimae*, *s_Verticillium_dahliae*, *g_Verticillium*, *g_Glomeraceae_gen_incertae_sedis*, *g_Sporormiella*, *s_Botryotrichum_atrogriseum*, *g_Botryotrichum*, and *f_Clavicipitaceae* were identified as the biomarkers of soil fungal communities in the CK, according to LEfse analysis. The *f_Clavicipitaceae*, *c_Dothideomycetes*, *o_Botryosphaeriales*, *f_Botryosphaeriaceae*, *f_Didymellaceae*, *g_Volutellas*, and *s_Volutella_ciliata* were identified as the biomarkers of soil fungal communities in the CS. In the CF treatment, *s_Solicoccozyma_terricola* served as the biomarker for soil fungal communities. In the BS treatment, *s_Biappendiculispora_sp*, *g_Biappendiculispora*, and *s_Preussia_lignicola* are biomarkers for soil fungal communities. In the FI treatment, *f_Nectriaceae*, *g_Fusarium*, *o_Hypocreales*, *c_Sordariomycetes*, *c_Tremellomycetes*, *s_Saitozyma_podzolica*, *f_Trimorphomycetaceae*, *g_Saitozyma*, *g_Mortierella*, and *o_Tremellales* were the biomarkers for soil fungal communities (Figure 3B).

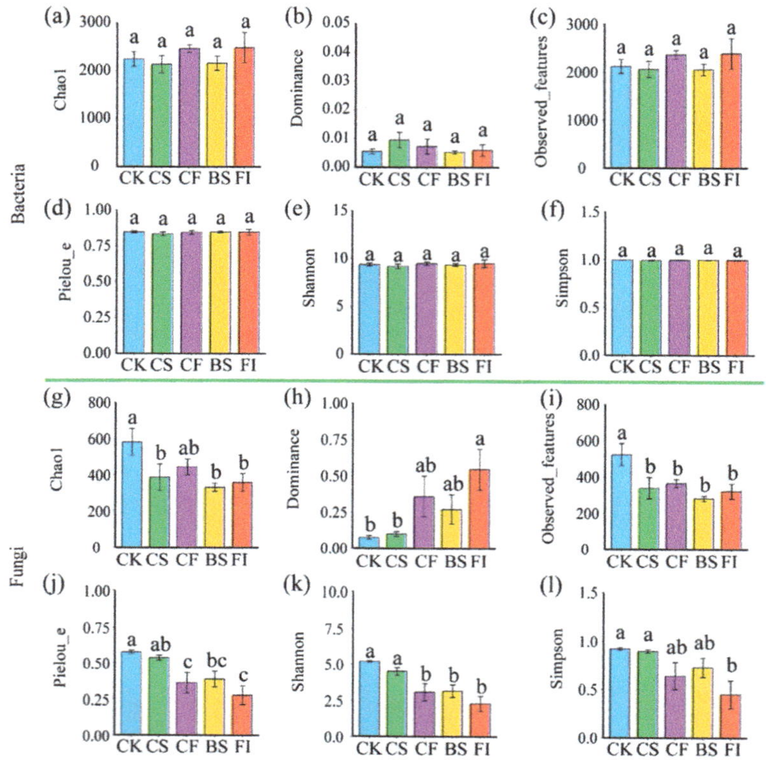

Figure 2. Effects of soil bacterial and fungal alpha diversity under different artificial long-term disturbance treatments. (**a–f**) represents the changes of soil bacterial diversity (Chao1, Dominance, Observed_features, Pielou_e, Shannon, and Simpson index, respectively) under different artificial long-term disturbance treatments. (**g–l**) represents the changes of soil fungal diversity (Chao1, Dominance, Observed_features, Pielou_e, Shannon, and Simpson index, respectively) under different artificial long-term disturbance treatments. Different lowercase letters (a–c) indicate significant differences among different artificial long-term disturbance treatments at the $p < 0.05$ level (ANOVA and Duncan's test).

3.4. *Multivariate Analyses (NMDS, PCA, and RDA) to Understand Key Interactions Among Soil Microbial Communities, Physicochemical Properties, Microbial Biomass, and Enzyme Activity*

We found that the linear fit between the observed dissimilarity and ordination distance of the bacterial and fungal communities was higher than 0.8 (Figure 4a,d). Under different artificial long-term disturbance treatments, NMDS revealed that bacterial community variations were predominantly localized, while fungal community variations were comparatively scattered (Figure 4b,e). In the first principal component, bacterial communities were notably reduced in BS compared to FI, while fungal communities in CK and CS were notably lower than those in BS, CF, and FI (Figure 4c,f). The variance in soil bacterial communities was primarily influenced by soil SMC, EC, SOC, and CBH (Figure 5a,b). Conversely, soil SOC and BG were the primary determinants of fungal community variation (Figure 5c,d).

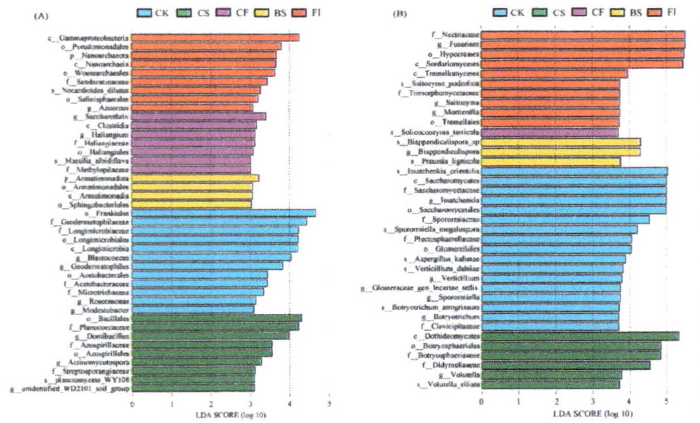

Figure 3. Biomakes screened by linear discriminant analysis effect size (LEfSe) analysis based on the community level. (**A**,**B**) LEfSe analysis of soil bacterial and fungal communities under different artificial long-term disturbance treatments, respectively. The figure shows the biomarkers with statistical differences between groups, and the horizontal axis is the LDA score. The bacterial communities with LDA scores greater than 3.01 are shown. The fungal communities with LDA scores greater than 3.69 are shown.

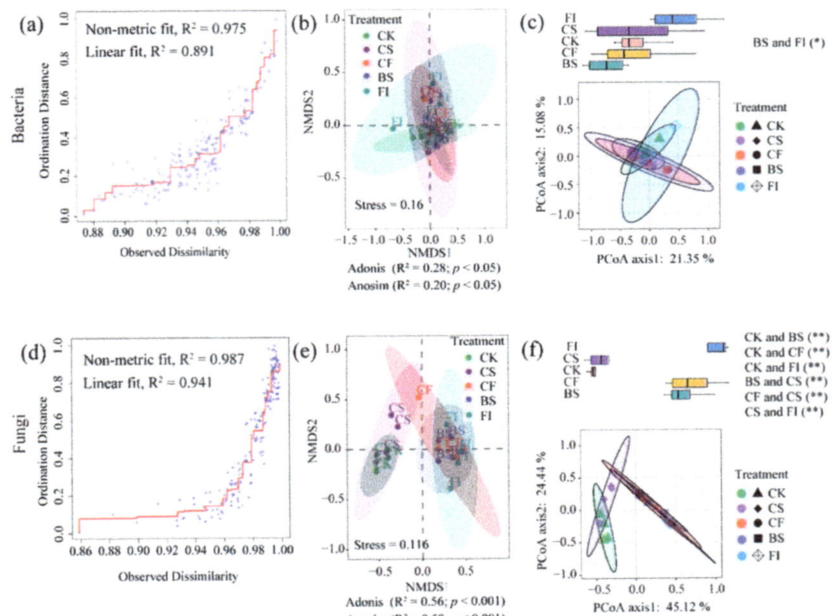

Figure 4. Principal component analysis (PCA) and non-metric multidimensional scaling analysis (NMDS) for soil bacterial and fungal communities (based on OTUs levels) under different artificial long-term disturbance treatments. (**a**,**d**) represent the relationship between observed dissimilarity and ordination distance of soil bacterial and fungal communities, respectively. (**b**,**e**) represent the non-metric multidimensional scaling analysis of soil bacterial and fungal communities under different artificial long-term disturbance treatments, respectively. (**c**,**f**) represent the principal component analysis of soil bacterial and fungal communities under different artificial long-term disturbance treatments, respectively. The data of the first axis of the principal component are selected in the histogram to compare the differences between different artificial long-term disturbance treatments. Significance codes, "*", $p < 0.05$; "**", $p < 0.01$.

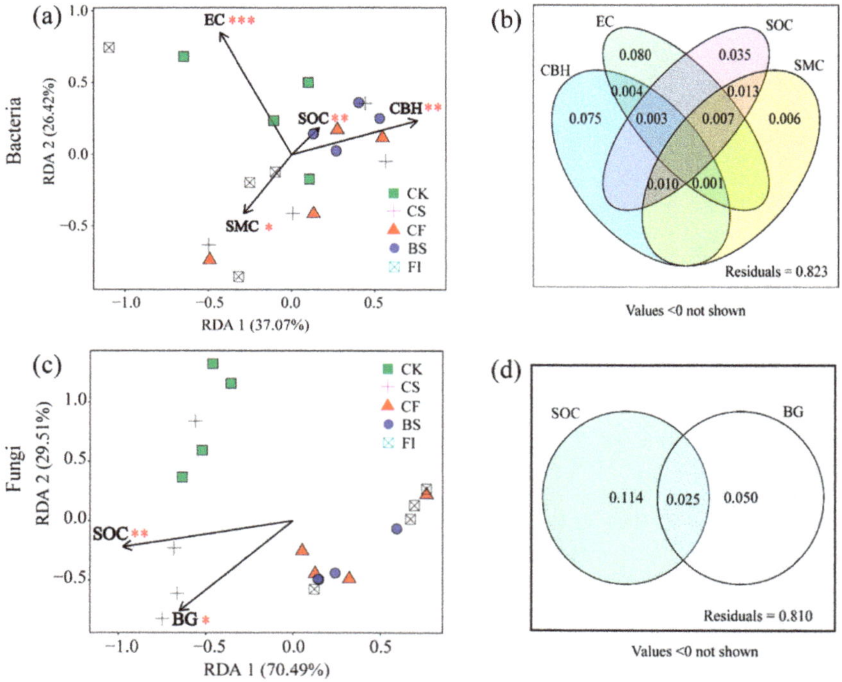

Figure 5. Redundancy analysis (RDA) and Venn diagrams among soil properties and bacterial and fungal communities (based on OTUs levels). (**a**,**c**) represent the effects of soil factors on soil bacterial and fungal microbial communities (based on OTUs levels), respectively. The screening indicators were screened by ordistep, and the significance test (ANOVA) was performed. (**b**,**d**) represent the percentage of individual and joint interpretations of soil factors. Significance codes, "*", $p < 0.05$; "**", $p < 0.01$; "***", $p < 0.001$.

3.5. Correlation of Soil Microbial Diversity with Soil Physicochemical Properties, Microbial Biomass, and Enzyme Activity

The CK demonstrated a significant correlation between the bacterial Pielou_e and soil AP, along with a notable relationship between the fungal alpha diversity (Chao1, Observed_features, and Shannon index) and soil pH, TK, MBC, MBN, MBP, BG, and CBH (Figure 6a,b). In CS, the bacterial alpha diversity showed a strong connection with soil TP, CBH, and LAP. Fungal Dominance and Shannon index showed a significant correlation with soil EC, TK, and AP (Figure 6c,d). In CF treatment, soil TK and AK were significantly correlated with the fungal alpha diversity (Chao1, Dominance, Pielou_e, Shannon, and Simpson index) (Figure 6f). There was no correlation between the soil's physical and chemical properties and soil bacteria (in CF, BS, and FI treatments) and fungal alpha diversity (in BS treatments) (Figure 6e,g–i). Conversely, In the case of FI, the fungal alpha diversity (Dominance, Pielou_e, Shannon, and Simpson index) was found to be significantly correlated with soil TP, BG, and CBH (Figure 6j).

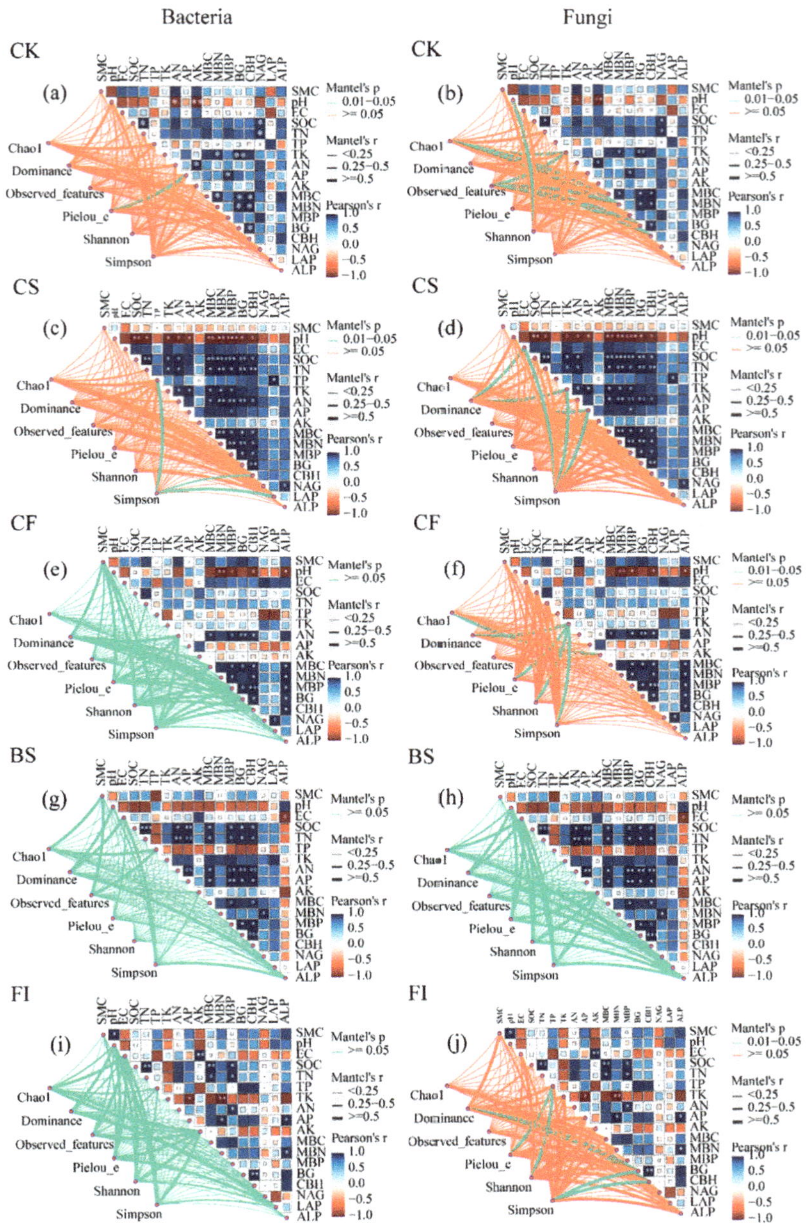

Figure 6. Mantel test analysis of soil bacterial and fungal alpha diversity and soil properties under different artificial long-term disturbance treatments. The size and color of the square in the figure represent the significant difference in soil properties. The greater the significant difference, the larger the square. The color of the square is blue, indicating a positive correlation, and the color of the square is red, indicating a negative correlation. The orange and green indicate the correlation between alpha diversity and soil factors. In (**a–d**,**f**,**j**), green indicates that the significance range is between 0.01 and 0.05, and orange indicates that the significance is greater than or equal to 0.05. In (**e**,**g–i**), green indicates that the significance is greater than or equal to 0.05. Significance codes, "*", $p < 0.05$; "**", $p < 0.01$; "***", $p < 0.001$.

3.6. Influence of Soil Physicochemical Properties, Microbial Biomass, and Enzyme Activity on Microbial Diversity Under Various Artificial Long-Term Disturbance Treatments

The analysis using RF indicated that the bacterial Dominance index was greatly influenced by soil AP and ALP, whereas the Simpson index was notably impacted by soil SMC and MBN (Figure 7b,f). In contrast, the fungal Dominance index was significantly influenced by soil AP, while the Pielou_e index was affected by soil EC, SMC, and AP (Figure 7h,j). Additionally, the fungal Shannon index showed significant associations with soil TP, AP, LAP, and NAG, and the Simpson index was influenced by SMC, AP, and TN (Figure 7k,l).

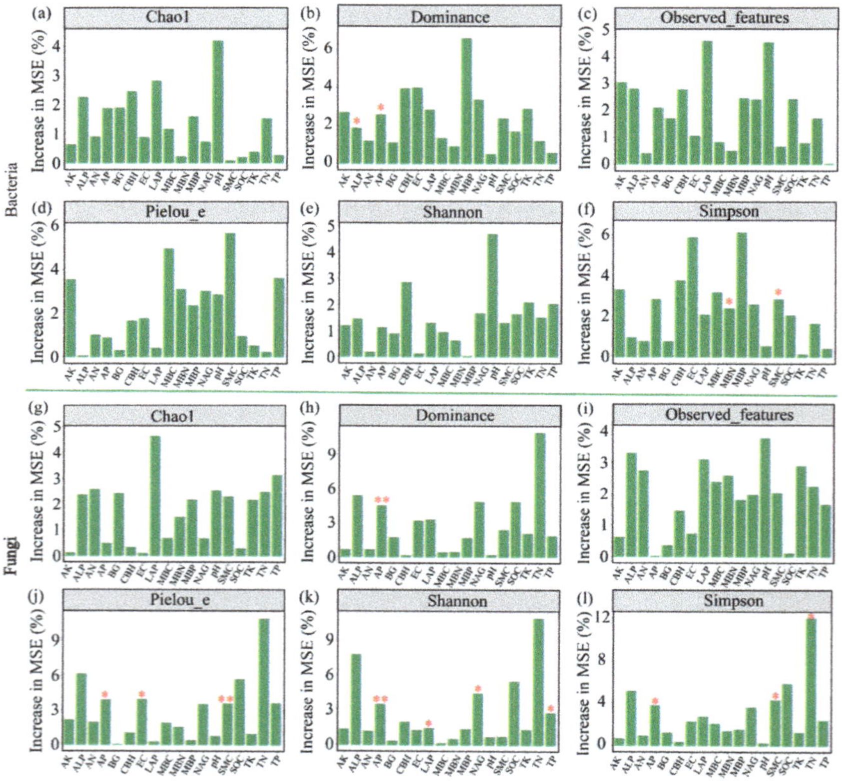

Figure 7. Random forest analysis of soil physical and chemical properties affecting soil bacterial and fungal alpha diversity under different artificial long-term disturbance treatments. **(a–f)** represent the effects of soil properties on bacterial diversity (Chao1, Dominance, Observed_features, Pielou_e, Shannon, and Simpson index) under different artificial long-term disturbance treatments, respectively. **(g–l)** represent the effects of soil properties on fungal diversity (Chao1, Dominance, Observed_features, Pielou_e, Shannon, and Simpson index) under different artificial long-term disturbance treatments, respectively. Significance codes, "*", $p < 0.05$; "**", $p < 0.01$.

4. Discussion

4.1. Effects of Various Management Patterns on Soil Microbial Communities and Diversity

In this study, *Actinomycetota*, *Pseudomonadota*, and *Ascomycota* emerged as the most prevalent bacterial and fungal phylum in desert soil across various management patterns. These findings are consistent with previous research, which identified *Actinomycetota*, *Pseudomonadota*, *Firmicuteota*, and *Bacteroidetes* as the dominant bacterial taxa in desert soils [53]. This consistency can be attributed to the nutrient-deficient conditions typical of desert ecosystems, where soil microbes predominantly depend on plant-derived organic matter for sustenance [13,54]. These microbes play a critical role in decomposing plant litter, thereby enhancing the soil's nutrient content [9,55]. *Actinomycetota* is known to dominate desert regions due to their ability to degrade cellulose and lignin in plant residues, thus providing essential nutrients for the soil [53,54,56]. Similarly, *Firmicuteota* can produce endospores, enabling them to withstand desiccation and environmental stress while stabilizing substrates and introducing nutrients, which allows them to thrive in nutrient-deficient sandy soils [57–59]. Additionally, *Ascomycota* is vital in arid regions with limited nutrients and rainfall, forming symbiotic relationships with other organisms to decompose organic material, improve nutrient availability, and promote plant growth under challenging environmental conditions [60–62]. The LEfSe analysis revealed that during the cutting treatment in spring, *f_Planococcaceae* and *f_Azospirillaceae* emerged as biomarkers for soil bacterial communities, whereas *f_Botryosphaeriaceae* and *f_Didymellaceae* were linked to soil fungal communities. In this context, *A. sparsifolia*, a leguminous plant, may establish a mutually beneficial relationship with rhizobia, facilitating nitrogen fixation and enhancing the nutrient-poor sandy soil affected by wind erosion [29,63]. This suggests that desert soils are nutrient-deficient, with bacteria playing a significant role in alleviating nitrogen limitations through ammonification of easily decomposed substances [2,64]. Changes in fungal diversity may also be influenced by geographic ecological conditions and varying tolerances to environmental factors [29,64].

Physical disturbances significantly impact the spatial distribution and temporal dynamics of soil microbial communities in arid environments, including sand covering, tree removal, and wildfires [13,65]. For instance, fire and post-fire management practices have been shown to alter soil organic carbon, pH, and enzyme levels, which in turn influence the diversity and abundance of soil bacteria and fungi [61,66]. This study revealed that the spring burning treatment significantly increased soil fungal alpha diversity. This effect may stem from the retention of ash, unburned residues, and root materials in the soil, which provide carbon and nitrogen sources for microbial decomposition [9,13,29]. This process contributes essential nutrients for soil fungi, promoting the stability of microbial communities and enhancing energy substrates in the soil, thereby affecting the nutrient dynamics of various plant communities in desert ecosystems [67,68]. Moreover, the spring-cutting treatment significantly improved the Shannon index for soil fungi. Additionally, soil organic carbon and available potassium contents were notably higher under the spring-cutting treatment compared to the floodwater irrigation treatment. The processes of cutting and burning may lead to a reduction in readily available nutrients, such as carbon and nitrogen, for soil microbes [23,69]. Furthermore, cutting and burning facilitate the breakdown, uptake, and alteration of nutrients and root exudates by soil microbes, ultimately creating a favorable environment for microbial growth and development [70,71].

4.2. Effects of Environmental Factors on Soil Microbial Communities and Diversity

The distribution, diversity, and composition of soil microbial communities are predominantly influenced by the interplay of biological and abiotic factors [72,73]. In arid ecosystems, microbial diversity is significantly influenced by various abiotic factors, including fluctuating temperatures, intense ultraviolet radiation, soil moisture levels, and nutrient availability. Additionally, biotic factors, such as plant species diversity and abundance, play a crucial role in shaping microbial communities [73,74]. The findings of this research indicate that the Dominance index for both bacterial and fungal communities was notably influenced by soil-available phosphorus, while the Simpson index for these communities was significantly impacted by SMC. Previous studies have demonstrated that variations in temperature and moisture can influence the decomposition and transformation of organic substances, thereby affecting the utilization of organic matter and nitrogen by soil microorganisms [75,76]. In this study, key factors affecting soil bacterial communities included EC, SMC, SOC, and CBH. In contrast, SOC and BG were identified as major drivers of variation in soil fungal communities. Cutting treatments significantly increased soil β-glucosidase activity, urease activity, phosphatase activity, moisture content, total nitrogen, total phosphorus, and hydrolyzable nitrogen, thus enriching the soil with nutrients for microorganisms [25,77] and promoting microbial reproduction [54,77,78]. These findings suggest that the rational utilization of desert plants can provide additional energy resources to the desert ecosystem, with cutting and burning serving as effective methods to achieve this goal [79]. In this study, no significant changes were observed in soil enzyme activities (Cellobiohydrolase, β-1,4-glucosidase, β-1,4-N-acetylglucosaminidase, Lleucine aminopeptidase, and Alkaline phosphatase) related to the cycling of carbon, nitrogen, and phosphorus under different management practices. However, other studies suggest that burning can influence soil pH and alter the availability of carbon and nitrogen, leading to nutrient losses that impact the metabolic activities of soil microbes [26,27]. While burning has been shown to significantly reduce extracellular enzyme activity associated with carbon, nitrogen, and phosphorus acquisition, it can simultaneously enhance oxidative enzyme activity [28]. The effects of cutting on soil enzyme activity and microbial biomass, however, largely depend on the intensity and duration of the treatment. Moderate cutting may have minimal impact on soil nutrient dynamics and microbial activity, whereas intensive or prolonged cutting can lead to substantial changes [80,81]. Additionally, factors such as soil type, climate conditions, and vegetation composition may buffer the effects of different management practices on soil enzyme activity and microbial biomass [82,83]. Soils with high organic matter content or well-established microbial communities may exhibit greater resilience to disturbances like cutting, thereby maintaining stable enzyme activities and nutrient levels [81,84]. This suggests that burning may contribute to a reduction in readily available nutrients such as carbon and nitrogen for soil microbes [23,69], which is inconsistent with the results of this study. These findings indicate that changes in soil organic carbon and available nutrients did not alter soil extracellular enzyme activity under different management measures.

Furthermore, the study found that the alpha diversity of fungi following floodwater irrigation was significantly lower than that of those in the control group. Additionally, available nitrogen in the control group was notably higher compared to the floodwater irrigation treatment. This suggests that long-term floodwater irrigation may lead to nutrient leaching from the surface to deeper soil layers, thereby addressing the nutrient requirements of the deeper soil [85,86]. Field experiments revealed an increase in total nitrogen content in the topsoil following floodwater irrigation, with available soil nutrients showing a tendency to migrate toward lower soil layers [87]. This indicates that the combined presence of water and heat after irrigation enhances microbial activity, promoting the decomposition of organic matter and improving nutrient availability in the topsoil [88]. The future direction of this research will focus on continued long-term monitoring to provide policymakers with insights into the rational use of desert plants and land management strategies.

4.3. Limitations and Considerations of the Study

This study provides valuable insights into the effects of management practices on soil microbial communities in desert ecosystems. However, several limitations should be acknowledged. One key limitation is the lack of long-term data, which is essential for assessing the sustained impacts of different management strategies on soil microbial diversity and function over time. Additionally, while our findings are relevant to desert ecosystems, they may not be directly generalizable to other environments with distinct ecological conditions, such as temperate forests or grasslands. Moreover, although the study incorporates multiple abiotic and biotic factors, it may not account for all potential environmental variables that influence microbial communities, including anthropogenic disturbances, extreme weather events, or broader climatic changes. While cutting and burning were identified as effective interventions, this study does not fully address their long-term ecological tradeoffs, such as potential soil degradation, shifts in microbial functional potential, or declines in biodiversity. To address these limitations, future research should prioritize long-term field experiments and monitoring programs to evaluate the prolonged effects of management practices on soil microbial dynamics. Additionally, employing advanced molecular techniques, such as metagenomics, stable isotope probing, and transcriptomics, would help elucidate the intricate relationships between soil microbes, nutrient cycling, and plant-soil feedback mechanisms. By integrating these approaches, future studies can enhance our understanding of sustainable land management in arid environments, ensuring that interventions optimize soil health while minimizing ecological risks.

5. Conclusions

Our study demonstrates that long-term artificial disturbances—including cutting in spring, cutting in fall, burning in spring, and flood water irrigation—significantly shape soil microbial communities in desert ecosystems. *Actinomycetota* and *Pseudomonadota* emerged as the dominant bacterial groups, while Ascomycota was the most abundant fungal group across all treatments. Notably, the control group exhibited the highest fungal alpha diversity (Chao1, Observed Features, Pielou's Evenness, Shannon, and Simpson indices), whereas floodwater irrigation resulted in the highest Dominance index, indicating shifts in microbial community structure. Our findings underscore the pivotal role of soil organic carbon (SOC) in influencing both bacterial and fungal communities. Additionally, cellobiohydrolase (CBH) was identified as a key factor shaping bacterial communities, while β-glucosidase (BG) played a crucial role in determining fungal community variations. These results provide valuable insights into the sustainable management of desert ecosystems, emphasizing the importance of responsible interventions. We advocate for carefully regulated cutting and burning practices, as these methods promote plant regeneration and microbial-driven organic matter decomposition, thereby enriching desert soils with essential nutrients. This research highlights the need for evidence-based management strategies to maintain soil health and microbial diversity, ensuring the long-term resilience and productivity of desert ecosystems.

Author Contributions: Conceptualization, Y.Z., W.I., Z.M. and F.Z.; methodology, Y.Z., Y.D., Z.Z., Z.M. and N.C.T.G.; software, Y.Z. and Z.Z.; validation, W.I. and F.Z.; formal analysis, Y.Z., Y.D., Z.Z., Z.M. and N.C.T.G.; investigation, Y.Z., Y.D., Z.M. and Z.Z.; writing—original draft preparation, Y.Z. and Y.D.; writing—review and editing, W.I., Z.M. and F.Z.; project administration, Z.M. and F.Z.; funding acquisition, Z.M. and F.Z. All authors have read and agreed to the published version of the manuscript.

Funding: This research was funded by the Natural Science Foundation of Xinjiang Uygur Autonomous Region, grant number 2024D01B86; National Natural Science Foundation of China, grant number 42271071; Startup Foundation for Introducing Talent of Xinjiang Institute of Ecology and Geography, CAS, grant number E4500109; National Key Research and Development Project of China, grant number 2022YFF1302504; and Tianshan Talents Program of Xinjiang Autonomous Region, grant number 2023TSYCLJ0046.

Data Availability Statement: Data will be made available on request. If requested, please contact the first author.

Acknowledgments: The authors are thankful to Susan Sen (native English speaker from Edinburgh, United Kingdom), English Lecturer at Rongqiao International School, Fuzhou, for her great support and help in revising the English writing of the article. Special thanks also go to Norela C. T. Gonzalez from Panama, Latin America, for her assistance with the English language and manuscript preparation. Norela C. T. Gonzalez is affiliated with the College of Forestry, Central South University of Forestry and Technology, Changsha 410004, Hunan, China. We further thank all the departmental colleagues for their help and support. We would like to thank all the reviewers who participated in the review during the preparation of this manuscript.

Conflicts of Interest: The authors declare no conflicts of interest.

Abbreviations

The following abbreviations are used in this manuscript:

CK	Control group
CS	Cutting in spring
CF	Cutting in fall
BS	Burning in spring
FI	Flood water irrigation
SMC	Soil moisture content
EC	Electrical conductivity
SOC	Soil organic carbon
TN	Total nitrogen
TP	Total phosphorus
TK	Total potassium
AN	Available nitrogen
AP	Available phosphorus
AK	Available potassium
MBC	Microbial biomass carbon
MBN	Microbial biomass nitrogen
MBP	Microbial biomass phosphorus

References

1. Tedersoo, L.; Bahram, M.; Polme, S.; Kõljalg, U.; Yorou, N.S.; Wijesundera, R.; Ruiz, L.V.; Vasco-Palacios, A.M.; Thu, P.Q.; Suija, A.; et al. Global diversity and geography of soil fungi. *Science* **2014**, *346*, 1256688. [CrossRef] [PubMed]
2. Zhang, H.J.; Dong, L.Z.; Yao, X.D.; Wang, W. Soil fertility shifts the relative importance of saprotrophic and mycorrhizal fungi for maintaining ecosystem stability. *Glob. Chang. Biol.* **2023**, *29*, 1206–1216. [CrossRef]
3. Bahram, M.; Hildebrand, F.; Forslund, S.K.; Anderson, J.L.; Soudzilovskaia, N.A.; Bodegom, P.M.; Bengtsson-Palme, J.; Anslan, S.; Coelho, L.P.; Harend, H.; et al. Structure and function of the global topsoil microbiome. *Nature* **2018**, *560*, 233–237. [CrossRef]
4. Zhang, Z.H.; Tariq, A.; Zeng, F.J.; Graciano, C.; Zhang, B. Nitrogen application mitigates drought-induced metabolic changes in *Alhagi sparsifolia* seedlings by regulating nutrient and biomass allocation patterns. *Plant Physiol. Biochem.* **2020**, *155*, 828–841. [CrossRef] [PubMed]
5. Jansson, J.K.; Hofmockel, K.S. Soil microbiomes and climate change. *Nat. Rev. Microbiol.* **2020**, *18*, 35–46. [CrossRef] [PubMed]
6. Trivedi, P.; Batista, B.D.; Bazany, K.E.; Singh, B.K. Plant–microbiome interactions under a changing world: Responses, consequences and perspectives. *New Phytol.* **2022**, *234*, 1951–1959. [CrossRef] [PubMed]
7. Geisen, S.; Heinen, R.; Andreou, E.; van Lent, T.; ten Hooven, F.C.; Thakur, M.P. Contrasting effects of soil microbial interactions on growth–defence relationships between early- and mid-successional plant communities. *New Phytol.* **2022**, *233*, 1345–1357. [CrossRef]
8. Islam, W.; Zeng, F.J.; Alotaibi, M.O.; Khan, K.A. Unlocking the potential of soil microbes for sustainable desertification management. *Earth-Sci. Rev.* **2024**, *252*, 104738. [CrossRef]
9. Ma, Y.; Dias, M.C.; Freitas, H. Drought and salinity stress responses and microbe-induced tolerance in plants. *Front. Plant Sci.* **2020**, *11*, 591911. [CrossRef]
10. Li, C.J.; Han, H.; Ablimiti, M.; Liu, R.; Zhang, H.; Fan, J.L. Morphological and physiological responses of desert plants to drought stress in a man-made landscape of the Taklimakan desert shelter belt. *Ecol. Indic.* **2022**, *140*, 109037. [CrossRef]

11. Islam, W.; Ullah, A.; Zeng, F.J. Response of total belowground soil biota in *Alhagi sparsifolia* monoculture at different soil vertical profiles in desert ecosystem. *Sci. Total Environ.* **2023**, *901*, 166027. [CrossRef] [PubMed]
12. Liu, H.; Li, J.; Singh, B.K. Harnessing co-evolutionary interactions between plants and Streptomyces to combat drought stress. *Nat. Plants* **2024**, *10*, 1159–1171. [CrossRef]
13. Gao, Y.J.; Akash, T.; Zeng, F.J.; Jordi, S.; Josep, P.; Zhang, Z.H.; Waqar, I.; Xu, M.Q. "Fertile islands" beneath three desert vegetation on soil phosphorus fractions, enzymatic activities, and microbial biomass in the desert-oasis transition zone. *Catena* **2022**, *212*, 106090. [CrossRef]
14. Feng, J.G.; He, K.Y.; Zhang, Q.F.; Han, M.G.; Zhu, B. Changes in plant inputs alter soil carbon and microbial communities in forest ecosystems. *Glob. Change Biol.* **2022**, *28*, 3426–3440. [CrossRef] [PubMed]
15. Yang, Y.; Zhang, Y.G.; Yu, X.X.; Jia, G.D. Soil microorganism regulated aggregate stability and rill erosion resistance under different land uses. *Catena* **2023**, *228*, 107176. [CrossRef]
16. Ren, C.J.; Wang, T.; Xu, Y.D.; Deng, J.; Zhao, F.Z.; Yang, G.H.; Han, X.H.; Feng, Y.Z.; Ren, G.X. Differential soil microbial community responses to the linkage of soil organic carbon fractions with respiration across land-use changes. *Forest Ecol. Manag.* **2018**, *409*, 170–178. [CrossRef]
17. Muyzer, G.; Stams, A.J.M. The ecology and biotechnology of sulphate-reducing bacteria. *Nat. Rev. Microbiol.* **2008**, *6*, 441–454. [CrossRef]
18. Coban, O.; de Deyn, G.B.; van der Ploeg, M. Soil microbiota as game-changers in restoration of degraded lands. *Science* **2022**, *375*, abe0725. [CrossRef]
19. Peng, Z.H.; Qian, X.; Liu, Y.; Li, X.M.; Gao, H.; An, Y.N.; Qi, J.J.; Jiang, L.; Zhang, Y.R.; Chen, S.; et al. Land conversion to agriculture induces taxonomic homogenization of soil microbial communities globally. *Nat. Commun.* **2014**, *15*, 3624. [CrossRef] [PubMed]
20. Barnard, R.L.; Osborne, C.A.; Firestone, M.K. Responses of soil bacterial and fungal communities to extreme desiccation and rewetting. *ISME J.* **2013**, *7*, 2229–2241. [CrossRef] [PubMed]
21. Worrich, A.; Stryhanyuk, H.; Musat, N.; König, S.; Banitz, T.; Centler, F.; Frank, K.; Thullner, M.; Harms, H.; Richnow, H.H.; et al. Mycelium-mediated transfer of water and nutrients stimulates bacterial activity in dry and oligotrophic environments. *Nat. Commun.* **2017**, *8*, 15472. [CrossRef] [PubMed]
22. Gruber, N.; Galloway, J.N. An Earth-system perspective of the global nitrogen cycle. *Nature* **2008**, *451*, 293–296. [CrossRef] [PubMed]
23. Högberg, P.; Nordgren, A.; Buchmann, N.; Taylor, A.F.; Ekblad, A.; Högberg, M.N.; Nyberg, G.; Ottosson-Löfvenius, M.; Read, D.J. Large-scale forest girdling shows that current photosynthesis drives soil respiration. *Nature* **2001**, *411*, 789–792. [CrossRef] [PubMed]
24. Kim, Y.S.; Makoto, K.; Takakai, F.; Shibata, H.; Satomura, T.; Takagi, K.; Hatano, R.; Koike, T. Greenhouse gas emissions after a prescribed fire in white birch-dwarf bamboo stands in northern Japan, focusing on the role of charcoal. *Eur. J. Forest Res.* **2011**, *130*, 1031–1044. [CrossRef]
25. Huang, X.B.; Chen, J.Y.; Li, S.F.; Su, J.R. Selective logging effects on plant functional traits depend on soil enzyme activity and nutrient cycling in a Pinus yunnanensis forest. *Forest Ecol. Manag.* **2023**, *545*, 121284. [CrossRef]
26. Zhuo, C.X.; Tong, X.; Hao, B.E.; Chen, Z.; Liu, H.Y. Feasibility analysis using a porous media model to simulate the wind protection effect of windbreak forests. *Land. Degra Dev.* **2022**, *34*, 207–220. [CrossRef]
27. Arunrat, N.; Sereenonchai, S.; Kongsurakan, P.; Iwai, C.B.; Yuttitham, M.; Hatano, R. Post-fire recovery of soil organic carbon, soil total nitrogen, soil nutrients, and soil erodibility in rotational shifting cultivation in Northern Thailand. *Front. Environ. Sci.* **2023**, *11*, 1117427. [CrossRef]
28. Pei, J.M.; Wan, J.R.; Wang, H.; Fang, C.M.; Nie, M.; Li, J.Q. Changes in the activity of soil enzymes after fire. *Geoderma* **2023**, *437*, 116599. [CrossRef]
29. Li, M.M.; Petrie, M.D.; Chen, H.; Zeng, F.J.; Zeeshan, A.; Sun, X.B. Effects of groundwater and seasonal streamflow on the symbiotic nitrogen fixation of deep-rooted legumes in a dryland floodplain. *Geoderma* **2023**, *434*, 116490. [CrossRef]
30. Wang, Y.H.; Shi, F.Z.; Yao, P.; Sheng, Y.; Zhao, C.Y. Assessing the evolution and attribution of watershed resilience in arid inland river basins, Northwest China. *Sci. Total Environ.* **2024**, *906*, 167534. [CrossRef] [PubMed]
31. Li, J.Y.; Zhang, D.Y.; Liu, M. Factors controlling the spatial distribution of soil organic carbon in Daxing'anling Mountain. *Sci. Rep.* **2020**, *10*, 12659. [CrossRef] [PubMed]
32. Lu, R.K. *Soil Agrochemical Analysis*; China Agricultural Science and Technology Press: Beijing, China, 1999; pp. 80–89.
33. Warra, H.H.; Ahmed, M.A.; Nicolau, M.D. Impact of land cover changes and topography on soil quality in the Kasso catchment, Bale Mountains of southeastern Ethiopia. *Singapore J. Trop. Geogr.* **2015**, *36*, 357–375. [CrossRef]
34. Bao, S.D. *Analysis of Soil and Agrochemistry*, 3rd ed.; China Agriculture Press: Beijing, China, 2000; pp. 25–200.
35. Steinweg, J.M.; Dukes, J.S.; Wallenstein, M.D. Modeling the effects of temperature and moisture on soil enzyme activity: Linking laboratory assays to continuous field data. *Soil Biol. Biochem.* **2012**, *55*, 85–92. [CrossRef]

36. Qi, R.M.; Li, J.; Lin, Z.A.; Li, Z.J.; Li, Y.T.; Yang, X.D.; Zhang, J.J.; Zhao, B.Q. Temperature effects on soil organic carbon, soil labile organic carbon fractions, and soil enzyme activities under long-term fertilization regimes. *Appl. Soil Ecol.* **2016**, *102*, 36–45. [CrossRef]
37. Vance, E.D.; Brookes, P.C.; Jenkinson, D.S. An extraction method for measuring soil microbial biomass C. *Soil Biol. Biochem.* **1987**, *19*, 703–707. [CrossRef]
38. Jenkinson, D.S.; Brookes, P.C.; Powlson, D.S. Measuring soil microbial biomass. *Soil Biol. Biochem.* **2004**, *36*, 5–7. [CrossRef]
39. Magoč, T.; Steven, S.L. FLASH: Fast length adjustment of short reads to improve genome assemblies. *Bioinformatics* **2011**, *27*, 2957–2963. [CrossRef] [PubMed]
40. Bokulich, N.A.; Subramanian, S.; Faith, J.J.; Gevers, D.; Gordon, J.I.; Knight, R.; Mills, D.A.; Caporaso, J.G. Quality-filtering vastly improves diversity estimates from Illumina amplicon sequencing. *Nat. Methods* **2013**, *10*, 57–59. [CrossRef] [PubMed]
41. Rognes, T.; Flouri, T.; Nichols, B.; Quince, C.; Mahé, F. VSEARCH: A versatile open source tool for metagenomics. *PeerJ* **2016**, *4*, e2584. [CrossRef] [PubMed]
42. Caporaso, J.G.; Kuczynski, J.; Stombaugh, J.; Bittinger, K.; Bushman, F.D.; Costello, E.K.; Fierer, N.; Peña, A.G.; Goodrich, J.K.; Gordon, J.I.; et al. QIIME allows analysis of high-throughput community sequencing data. *Nat. Methods* **2010**, *7*, 335–336. [CrossRef] [PubMed]
43. Edgar, R.C. UPARSE: Highly accurate OTU sequences from microbial amplicon reads. *Nat. Methods* **2013**, *10*, 996–998. [CrossRef]
44. Wang, Q.; Garrity, G.M.; Tiedje, J.M.; Cole, J.R. Naive Bayesian classifier for rapid assignment of rRNA sequences into the new bacterial taxonomy. *Appl. Environ. Microbiol.* **2007**, *73*, 5261–5267. [CrossRef]
45. Quast, C.; Pruesse, E.; Yilmaz, P.; Gerken, J.; Schweer, T.; Yarza, P.; Peplies, J.; Glöckner, F.O. The SILVA ribosomal RNA gene database project: Improved data processing and webbased tools. *Nucleic Acids Res.* **2013**, *41*, D590–D596. [CrossRef]
46. Altschul, S.F.; Gish, W.; Miller, W.; Myers, E.W.; Lipman, D.J. Basic local alignment search tool. *J. Mol. Biol.* **1990**, *215*, 403–410. [CrossRef]
47. Hahsler, M.; Hornik, K.; Buchta, C. Getting things in order: An introduction to the R package seriation. *J. Stat. Softw.* **2008**, *25*, 1–34. [CrossRef]
48. Wei, T.; Simko, V. Corrplot: Visualization of a correlation matrix. *MMWR Morb. Mortal. Wkly. Rep.* **2013**, *52*, 145–151.
49. Archer, E. RfPermute: Estimate Permutation p-Values for Random Forest Importance Metrics. R Package Version 1. 2016. Available online: https://github.com/EricArcher/rfPermute (accessed on 5 February 2025).
50. Oksanen, J.; Blanchet, F.G.; Friendly, M.; O'Hara, B.; Stevens, M.H.H.; Oksanen, M.J.; Suggests, M. Vegan: Community Ecology Package. Ordination Methods, Diversity Analysis and Other Functions for Community and Vegetation Ecologists. Version 2.4-2. 2017. Available online: https://github.com/vegandevs/vegan/issues (accessed on 5 February 2025).
51. Zhang, Y.L.; Du, Y.; Chai, X.T.; Li, X.Y.; Zhang, Z.H.; Islam, W.; Zeng, F.J. Combined effects of planting patterns and mowing time on different organs and soil stoichiometry of *Cyperus esculentus* in desert oasis transition zone. *J. Plant Physiol.* **2023**, *287*, 154033. [CrossRef]
52. Segata, N.; Izard, J.; Waldron, L.; Gevers, D.; Miropolsky, L.; Garrett, W.S.; Huttenhower, C. Metagenomic biomarker discovery and explanation. *Genome Biol.* **2011**, *12*, R60. [CrossRef] [PubMed]
53. Crits-Christoph, A.; Robinson, C.K.; Barnum, T.; Fricke, W.F.; Davila, A.F.; Jedynak, B.; McKay, C.P.; Diruggiero, J. Colonization patterns of soil microbial communities in the Atacama Desert. *Microbiome* **2013**, *1*, 28. [CrossRef]
54. Dong, L.; Li, M.X.; Li, S.; Yue, L.X.; Ali, M.; Han, J.R.; Lian, W.H.; Hu, C.J.; Lin, Z.L.; Shi, G.Y.; et al. Aridity drives the variability of desert soil microbiomes across north-western China. *Sci. Total Environ.* **2024**, *907*, 168048. [CrossRef] [PubMed]
55. Zhu, M.H.; Fanin, N.; Wang, Q.K.; Xu, Z.C.; Liang, S.; Ye, J.; Lin, F.; Yuan, Z.Q.; Mao, Z.K.; Wang, X.G.; et al. High functional breadth of microbial communities decreases home-field advantage of litter decomposition. *Soil Biol. Biochem.* **2024**, *188*, 109232. [CrossRef]
56. Bao, Y.Y.; Dolfing, J.; Guo, Z.Y.; Chen, R.R.; Wu, M.; Li, Z.P.; Lin, X.G.; Feng, Y.Z. Important ecophysiological roles of non-dominant Actinobacteria in plant residue decomposition, especially in less fertile soils. *Microbiome* **2021**, *9*, 84. [CrossRef] [PubMed]
57. Mandic-Mulec, I.; Stefanic, P.; van Elsas, J.D. Ecology of bacillaceae. *Microbiol. Spectr.* **2015**, *3*, TBS-0017-2013. [CrossRef] [PubMed]
58. Maier, S.; Tamm, A.; Wu, D.; Caesar, J.; Grube, M.; Weber, B. Photoautotrophic organisms control microbial abundance, diversity, and physiology in different types of biological soil crusts. *ISME J.* **2018**, *12*, 1032–1046. [CrossRef] [PubMed]
59. Zhou, H.; Gao, Y.; Jia, X.H.; Wang, M.M.; Ding, J.J.; Cheng, L.; Bao, F.; Wu, B. Network analysis reveals the strengthening of microbial interaction in biological soil crust development in the Mu Us Sandy Land, northwestern China. *Soil Biol. Biochem.* **2020**, *144*, 107782. [CrossRef]
60. Fierer, N.; Jackson, R.B. The diversity and biogeography of soil bacterial communities. *Proc. Natl. Acad. Sci. USA* **2006**, *103*, 626–631. [CrossRef] [PubMed]
61. Lauber, C.L.; Micah, H.; Rob, K.; Fierer, N. Pyrosequencing-based assessment of soil pH as a predictor of soil bacterial community structure at the continental scale. *Appl. Environ. Microb.* **2009**, *75*, 5111–5120. [CrossRef] [PubMed]

62. Glassman, S.I.; Wang, I.J.; Bruns, T.D. Environmental filtering by pH and soil nutrients drives community assembly in fungi at fine spatial scales. *Mol. Ecol.* **2017**, *26*, 6960–6973. [CrossRef] [PubMed]
63. Reeve, W.; Ardley, J.; Tian, R.; De Meyer, S.; Terpolilli, J.; Melino, V.; Tiwari, R.; Yates, R.; O'Hara, G.; Howieson, J.; et al. Genome sequence of the Listia angolensis microsymbiont Microvirga lotononidis strain WSM3557(T.). *Stand Genomic Sci.* **2013**, *9*, 540–550. [CrossRef] [PubMed]
64. Zhao, S.; Liu, J.J.; Banerjee, S.; Zhou, N.; Zhao, Z.Y.; Zhang, K.; Hu, M.F.; Tian, C.Y. Biogeographical distribution of bacterial communities in saline agricultural soil. *Geoderma* **2020**, *361*, 114095. [CrossRef]
65. Liu, R.Y.; Li, K.; Zhang, H.X.; Zhu, J.G.; Joshi, D. Spatial distribution of microbial communities associated with dune landform in the Gurbantunggut Desert, China. *J. Microbiol.* **2014**, *52*, 898–907. [CrossRef] [PubMed]
66. Yang, M.; Luo, X.; Cai, Y.; Mwangi, B.N.; Khan, M.S.; Haider, F.U.; Huang, W.; Cheng, X.; Yang, Z.; Zhou, H.; et al. Effect of fire and post-fire management on soil microbial communities in a lower subtropical forest ecosystem after a mountain fire. *J. Environ. Manag.* **2024**, *351*, 119885. [CrossRef]
67. Fuentes-Ramirez, A.; Veldman, J.W.; Holzapfel, C.; Moloney, K.A. Spreaders, igniters, and burning shrubs: Plant flammability explains novel fire dynamics in grass-invaded deserts. *Ecol. Appl.* **2016**, *26*, 2311–2322. [CrossRef]
68. Pilliod, D.S.; Welty, J.L.; Arkle, R.S. Refining the cheatgrass–fire cycle in the Great Basin: Precipitation timing and fine fuel composition predict wildfire trends. *Ecol. Evol.* **2017**, *7*, 8126–8151. [CrossRef] [PubMed]
69. Morgan, J.A.; Bending, G.D.; White, P.J. Biological costs and benefits to plant-microbe interactions in the rhizosphere. *J. Exp. Bot* **2005**, *56*, 1729–1739. [CrossRef] [PubMed]
70. Pellegrini, A.F.A.; Caprio, A.C.; Georgiou, K.; Finnegan, C.; Hobbie, S.E.; Hatten, J.A.; Jackson, R.B. Low-intensity frequent fires in coniferous forests transform soil organic matter in ways that may offset ecosystem carbon losses. *Glob. Chang. Biol.* **2021**, *27*, 3810–3823. [CrossRef] [PubMed]
71. Riutta, T.; Kho, L.K.; Teh, Y.A.; Ewers, R.; Majalap, N.; Malhi, Y. Major and persistent shifts in below-ground carbon dynamics and soil respiration following logging in tropical forests. *Glob. Chang. Biol.* **2021**, *27*, 2225–2240. [CrossRef]
72. Zheng, Q.; Hu, Y.T.; Zhang, S.S.; Noll, L.; Böckle, T.; Dietrich, M.; Herbold, C.W.; Eichorst, S.A.; Woebken, D.; Richter, A.; et al. Soil multifunctionality is affected by the soil environment and by microbial community composition and diversity. *Soil Biol. Biochem.* **2019**, *136*, 107521. [CrossRef] [PubMed]
73. Bastida, F.; Eldridge, D.J.; García, C.; Kenny Png, G.; Bardgett, R.D.; Delgado-Baquerizo, M. Soil microbial diversity-biomass relationships are driven by soil carbon content across global biomes. *ISME J.* **2021**, *15*, 2081–2091. [CrossRef]
74. Andrew, D.R.; Fitak, R.R.; Munguia-Vega, A.; Racolta, A.; Martinson, V.G.; Dontsova, K. Abiotic factors shape microbial diversity in Sonoran Desert soils. *Appl. Environ. Microbiol.* **2012**, *78*, 7527–7537. [CrossRef]
75. Zhao, J.Y.; Xie, X.; Jiang, Y.Y.; Li, J.X.; Fu, Q.; Qiu, Y.B.; Fu, X.H.; Yao, Z.Y.; Dai, Z.M.; Qiu, Y.P.; et al. Effects of simulated warming on soil microbial community diversity and composition across diverse ecosystems. *Sci. Total Environ.* **2024**, *911*, 168793. [CrossRef] [PubMed]
76. Qu, L.R.; Wang, C.; Manzoni, S.; Dacal, M.; Maestre, F.T.; Bai, E. Stronger compensatory thermal adaptation of soil microbial respiration with higher substrate availability. *ISME J.* **2024**, *18*, wrae025. [CrossRef] [PubMed]
77. Chen, Y.J.; Neilson, J.W.; Kushwaha, P.; Maier, R.M.; Barberán, A. Life-history strategies of soil microbial communities in an arid ecosystem. *ISME J.* **2021**, *15*, 649–657. [CrossRef] [PubMed]
78. Hakobyan, A.; Velte, S.; Sickel, W.; Quandt, D.; Stoll, A.; Knief, C. Tillandsia landbeckii phyllosphere and laimosphere as refugia for bacterial life in a hyperarid desert environment. *Microbiome* **2023**, *11*, 246. [CrossRef] [PubMed]
79. Kulmala, L.; Aaltonen, H.; Berninger, F.; Antti-Jussi, K.; Levula, J.; Bäck, J.; Hari, P.; Kolari, P.; Korhonen, J.F.J.; Kulmala, M.; et al. Changes in biogeochemistry and carbon fluxes in a boreal forest after the clear-cutting and partial burning of slash. *Agr. Forest Meteorol.* **2014**, *188*, 33–44. [CrossRef]
80. Chen, J.; Luo, Y.Q.; Li, J.W.; Zhou, X.H.; Cao, J.J.; Wang, R.W.; Wang, Y.Q.; Shelton, S.; Jin, Z.; Walker, L.M.; et al. Costimulation of soil glycosidase activity and soil respiration by nitrogen addition. *Glob. Chang. Biol.* **2017**, *23*, 1328–1337. [CrossRef] [PubMed]
81. Jessica, L.M.; Gutknecht, H.A.L.; Henry, T.C.B. Inter-annual variation in soil extra-cellular enzyme activity in response to simulated global change and fire disturbance. *Pedobiologia* **2010**, *53*, 283–293.
82. Cheng, J.M.; Jing, G.H.; Wei, L.; Jing, Z.B. Long-term grazing exclusion effects on vegetation characteristics, soil properties and bacterial communities in the semi-arid grasslands of China. *Ecol. Eng.* **2016**, *97*, 170–178. [CrossRef]
83. Zhang, X.W.; Han, X.Z.; Yu, W.T.; Wang, P.; Cheng, W.X. Priming effects on labile and stable soil organic carbon decomposition: Pulse dynamics over two years. *PLoS ONE* **2017**, *12*, e0184978 [CrossRef] [PubMed]
84. Zhou, Z.H.; Wang, C.K.; Luo, Y.Q. Effects of forest degradation on microbial communities and soil carbon cycling: A global meta-analysis. *Glob. Ecol. Biogeogr.* **2017**, *27*, 110–124. [CrossRef]
85. Li, H.F.; Zeng, F.J.; Gui, D.W.; Zhang, L.G.; Song, C.; Luo, W.C.; Peng, S.L. Effects of cutting and burning on regeneration of *Alhagi sparsifolia* Shap. on the southern fringe of the Taklamakan Desert, North-west China. *Rangel. J.* **2012**, *34*, 389. [CrossRef]

86. Losada, J.M.; Blanco-Moure, N.; Fonollá, A.; Martínez-Ferrí, E.; Hormaza, J.I. Hydraulic tradeoffs underlie enhanced performance of polyploid trees under soil water deficit. *Plant Physiol.* **2023**, *192*, 1821–1835. [CrossRef]
87. Xu, J.T.; Mu, Q.; Sun, S.K.; Yang, N.; Peng, X.B.; Zou, Y.F.; Feng, S.Y.; Cai, H.J. Evaluation of long-term spatio-temporal characteristics of water and nitrogen in the vadose zone and potential threats to groundwater under different irrigation strategies. *J. Clean Prod.* **2023**, *429*, 139495. [CrossRef]
88. Gerlach, R.; Buhk, C.; Schirmel, J.; Jungkunst, H.F. Exploration of the response of soil properties and plant species composition to centuries-old traditional meadow irrigation. *J. Plant Ecol.* **2023**, *16*, rtac072. [CrossRef]

forests

MDPI

Article

Alleviating Microbial Carbon Limitation in *Pinus armandii* Forests Through *Panax notoginseng* Cultivation

Rui Rui [1,†], Jingying Hei [1,†], Yue Li [1], Xiaoli Wan [2], Shu Wang [1,*] and Xiahong He [3,*]

1 Yunnan Provincial Key Laboratory for Conservation and Utilization of In-Forest Resource, Southwest Forestry University, Kunming 650224, China; ruirui@swfu.edu.cn (R.R.); jyhei@swfu.edu.cn (J.H.); liyue_421@swfu.edu.cn (Y.L.)

2 Lincang Academy of Forestry Sciences, Lincang 677099, China; wanli82127@swfu.edu.cn

3 Ministry of Education Key Laboratory for Forest Resources Conservation and Utilization in the Southwest Mountains of China, Southwest Forestry University, Kunming 650224, China

* Correspondence: wangshu@swfu.edu.cn (S.W.); hxh@swfu.edu.cn (X.H.)

† These authors contributed equally to this work.

Abstract: The cultivation of *Panax notoginseng* (Sanqi) within the *Pinus armandii* forest understory has been widely promoted in Yunnan, China. However, relatively little is known about how Sanqi cultivation influences microbial metabolic limitations and their driving factors in *P. armandii* ecosystems in terms of soil extracellular enzyme activity (EEA) and stoichiometry (EES). In this study, we established monoculture *P. armandii* (MPA) and Sanqi–*P. armandii* agroforestry (SPA) systems to investigate microbial resource limitations in *P. armandii* soils over 12 months (semi-monthly sampling). Sanqi cultivation decreased EEAs in *P. armandii* soils in the SPA system. Moreover, the vector length in both the bulk and rhizosphere soils of *P. armandii* decreased significantly from 1.31 to 1.12 and 1.29 to 1.21, respectively, indicating a decrease in the microbial C limitation of *P. armandii* soils. A vector angle < 45° in both systems revealed that N, rather than P, predominantly restricted microbial metabolism. The most influential factors affecting vector length and angle were Sanqi cultivation and seasonal dynamics. Structural equation modelling (SEM) revealed that fungi-to-bacteria ratios and soil chemical properties were direct factors positively affecting vector length. Overall, our findings suggest that Sanqi cultivation benefited soil microorganisms in *P. armandii* soils and should be encouraged to supply N to promote the sustainable development of *P. armandii*.

Keywords: *Pinus armandii*; Sanqi cultivation; enzymatic stoichiometry; fungi-to-bacteria ratios; metabolism; microbial limitation

1. Introduction

Soil extracellular enzymes secreted by microbes catalyse the decomposition of soil organic matter (SOM), releasing energy and mineral nutrients for plant and microorganismal growth [1]. This process signifies an equilibrium between nutrient and energy supply and demand [2], with extracellular enzyme activity (EEA) serving as a robust indicator of microbial metabolic activity and soil quality [3]. C-acquiring enzymes, including xylosidase (XYL), β-1,4 glucosidase (BG), α-glucosidase (AG), and cellulosic disaccharide hydrolase (CBH), break down complex carbohydrates into simpler sugar molecules, thereby enhancing the uptake and utilisation of carbon by plants. Leucine aminopeptidase (LAP) and β-1,4-N-acetylaminoglucosidase (NAG) primarily degrade proteins and polysaccharides, providing a nitrogen source for microorganisms by releasing amino acids and N-acetylglucosamine. Acid phosphatase (AP) is chiefly involved in the phosphorus cycle, catalysing the hydrolysis of organic phosphoric acid compounds to release phosphorus [4]. Thus, the ratios of soil hydrolytic enzyme activities can serve as quantitative indicators of microbial resource allocation to carbon, nitrogen, and phosphorus acquisition, as well as soil organic carbon stability [5]. Soil ecoenzymatic stoichiometry (EES) reflects the restriction of soil microbial energy and nutrient resources compared with soil extracellular enzyme chemometrics on a global scale (C:N:P = 1:1:1) [6] or vector length and angle values [7–9]. Numerous factors, including introducing plants [10], seasonal dynamics [11], and soil compartments [12] modulate biological and abiotic factors, thereby affecting soil EEA and EES. Therefore, understanding the relationships among environmental (abiotic) factors, biological factors, EEA, and EES is pivotal for elucidating the investment priority of resource acquisition [13], and providing a theoretical basis for production practice [14], plant growth [15], productivity [16], soil characteristics [17], and responses to climate change in terrestrial ecosystems.

Sustainable development land, agroforestry systems that organically combine agricultural plants (introduced) with forestry plants (in situ), have a strong influence on soil EEAs [18–20]. Previous studies have demonstrated that plant species introductions either have positive, negative, or no effects on EEA related to C, N, and P in plants in situ. For example, after introducing plants, the soil EEA of in situ plants such as kiwifruit (kiwifruit/alfalfa) (*Actinidia chinensis/Medicago sativa* L.) [21], apple (*Malus pumila* Mill.) (apple/corn) [22], hickory (hickory/peony or kula clover) (*Carya cathayensis* Sarg./*Paeonia* L. or *Trifolium* L.) [23,24], and peach (peach/morel) (*Amygdalus persica* Linn./*Morchella esculenta* L.) [25] increased significantly, but the degree of EEA increase involved in C, N, and P was inconsistent. In contrast, C-, N-, and P-acquiring enzyme activities in alfalfa (alfalfa/poplar) (*Medicago sativa* L./*Populus tomentosa* Carr.), coffee (*Coffea arabica* Linn.) (coffee/thorn), and pear (pear/konjac) (*Pyrus* L./*Amorphophallus konjac* K. Koch) soils were significantly reduced or unchanged by the introduced plants [20,26,27]. In addition, changes in enzyme activities related to C, N, and P vary (increases, decreases, or no changes) in soils with different plants [28–30]. This variability can be attributed to differences in the soil extracellular enzyme species [31], introduced and in situ plant species [24], and environmental factors [20]. In summary, a deep understanding of how introduced plants cause changes in the soil EEA of in situ plants is crucial for evaluating the sustainable development of agroforestry systems.

Seasonal variations (e.g., temperature and precipitation) significantly influence biotic and abiotic factors [32–35], which further lead to alterations in soil EEA and EES [36,37]. Seasonal changes can alter soil microbial biomass and the quality and quantity of soil input by plants, thereby affecting EEA and EES. For example, EEAs are usually higher in summer than in other seasons because of the higher microbial biomass [11,34,36]. Moreover, changes in the quality and quantity of soil input by plants (e.g., lignin-to-nitrogen ratio) affect microbial nutrient utilisation efficiency, leading to changes in EEA and EES [2,13]. In addition, seasonal variations in temperature and rainfall cause variations in edaphic factors, microbial growth, and available soil nutrients, which can directly or indirectly affect EEA and EES [35,37–39]. Although distinguishing the individual effects of seasonal dynamics, plants, and the soil environment on EEA is challenging, a deeper understanding of the relationship between EEA/EES and biological/abiotic factors under different seasonal conditions [4,35] is beneficial for elucidating microbial responses to climate change [40].

Different soil compartments (rhizosphere and bulk) significantly affect soil EEAs and EES [12,41], and rhizosphere rather than bulk soils, had higher soil enzyme activities [12,42]. On the one hand, rhizosphere soil exhibits higher microbial activity and more frequent nutrient flow [12]. In contrast, soil physicochemical properties are influenced by plant root growth, further altering the rhizosphere soil microbial activity [8]. Furthermore, root-derived carbon compounds [43], mainly derived from root secretions, stimulate microbial activity, leading to higher rhizosphere soil enzyme activities [41]. Plant root biomass also increases rhizosphere soil enzyme activities (proteins and gibberellins) to address N limitations [44]. As a physiological hub of the arable crops in agroforestry systems, the rhizosphere significantly affects soil microorganisms, root exudates, soil physicochemical properties, and plant root biomass [41,44,45]. Nevertheless, the relationship between the trees and arable crops in agroforestry systems is more intricate [18], with nutrient competition among different plants [46], the mutual transmission of soil microorganisms [47], and the diffusion of root exudates [48] significantly impacting EEA. Therefore, an in-depth analysis of the changes in EEA/EES in different soil compartments can reveal the nutrient requirements, acquisition strategies, and metabolic models of microbes in various soils [8,10,12], providing a basis for establishing a rational agroforestry system.

The long-term monoculture management of pine forests can trigger ecological issues such as soil degradation, biodiversity loss, and environmental disturbance [49,50]. Conversely, the SPA system, as sustainable land-use management [47], has been widely adopted in the Yunnan province, covering approximately 666.67 ha [51]. Sanqi is a rare herbal medicine, which belongs to the family Araliaceae [52,53]. Simulative habitat cultivation involves simulating the original habitat and site environments of Chinese medicinal plants to preserve their quality [54]. The simulative cultivation of organic Sanqi in *P. armandii* forests has been shown to improve Sanqi quality, prevent obstacles to continuous cropping, and increase the economic benefits for producers [51–53]. However, attention should be given not only to Sanqi but also to pine trees in the SPA system. This is because pine forests have been deficient in carbon, and the organically managed Sanqi with no application of pesticides and fertilisers potentially affects the carbon demand in the SPA agroforestry system [51]. Reducing microbial carbon restriction can enhance soil organic carbon accumulation [35], bolster ecosystem stability and resilience, improve soil fertility and structure, mitigate climate change [13], and foster the sustainable development of agricultural systems. Consequently, investigating the impact of Sanqi cultivation on microbial carbon restriction in pine forests holds not only practical application value for local regions but also offers theoretical significance and practical guidance for other agroforestry ecosystems globally. By illustrating how to optimize ecosystem functions through plant–microbial interactions, it presents innovative concepts for the design and sustainable advancement of agroforestry ecosystems on a worldwide scale.

Previous studies have primarily focused on the planting methods [51], quality analysis [55], and pest control [51] of Sanqi. However, less attention has been paid to the effects of Sanqi planting on Pinus soil quality, particularly on the soil EEA. Therefore, we designed monoculture *P. armandii* (MPA) and Sanqi–*P. armandii* agroforestry (SPA) systems to compare and analyse the changes in EEA (CBH, AG, XYL, BG, NAG, LAP, and AP) and EES in *P. armandii* soils. Furthermore, we assessed the interactive effects of season, plant introduction, and soil compartments on EEA and EES in *P. armandii* soils. We also conducted a one-year multi-frequency (twice per month) sampling to accurately analyse the nutrient limitations of soil microorganisms in different *P. armandii* systems. We tested the following hypotheses: (1) Sanqi planting significantly reduces the ecoenzymatic C:N and C:P and vector length and alleviates the C limitations of microbial resources in *P. armandii* soils; and (2) plant introduction, seasons, and soil compartments (rhizosphere/bulk) exert differential effects on EEA and EES in *P. armandii* soils.

2. Materials and Methods

2.1. Study Area

The base of Sanqi cultivation is mainly located in Xundian Hui and Yi autonomous county, Kunming city, Yunnan province (103.25° E, 25.55° N, 2247.78 m a.s.l.), with an average annual temperature of 14.7 °C and precipitation of 1624.0 mm. This region experiences a subtropical plateau monsoon climate with distinct dry (November–April) and rainy seasons (May–October). The forest soil was mountainous red soil, and the vegetation was dominated by *P. armandii*, with an average diameter at a breast height of 18 cm and an average tree height of 9.5 m.

2.2. Experimental Design and Sampling

The MPA and SPA systems were established under similar conditions, including vegetation, soil types, aspects, and angles. The planting of Sanqi involved several steps. First, weeds and shrubs in the forest understory were removed and *P. armandii* branches were trimmed up to a height of ~3 m. Second, the soils (0–20 cm) were ploughed deeply 2–3 times using a small rotary tiller. Moreover, a soil bed was set up along the contour line of the *P. armandii* forest (10–20 m, 1–1.5 m, and 0.3–0.4 m in length, width, and height), and a drainage outlet was left between two adjacent plots (1.5–2.0 m). Subsequently, 1 year old Sanqi seedlings were transplanted into the soil bed with intra- and inter-row spacings of 10 and 15 cm, respectively, and covered with *P. armandii* needles (3–5 cm). In addition, a circular arched rain shelter was built to prevent water accumulation during the rainy season, drip pipes were installed for irrigation, and no pesticides, fertilisers, or artificial rodent control measures were applied (Figure 1).

Figure 1. Monoculture *P. armandii* (MPA) and Sanqi–*P. armandii* agroforestry (SPA) systems.

Six plots of 10 m × 10 m were randomly set up in both the MPA and SPA system. Sanqi was planted in January 2020, but due to the influence of COVID-19, 288 soil samples were collected from 10 September 2020 to 20 August 2021, including two soil compartments (rhizosphere and bulk) × 2 systems × 12 months × 2 times/month × 3 replicates. Semi-monthly sampling was applied to accurately analyse the seasonal dynamic of EEA and EES. The soil sampling method used the 5-point method, in which three soil samples were collected at each point and mixed as a composite sample. Soils attached to the root surface at a thickness of 0–2 mm and 20 cm away from the *P. armandii* tree (depth of 0–20 cm) were considered as the rhizosphere and bulk soil, respectively. Soil samples were stored in self-sealing bags and returned to the laboratory in liquid nitrogen. After removing litter and impurities from the soil sample, the soil was sieved through a 4 mm sieve. The processed soil samples were separated into two subsamples and stored at 4 °C and −80 °C for subsequent analyses of soil EEA, edaphic factors, and microbes, respectively.

2.3. Analysis of Edaphic Factors and Soil EEA

Fresh soils underwent oven-drying at 100 °C for 24 h to determine the water content (WC) by measuring the weight variation. A pH meter was used to determine the soil pH (soil/water = 1:5). After the soils were treated (H_2SO_4 digestion and KCl extraction), the total nitrogen (TN), total phosphorus (TP), nitrate nitrogen (NO_3^-−N), and ammonium nitrogen (NH_4^+−N) were determined using a flow analyser (SEAL Analytical AA3, Mequon, WI, USA). Soil organic carbon (SOC) was measured using the $K_2Cr_2O_7$ oxidation method. The total potassium (TK) was measured using atomic emission spectrometry and an AA-6300C flame photometer (Shimadzu Corporation, Kyoto, Japan).

Four C-acquiring enzymes (xylosidase, XYL; β-1,4 glucosidase, BG; α-glucosidase, AG; cellulosic disaccharide hydrolase, CBH), two N-acquiring enzymes (leucine aminopeptidase, LAP; β-1,4-N-acetylaminoglucosidase, NAG), and one P-acquiring enzyme (acid phosphatase, AP) were assayed using fluorometric protocols. Briefly, 4-methylumbelliferone (MUB) and 7-amino-4-methylcoumarin (MUC) solutions with varying concentrations (0, 2.5, 5, 10, 25, 50, and 100 μM, 150 μL of each concentration) were prepared and added to enzyme-labelled plates. A standard curve was generated after incubating the standard solutions in the dark at 25 °C for 3 h. Furthermore, 1 g of fresh soil was mixed with 125 mL of acetate buffer (50 mM, pH 5.0) in a 250 mL conical flask and stirred thoroughly with a magnetic stirrer. Subsequently, 150 μL of the suspension was pipetted to a black polystyrene 96-well microtiter plate, to which an enzyme–substrate solution (50 μL) was added. After incubation, the catalytic reaction was terminated by adding 10 μL of NaOH solution (1 M, termination solution) to each well. Finally, fluorescence intensity was determined using a microplate reader (Tecan Infinite M200, Salzburg, Austria) at excitation and emission wavelengths of 365 and 450 nm, respectively. Eight replicates were analysed from each soil sample.

2.4. DNA Extraction and Quantitative Polymerase Chain Reaction (qPCR) Assays

Total DNA was extracted from soil samples (0.5 g) according to the instructions of the Qiagen DNeasy PowerSoil Kit (MP Biomedicals, Solon, CA, USA), and DNA quality and concentration were assessed using agarose gel electrophoresis and a NanoDrop2000 (Thermo Fisher Scientific, Waltham, MA, USA). Absolute quantification of bacterial and fungal genes was carried out on a LightCycler 480 II system (Roche, Basel, Switzerland), with the amplification primers (338F/806R and ITS1F/ITS2R) shown in Table S1 [47,51]. The PCR reaction system totalled 20 μL and consisted of 1 μL of DNA, 0.5 μL of forward and reverse primers, 10 μL of TB Green Premix Ex Taq (TaKaRa, Kyoto, Japan), and 8 μL of ddH_2O. Standard curves were generated after PCR reaction using a tenfold gradient dilution of plasmid DNA concentration (104.1 ng/μL for bacteria and 130.6 ng/μL for fungi). The amplification efficiencies and slopes for bacterial and fungal genes were 93.43%/95.30% ($R^2 > 0.991$) and −3.49/−3.44, respectively. All samples were analysed in triplicate for qPCR.

2.5. Determination of Soil EEA

Ln (CBH+BG+XYL+AG):Ln (NAG+LAP), Ln (CBH+BG+XYL+AG):Ln (AP), and Ln (NAG+LAP):Ln (AP) denote the ecoenzymatic C:N, C:P, and N:P ratios, respectively. Moreover, the vector length and angle of the soil EEA quantify nutrient limitations for soil microorganisms in ecosystems [44]. Specifically, the vector length signifies the carbon limitation of microbial metabolism, with a longer length indicating stronger carbon limitation [56].

The vector angle represents the nitrogen or phosphorus limitation of microbial metabolism, with a vector angle greater or less than 45° indicating significant P and N limitations, respectively. The vector length and angle were calculated as follows:

$$\textit{Vector length}\ (\text{unitless}) = SQRT(x^2 + y^2) \tag{1}$$

$$\textit{Vector angle}\ (\text{degrees}) = Degrees\ [Atan2(x,y)] \tag{2}$$

where x denotes the EES related to C, N (Ln (CBH + BG + XYL + AG):Ln (AP)), y denotes the EES related to P (Ln (CBH + BG + XYL + AG):Ln (NAG + LAP)), and *Atan2* denotes the four-quadrant inverse tangent (tan-1 (y/x)), which serves as the reference angle for the positive x-axis of all plots.

2.6. Statistical Analysis

One- and two-way ANOVA, correlation analysis and principal effects analysis were performed using SPSS 25.0. Each ANOVA used LSD and Duncan for multiple post-hoc testing. Graphing and redundancy analyses (RDA) were performed using GraphPad Prism 10 and Canoco 5, respectively. In addition, structural equation modelling (SEM) analysis was carried out using IBM SPSS Amos v28.0.0 to determine the direct and indirect impacts of edaphic factors, fungal gene abundance, bacterial gene abundance, and fungi-to-bacteria (F:B) ratios on microbial metabolic restriction. Each structural model for the overall goodness of fit analysis was assessed using the chi-square (χ^2) test, distribution, degrees of freedom (df), p-values, approximate root mean square error (RMSEA), and goodness of fit index (GFI).

3. Results

3.1. Soil Physicochemical Properties

Sanqi cultivation dramatically affected various *P. armandii* soil properties. Specifically, fungal gene abundance and SOC, TN, TP, and TK, and soil C:P and N:P values decreased significantly (ANOVA, $p < 0.05$) in both the rhizosphere and bulk soils of *P. armandii*. Conversely, the NO_3^-—N contents, pH, WC, and soil C:N increased, whereas bacterial gene abundance, ST, and F:B content did not show significant differences. In the monoculture *P. armandii* system, rhizosphere soils had a higher SOC, whereas they had a lower NH_4^+—N content and WC compared to bulk soil. In the Sanqi–*P. armandii* agroforestry system, the SOC, NH_4^+—N, NO_3—N, and TN contents and soil C:N and C:P ratios in rhizospheric soil were higher than those in bulk soil, whereas the WC and pH values in rhizospheric soil were lower than those in bulk soil (Table S2 and Figure S1).

3.2. Soil EEA in the MPA and SPA Systems

The planting of Sanqi caused a decrease in EEA (NAG, BG, CBH, AP, XYL, and AG) in *P. armandii* soil in the SPA system compared to the MPA system, while LAP activity remained unchanged. Moreover, except for the highest LAP activity that was observed in autumn, all EEAs were higher in spring and summer than in autumn and winter. In addition, there was no difference in extracellular enzyme activity between *P. armandii* soils in the MPA and SPA systems for either the rhizosphere or bulk soils (Figures 2 and S2).

3.3. Variation in EES and EEA in the MPA and SPA Systems

Compared to the monoculture *P. armandii* system, the ecoenzymatic C:N and C:P ratios decreased significantly in the Sanqi–*P. armandii* system, whereas the ecoenzymatic N:P ratio remained unchanged. Moreover, the ecoenzymatic C:N ratio was consistently lower in autumn across all treatments than in the other seasons, whereas the enzymatic C:P ratio varied seasonally. The eco-enzymatic N:P ratio was lowest in summer and highest in autumn. In addition, the cultivation of Sanqi resulted in a significant decrease in the C:N and C:P ratios in the bulk soils rather than in the rhizosphere soils of *P. armandii*, whereas the N:P ratio increased dramatically in the bulk soil of *P. armandii* (Figure 3). Notably, all treatments exhibited C:N and C:P ratios of less than one, whereas the N:P ratios were greater than one (Figure 3). The C:N:P ratios of rhizospheric and bulk soils in the MPA system were 1:1.10:1.06 and 1:1.11:1.09, respectively, whereas those in the SPA system were 1:1.30:1.24 and 1:1.17:1.17, respectively (Table S3). Compared to the monoculture *P. armandii* system, the vector length decreased in the Sanqi–*P. armandii* system, whereas the vector angle remained unchanged, and both the vector length and angle were lowest in autumn. Moreover, there were no differences between the vector lengths and angles of the rhizosphere and bulk soils in the MPA system, whereas planting in Sanqi led to a significant decrease in the vector lengths and angles of the rhizosphere and bulk soils in autumn (Figure 3). These results indicate that the planting of Sanqi can alleviate carbon limitation, with the strongest effect observed in autumn. In the different systems, P was significantly restricted in summer, whereas N was significantly restricted in spring, autumn, and winter, with the strongest restriction observed in autumn (Table S4).

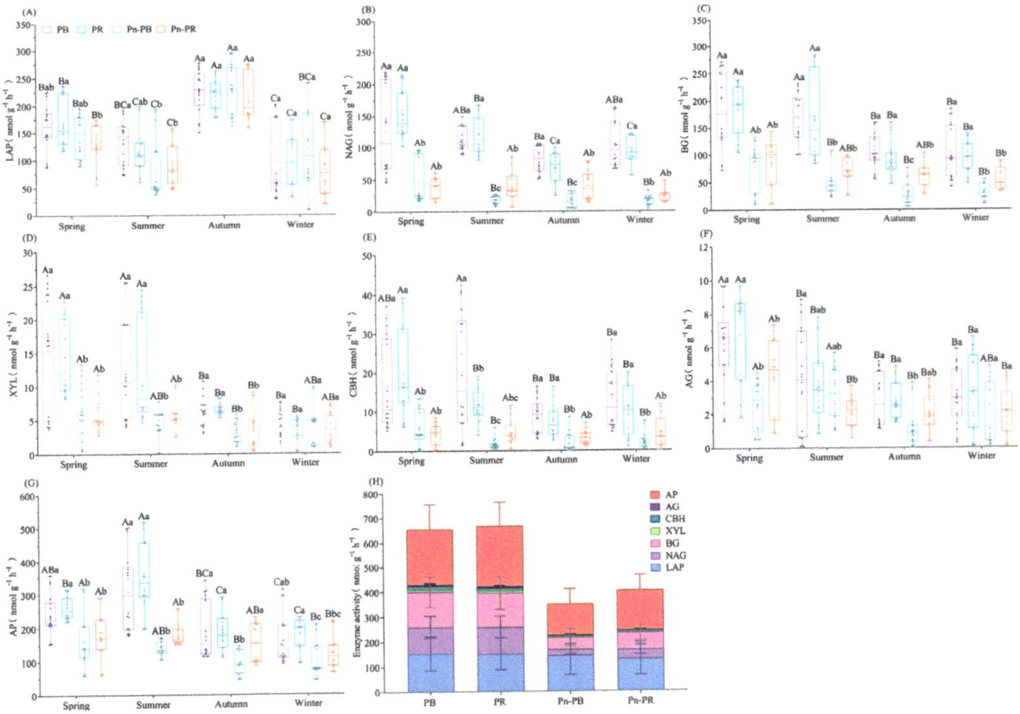

Figure 2. Seasonal dynamics of soil EEAs under different *P. armandii* systems. PB and PR: the bulk and rhizosphere soils of *P. armandii* in the MPA system; Pn-PB and Pn-PR: the bulk and rhizosphere soils of *P. armandii* in the SPA system. Different upper- and lower-case letters denote significant differences at $p < 0.05$. (**A–G**) the activity of leucine aminopeptidase (LAP), β-1,4-N-acetylaminoglucosidase (NAG), β-1,4 glucosidase (BG), xylosidase (XYL), cellulosic disaccharide hydrolase (CBH), α-glucosidase (AG), acid phosphatase (AP) under different seasons, respectively. (**H**) the activity of seven extracellular enzyme under different treatments.

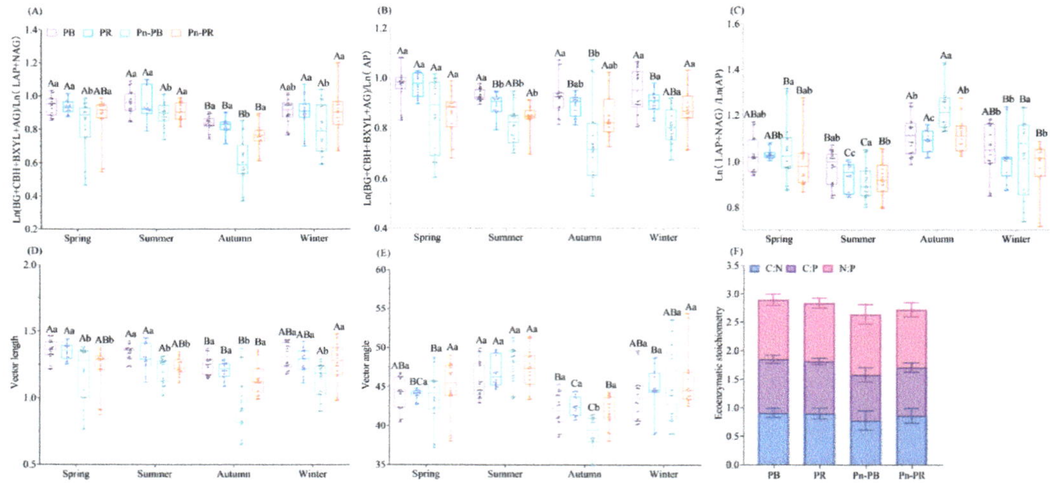

Figure 3. Seasonal dynamics of enzymatic ratios and vector length/angle in the *P. armandii* soils under the MPA and SPA systems. Different upper- and lower-case letters denote significant differences at $p < 0.05$. (**A**) extracellular enzyme C (XYL+BG+AG+CBH)/N (LAP+NAG). (**B**) extracellular enzyme C (XYL+BG+AG+CBH)/P (AP). (**C**) extracellular enzyme N (LAP+NAG)/P (AP). (**D**,**E**) vector length and vector angle of extracellular enzyme under different seasons. (**F**) eco-enzymatic stoichiometry under different treatments.

3.4. Influence of Environmental Factors on EEA and EES

Correlation analysis revealed significant positive correlations between SOC, TN, TP, TK, and fungal abundance and the activities of seven soil EEAs, as well as enzymatic C:N and C:P ratios. Moreover, NO_3^-−N, and ST exhibited significant negative and positive correlations with the economic N:P ratio, respectively. In addition, vector length displayed a notable positive correlation with SOC, TN, TK, and fungal gene abundance but a negative correlation with NO_3^-−N, WC, and pH. Vector angle exhibited a dramatically positive correlation with ST and a negative correlation with NH_4^+−N and TK (Table S5).

The RDA results show that the factors influencing soil EEA and EES varied among the different *P. armandii* systems. ST, TK, pH, WC, and soil C:P ratio were the most significant factors in the MPA system, whereas NO_3^-−N, ST, F:B, NH_4^+−N, and soil C:P were the most crucial factors in the agroforestry system (Figure 4, Table S6).

3.5. The Main Effect Analysis of EES

The main effect analysis evaluated the effects of plant introduction (planting), seasons, and soil compartments (rhizosphere and bulk) on soil EES (Figure 5, Table S7). We found that plant introduction, seasons, and soil compartments were important factors affecting soil EES, vector length, and vector angle. However, the degree of influence of these factors on EES was inconsistent. Notably, the season emerged as the strongest factor affecting C:N and N:P ratios ($\eta^2 = 0.296$, $p < 0.001$; $\eta^2 = 0.35$, $p < 0.001$, respectively), while plant introduction exerted the greatest impact on C:P ratio ($\eta^2 = 0.275$, $p < 0.001$). In addition, the factors affecting the vector length ranked in the following order: plant introduction ($\eta^2 = 0.248$, $p < 0.001$) > season ($\eta^2 = 0.218$, $p < 0.001$) > soil compartment ($\eta^2 = 0.017$, $p < 0.05$). Those affecting the vector angle ranked in the following order: season ($\eta^2 = 0.343$, $p < 0.001$) > soil compartment ($\eta^2 = 0.027$, $p < 0.01$) > plant introduction ($\eta^2 = 0.001$, $p > 0.05$) (Figure 5).

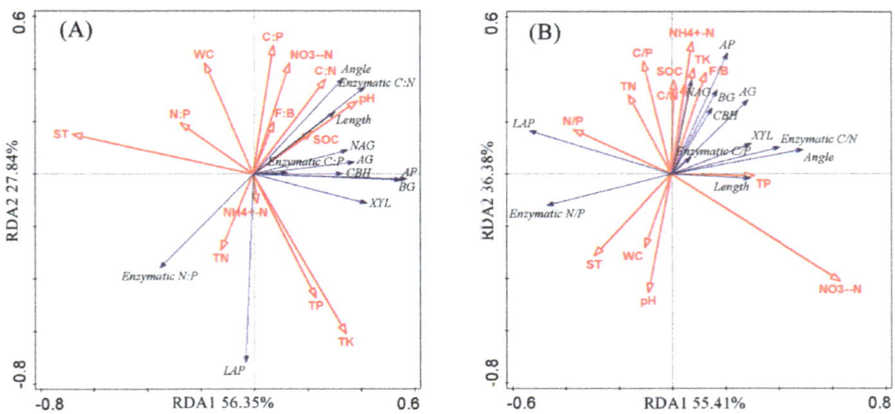

Figure 4. Redundancy analysis for the relationship between soil EEAs/EES and edaphic factors under the MPA (**A**) and SPA (**B**) systems.

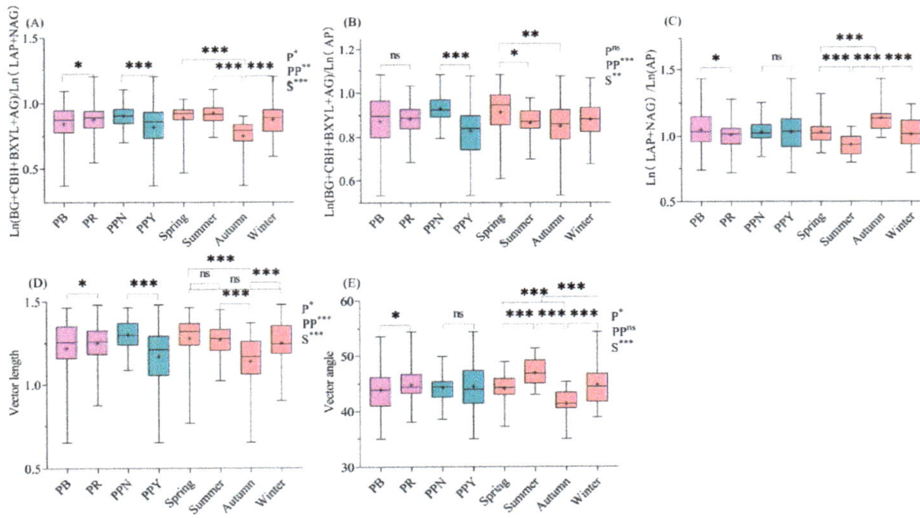

Figure 5. Main effect analysis of soil EES (**A–C**), vector length (**D**), and angle (**E**). P, PP, and S represent soil compartments factors, planting factors, and seasonal factors, respectively. PB and PR: *P. armandii* bulk or rhizosphere soils, respectively; PPY and PPN: *P. armandii* soils with or without Sanqi cultivation, respectively. "*", "**", and "***" indicate statistically significant differences at the $p < 0.05$, $p < 0.01$, and $p < 0.001$ levels, respectively, and "ns" denotes no statistical difference.

3.6. Structural Equation Modelling (SEM) Analysis in the MPA and SPA Systems

SEM explained the direct and indirect effects of the soil microenvironment, soil chemical properties, fungal and bacterial gene abundance, and F:B ratios on vector length and angle (Figure 6; Table S8). In both the monoculture *P. armandii* and Sanqi–*P. armandii* systems, F:B ratios and soil chemical properties as direct factors positively affected vector length, whereas bacterial and fungal abundances served as indirect factors that significantly affected vector length. Moreover, vector length, and soil microenvironment as direct factors had positive and negative effects on the vector angle, respectively. In addition, the chemical properties of SPA, rather than the MPA system, displayed a negative effect on the vector angle as a direct factor.

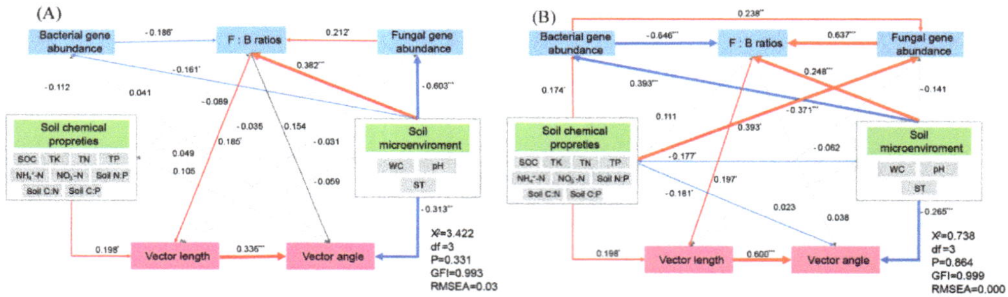

Figure 6. Analysis of the SEM paths under the MPA (**A**) and SPA (**B**) systems. The colour and width of the arrows denote the negative (blue) and positive (red) relationships and the coefficient strength, respectively. *, **, and *** indicate $p < 0.05$, 0.01, and 0.001 levels, respectively.

4. Discussion

The effects of introduced plant species in agroforestry systems on the activities of extracellular enzymes involved in C, N, and P in plant soils in situ have been inconsistent [20,21,27]. In the Sanqi–*P. armandii* agroforestry system, the cultivation of Sanqi resulted in a decrease in the EEA related to C, N, and P in *P. armandii* soils. Our findings diverged from those observed in the kiwi/alfalfa and apple/corn agroforestry systems [21,22] yet aligned with the outcomes in the alfalfa/poplar and coffee/bramble systems [20,26]. Additionally, the C:N and C:P ratios decreased, whereas the N:P ratio increased in bulk soils compared to rhizosphere soils. Furthermore, the EES (C:N, C:P, N:P, and C:N:P) in both the rhizosphere and bulk soils did not deviate from global scale EES benchmarks (C:N = 1.41; C:P = 0.62; N:P = 0.44; C:N:P = 1:1:1) [57]. In addition, vector lengths were significantly reduced in the SPA system, indicating that planting Sanqi could alleviate C limitation in *P. armandii* soils. This evidence supports our first hypothesis. Notably, the vector angles across the different systems were all less than 45°, suggesting an overall N limitation. Previous studies have demonstrated that a reduction in C limitation could potentially exacerbate N or P limitation in various terrestrial systems [13,58], owing to variations in climate and land-use conversion types, although this was not observed in our study. However, in agroforestry systems, plant introduction [10], seasonal changes [11], and soil compartments [12] can alter both biotic and abiotic factors such as soil microbes [25,58,59] and soil physicochemical properties [10,20,24,54], thereby influencing EEA and EES. These factors support our second hypothesis and significantly affect the C limitations in *P. armandii* soils within agroforestry systems.

Introducing plants has been identified as a significant factor influencing the EES [60], largely because of their ability to alter soil physicochemical properties [10,20,54] and soil microbial community structures [60]. In our study, Sanqi planting significantly reduced vector length rather than vector angle, implying a reduction in C limitation rather than constraints related to N and P. Correlation analyses showed that SOC, TN, TK, and fungal abundance were positively correlated with EEA vector length, all of which were significantly lower in the *P. armandii soils* of the SPA system. Therefore, we posit that introducing Sanqi-induced alterations in soil physicochemical properties and microbial structure serves as the primary driver mitigating C limitation in *P. armandii* soils.

Previous studies have consistently demonstrated positive correlations among SOC, TN, TK, and EEA/EES [10,11], which is consistent with our findings. The reduction in SOC, TN, and TK levels in soils could be attributed to nutrient competition between Sanqi and *P. armandii*, thereby diminishing C restrictions. However, soil tillage during Sanqi cultivation can influence soil physicochemical properties, reducing soil EES and vector length [61,62], which may be the main reason for the alleviation of C restriction in the SPA system. Furthermore, pine soil in both the MP and SPA systems is limited by N (vector angle less than 45°), although the cultivation of Sanqi can marginally alleviate this N limitation in the SPA system (with no significant difference). On one hand, Sanqi has a substantial nitrogen requirement and competes with pine trees for this nutrient [63], thereby reducing the soil's nitrogen content. The application of organic fertiliser during Sanqi cultivation can partially mitigate nitrogen deficiency, which explains the slight reduction in nitrogen limitation in the SPA system's pine soil. On the other hand, subtropical pine forests are inherently nitrogen-limited (Luo et al., 2020), and the organic management of Sanqi without pesticides and fertilisers [64] can intensify the nitrogen demand within the SPA system. The decline in nitrogen typically elevates the C/N ratio and accelerates organic matter decomposition, while carbon decomposition proceeds more slowly, potentially leading to rapid nitrogen depletion in the soil [65]. Our findings also indicate that pine soils in both the MP and SPA systems are not restricted by P (vector angle less than 45°), likely due to Sanqi's relatively low P demand and the high background P content in pine forest soil, which suffices for both Sanqi and pine requirements. Consequently, we recommend increasing the input of organic fertiliser during Sanqi cultivation to mitigate microbial N restriction. In addition, fungal diversity is more susceptible to the influence of EES than bacterial diversity [59]. Introducing Sanqi alters the fungal community structure in *P. armandii* soil significantly, potentially transmitting certain fungi to pine-associated soil microorganisms [47]. There is a positive correlation between fungal abundance and vector length. Thus, we conclude that the reduction in SOC and N content in *P. armandii* soils following Sanqi cultivation affects the EES and further decreases fungal abundance. In addition, fungi can assist plants in acquiring soil nitrogen, while nitrogen serves as the primary factor influencing the assembly of fungal communities in both pine and Sanqi soils [47]. Consequently, the decline in soil nitrogen levels directly contributes to the reduction in fungal abundance.

Seasonal variations are pivotal in shaping soil enzyme dynamics, as acknowledged in biogeochemical models [15,35]. The soil EEA in both *P. armandii* systems was significantly higher in spring/summer than in autumn/winter, which is inconsistent with other research results indicating that EEA peaks during summer in *Quercus*, wetland, and *Betula platyphylla* soil [36,66,67]. Seasonal variations can influence abiotic factors, such as temperature and moisture, leading to differences in soil enzyme acquisition activities [32–35,37]. In particular, the soil water content demonstrated a significant negative correlation with EEA, likely because of the distinct alternation of drought and rainy seasons in subtropical forests, resulting in alternating changes in soil moisture. Concurrently, alterations in the quality and quantity of plant inputs to the soil affect the nutrient utilisation efficiency of microorganisms, consequently modulating EEA and EES [2,13]. Therefore, the absence of a well-defined growing season in subtropical forests, which results in increased plant inputs, has emerged as a primary factor influencing soil enzyme activity.

P. armandii soils in the SPA system experienced a significant reduction in C limitation (indicated by decreases in C:N and C:P ratios and vector length) and an increase in N limitation (evidenced by decreases in the C:N ratio, increases in the N:P ratio, and decreases in vector angle) during autumn, which was inconsistent with previous studies [34,37,67]. This discrepancy can be attributed to seasonal shifts in the biological factors affecting EEA and EES [2,11,13,34,36]. Correlation analyses revealed a positive association between vector length and fungal abundance, suggesting that reduced fungal abundance in *P. armandii* soils during autumn, possibly due to Sanqi planting, plays a crucial role in alleviating carbon restriction. Moreover, soil water content was negatively correlated with vector length, whereas temperature was positively correlated with vector angle. Soil water can modulate soil extracellular enzyme activities by changing substrate diffusion rates (mainly C) and inhibitor levels [68], as well as the relative utilisation rates of different nutrient elements [69]. Furthermore, the soil temperature exerts substantial control over enzymatic activity, thereby affecting both EEA and EES [2,35,36,70]. Therefore, the observed alleviation of carbon limitations in autumn may be attributed to heavy rainfall and adequate water replenishment during Sanqi planting. In contrast, higher temperatures exacerbated N-limitation during this season. In addition, a strong interrelationship exists among the EEA related to C, N, and P in the soil; that is, restrictions on one element can affect the acquisition of another [13,71,72]. Therefore, the reduction in C limitation in autumn may exacerbate N limitation.

Compared to seasonal variations and plant introduction, the impact of the soil compartments on C:N, N:P, vector angle, and vector length was relatively minor (Figure 5). Rhizosphere soil exhibits higher soil enzyme activity [12,42], which is consistent with our findings. Moreover, compared to the rhizosphere soil, the bulk soil of *P. armandii* in agroforestry systems showed significant decreases in C:N, C:P, and vector length compared to the MPA system, indicating that the planting of Sanqi effectively reduced C limitation in bulk *P. armandii* soil. Our observations are consistent with those of a previous study, which noted a smaller C limit in *Pinus sylvestris* rhizosphere soil than in bulk soil [12]. In addition, the interplay of nutrient competition among different plants [46], the mutual transfer of soil microorganisms [47], and the diffusion of root exudates [48] could complicate the plant–soil–plant relationship in agroforestry systems [18], significantly affecting both rhizosphere and non-rhizosphere soils.

5. Conclusions

Our hypotheses regarding the effect of Sanqi planting on EEA in *P. armandii* soil in an agroforestry system were confirmed. The planting of Sanqi decreased the EEAs in *P. armandii* soil in the agroforestry system. However, the C:N:P ratios (1:1.30:1.24 and 1:1.17:1.17) did not significantly deviate from the global mean EES (1:1:1). After Sanqi cultivation, both the bulk and rhizosphere soils of *P. armandii* exhibited a significant decrease in vector length, indicating a reduction in microbial metabolic C limitations. Moreover, the vector angle <45° in *P. armandii* soils under both systems revealed that microbial metabolism was primarily restricted by N rather than P. Furthermore, the strongest factors affecting the vector length and vector angle were plant introduction ($\eta^2 = 0.248$, $p < 0.001$) and season ($\eta^2 = 0.343$, $p < 0.001$), respectively. Structural equation modelling (SEM) indicated that F:B (fungi/bacteria) ratios and soil chemical properties as direct factors positively influenced vector length. Therefore, we recommend increasing the application of organic fertiliser in the management of Sanqi cultivation to alleviate the microbial N limitations, thereby promoting the sustainable development of the SPA system. Concurrently, the introduction of plants into an agroforestry system significantly influences microbial nutrition limitations; thus, it is essential to ensure the judicious selection and integration of plants to maintain the optimal design of the agroforestry system. As a next step, we will employ multi-omics techniques to analyse the microbiome, thereby confirming the sustainability of the agroforestry system.

Supplementary Materials: The following supporting information can be downloaded at https://www.mdpi.com/article/10.3390/f16010158/s1. Table S1. The primer sequences and PCR conditions for bacteria and fungi; Table S2. Soil physicochemical properties and bacterial/fungal abundance in the MPA and SPA systems; Table S3. Seasonal changes in soil physicochemical properties and bacterial/fungal abundance in the MPA and SPA system; Table S4. EES, the vector length, and angle in the MPA and SPA systems; Table S5. Pearson correlation between soil extracellular enzyme activity and enzyme stoichiometry, soil properties, and fungal and bacterial abundance; Table S6. The permutation test of the redundancy analysis (RDA); Table S7. Analysis of the interaction between soil extracellular enzymes and stoichiometry with seasons and treatments; Table S8. Summary of standardized direct, indirect, and total effects of predictor variables on enzymatic vector length and angle in the SEM; Figure S1. One-year changes in soil physicochemical properties and bacterial/fungal abundance in the MPA and SPA systems; Figure S2. Annual changes in soil extracellular enzyme activity and stoichiometry in the MPA and SPA systems.

Author Contributions: R.R.: writing—original draft, methodology, and funding acquisition. J.H.: formal analysis and writing—original draft. Y.L.: formal analysis and methodology. X.W.: methodology and formal analysis. S.W.: conceptualization, and writing—review and editing. X.H.: funding acquisition, and writing—review and editing. All authors have read and agreed to the published version of the manuscript.

Funding: This work was supported by the Yunnan Fundamental Research Project (202401BD070001-122), Training Project of Technological Innovation Talents in Yunnan Province (202405AD350065), Yunnan Ten Thousand People Plan Youth Top Talent Project (YNWR-QNBJ-2019-028), China Agriculture Research System (CARS-21), Major Science and Technology Project of Yunnan Province (202102AE090042, 202202AE090036, 2019ZG0901, 2021Y250), and Kunming Science and Technology Bureau (2021JH002).

Data Availability Statement: The datasets generated and analysed in the current study are available from the corresponding author on reasonable request.

Conflicts of Interest: The authors declare no conflicts of interest.

References

1. Yao, X.H.; Min, H.; Lv, Z.H.; Yuan, H.P. Influence of acetamiprid on soil enzymatic activities and respiration. *Eur. J. Soil Biol.* **2006**, *42*, 120–126. [CrossRef]
2. Nannipieri, P.; Trasar-Cepeda, C.; Dick, R.P. Soil enzyme activity: A brief history and biochemistry as a basis for appropriate interpretations and meta-analysis. *Biol. Fert. Soils* **2018**, *54*, 11–19. [CrossRef]
3. Luo, L.; Meng, H.; Gu, J.D. Microbial extracellular enzymes in biogeochemical cycling of ecosystems. *J. Environ. Manag.* **2017**, *197*, 539–549. [CrossRef]
4. Lybrand, R.A.; Gallery, R.E.; Trahan, N.A.; Moore, D.J. Disturbance alters the relative importance of topographic and biogeochemical controls on microbial activity in temperate montane forests. *Forests* **2018**, *9*, 97. [CrossRef]
5. Zuccarini, P.; Asensio, D.; Ogaya, R.; Sardans, J.; Penuelas, J. Effects of seasonal and decadal warming on soil enzymatic activity in a P-deficient Mediterranean shrubland. *Glob. Chang. Biol.* **2020**, *26*, 3698–3714. [CrossRef] [PubMed]
6. Sinsabaugh, R.L.; Lauber, C.L.; Weintraub, M.N.; Ahmed, B.; Allison, S.D.; Crenshaw, C.; Contosta, A.R.; Cusack, D.; Frey, S.; Gallo, M.E.; et al. Stoichiometry of soil enzyme activity at global scale. *Ecol. Lett.* **2008**, *11*, 1252–1264. [CrossRef]
7. Peng, X.Q.; Wang, W. Stoichiometry of soil extracellular enzyme activity along a climatic transect in temperate grasslands of northern China. *Soil Biol. Biochem.* **2016**, *98*, 74–84. [CrossRef]
8. Cui, Y.X.; Fang, L.C.; Guo, X.B.; Wang, X.; Zhang, Y.J.; Li, P.F.; Zhang, X.C. Ecoenzymatic stoichiometry and microbial nutrient limitation in rhizosphere soil in the arid area of the northern Loess Plateau, China. *Soil Biol. Biochem.* **2018**, *116*, 11–21. [CrossRef]
9. He, Q.Q.; Wu, Y.H.; Bing, H.J.; Zhou, J.; Wang, J.P. Vegetation type rather than climate modulates the variation in soil enzyme activities and stoichiometry in subalpine forests in the eastern Tibetan Plateau. *Geoderma* **2020**, *374*, 114424. [CrossRef]
10. Gao, P.Z.; Zhu, J.J.; Yang, K.; Yan, Q.L.; Zhang, J.X.; Yu, L.Z.; Diao, M.M.; Xu, S. Can larch-*Aralia elata* agroforestry systems improve the soil chemical and microbial properties of larch plantations? *Agrofor. Syst.* **2022**, *96*, 885–896. [CrossRef]
11. Zhang, Q.; Zhang, D.D.; Wu, J.J.; Li, J.S.; Feng, J.; Cheng, X.L. Soil nitrogen-hydrolyzing enzyme activity and stoichiometry following a subtropical land use change. *Land Degrad. Dev.* **2021**, *32*, 4277–4287. [CrossRef]

12. Bi, B.; Wang, Y.; Wang, K.; Zhang, H.; Fei, H.; Pan, R.; Han, F. Changes in microbial metabolic C-and N-limitations in the rhizosphere and bulk soils along afforestation chronosequence in desertified ecosystems. *J. Environ. Manag.* **2022**, *303*, 114215. [CrossRef] [PubMed]
13. Feyissa, A.; Gurmesa, G.A.; Yang, F.; Long, C.Y.; Zhang, Q.; Cheng, X.L. Soil enzyme activity and stoichiometry in secondary grasslands along a climatic gradient of subtropical China. *Sci. Total Environ.* **2022**, *825*, 154019. [CrossRef]
14. Sinsabaugh, R.L.; Moorhead, D.L. Resource allocation to extracellular enzyme production: A model for nitrogen and phosphorus control of litter decomposition. *Soil Biol. Biochem.* **1994**, *26*, 1305–1311. [CrossRef]
15. Bell, C.; Carrillo, Y.; Boot, C.M.; Rocca, J.D.; Pendall, E.; Wallenstein, M.D. Rhizosphere stoichiometry: Are C: N: P ratios of plants, soils, and enzymes conserved at the plant species-level? *New Phytol.* **2014**, *201*, 505–517. [CrossRef]
16. Sinsabaugh, R.L.; Follstad, S.J.J. Ecoenzymatic stoichiometry and ecological theory. *Annu. Rev. Ecol. Evol. Syst.* **2012**, *43*, 313–343. [CrossRef]
17. Raiesi, F.; Salek-Gilani, S. The potential activity of soil extracellular enzymes as an indicator for ecological restoration of rangeland soils after agricultural abandonment. *Appl. Soil Ecol.* **2018**, *126*, 140–147. [CrossRef]
18. Ma, Y.H.; Fu, S.L.; Zhang, X.P.; Zhao, K.; Chen, H.Y.H. Intercropping improves soil nutrient availability, soil enzyme activity and tea quantity and quality. *Appl. Soil Ecol.* **2017**, *119*, 171–178. [CrossRef]
19. Dollinger, J.; Jose, S. Agroforestry for soil health. *Agrofor. Syst.* **2018**, *92*, 213–219. [CrossRef]
20. Clivot, H.; Petitjean, C.; Marron, N.; Erwin, D.; Genestier, J.; Blaszczyk, N.; Santenoise, P.; Laflotte, A.; Piutti, S. Early effects of temperate agroforestry practices on soil organic matter and microbial enzyme activity. *Plant Soil* **2020**, *453*, 189–207. [CrossRef]
21. Wang, Q.P.; Zhang, C.; Li, J.H.; Wu, X.M.; Long, Y.H.; Su, Y. Intercropping *Vicia sativa* L. improves the moisture, microbial community, enzyme activity and nutrient in rhizosphere soils of young kiwifruit plants and enhances plant growth. *Horticulturae* **2021**, *7*, 335. [CrossRef]
22. Fan, L.J.; Liu, Q.Z.; Wang, H.; Xu, Z.X.; Li, W.H. Effects of corn-apple rotation on enzyme activity and pH value in rhizosphere soil of apple trees. *Acta Agric. Zhejiangensis* **2017**, *29*, 2084–2090. [CrossRef]
23. Kremer, R.J.; Kussman, R.D. Soil quality in a pecan–kura clover alley cropping system in the Midwestern USA. *Agrofor. Syst.* **2011**, *83*, 213–223. [CrossRef]
24. Wang, Z.C.; Zhou, M.Y.; Liu, H.; Huang, C.; Ma, Y.H.; Ge, H.X.; Ge, X.; Fu, S.L. Pecan agroforestry systems improve soil quality by stimulating enzyme activity. *Peer J.* **2022**, *10*, 12663. [CrossRef] [PubMed]
25. Song, H.Y.; Chen, D.; Sun, S.X.; Li, J.; Tu, M.Y.; Xu, Z.H.; Gong, R.G.; Jiang, G.L. Peach-morchella intercropping mode affects soil properties and fungal composition. *Peer J.* **2021**, *9*, e11705. [CrossRef]
26. Dai, S.G.; Zhou, H.L.; Lan, J.P. Effects of konjac planting on soil enzyme activity of two kinds of pear plantations. *J. Sichuan For. Sci. Technol.* **2015**, *36*, 102–105. [CrossRef]
27. López, R.E.I.; Navarrete, E.F.C.; Rosado, J.T.P.; García, C.R.S.; Margenot, A.J. Soil nitrogen cycling under contrasting management systems in Amazon Coffea canephora agroecosystems. *Soil Sci. Soc. Am. J.* **2021**, *85*, 1634–1648. [CrossRef]
28. Mijangos, I.; Epelde, L.; Garbisu, C.; González-Oreja, J.A. Modification of soil enzyme activities as a consequence of replacing meadows by pine plantations under temperate climatei. *Pedobiologia* **2014**, *57*, 61–66. [CrossRef]
29. Wen, H. The Effect of Eucalyptus with Agroforestry Management Modes on Soil Physical and Chemical Properties and Enzyme Activity. Master's Thesis, Guizhou University, Guiyang, China, 2021; p. 25.
30. Bian, X.B.; Yang, X.H.; Li, Q.; Sun, X. Effects of planting of two common crops, *Allium fistulosum* and *Brassica napus*, on soil properties and microbial communities of ginseng cultivation in northeast china. *BMC Microbiol.* **2022**, *22*, 182. [CrossRef]
31. Curtright, A.J.; Tiemann, L.K. Intercropping increases soil extracellular enzyme activity: A meta-analysis. *Agric. Ecosyst. Environ.* **2021**, *319*, 107489. [CrossRef]
32. Boerner, R.E.J.; Brinkman, J.A.; Smith, A. Seasonal variations in enzyme activity and organic carbon in soil of a burned and unburned hardwood forest. *Soil Biol. Biochem.* **2005**, *37*, 1419–1426. [CrossRef]
33. Fenner, N.; Freeman, C.; Reynolds, B. Observations of a seasonally shifting thermal optimum in peatland carbon-cycling processes: Implications for the global carbon cycle and soil enzyme methodologies. *Soil Biol. Biochem.* **2005**, *37*, 1814–1821. [CrossRef]
34. Wallenstein, M.D.; McMahon, S.K.; Schimel, J.P. Seasonal variation in enzyme activities and temperature sensitivities in Arctic tundra soils. *Glob. Chang. Biol.* **2009**, *15*, 1631–1639. [CrossRef]
35. Zhang, Q.; Feng, J.; Wu, J.J.; Zhang, D.D.; Chen, Q.; Li, Q.X.; Long, C.Y.; Feyissa, A.; Cheng, X.L. Variations in carbon-decomposition enzyme activities respond differently to land use change in central china. *Land Degrad. Dev.* **2018**, *30*, 459–469. [CrossRef]
36. Bai, X.J.; Dippold, M.A.; An, S.S.; Wang, B.R.; Zhang, H.X.; Loeppmann, S. Extracellular enzyme activity and stoichiometry: The effect of soil microbial element limitation during leaf litter decomposition. *Ecol. Indic.* **2021**, *121*, 107200. [CrossRef]
37. Schaap, K.J.; Fuchslueger, L.; Quesada, C.A.; Hofhansl, F.; Valverde-Barrantes, O.; Camargo, P.B.; Hoosbeek, M.R. Seasonal fluctuations of extracellular enzyme activities are related to the biogeochemical cycling of C, N and P in a tropical terra-firme forest. *Biogeochemistry* **2023**, *163*, 1–15. [CrossRef]

38. Waring, B.G.; Weintraub, S.R.; Sinsabaugh, R.L. Ecoenzymatic stoichiometry of microbial nutrient acquisition in tropical soils. *Biogeochemistry* **2013**, *117*, 101–113. [CrossRef]
39. Kivlin, S.N.; Treseder, K.K. Soil extracellular enzyme activities correspond with abiotic factors more than fungal community composition. *Biogeochemistry* **2014**, *117*, 23–37. [CrossRef]
40. Chen, Y.; Han, M.G.; Yuan, X.; Cao, G.M.; Zhu, B. Seasonal changes in soil properties, microbial biomass and enzyme activities across the soil profile in two alpine ecosystems. *Soil Ecol. Lett.* **2021**, *3*, 383–394. [CrossRef]
41. Svenja, S.C.; Köster, M.; Dippold, M.A.; Nájera, F.; Merino, C.; Kuzyakov, Y. Environmental drivers and stoichiometric constraints on enzyme activities in soils from rhizosphere to continental scale. *Geoderma* **2019**, *337*, 973–982. [CrossRef]
42. Kuzyakov, Y.; Razavi, S.B. Rhizosphere size and shape: Temporal dynamics and spatial stationarity. *Soil Biol. Biochem.* **2019**, *135*, 343–360. [CrossRef]
43. Kuzyakov, Y. Factors affecting rhizosphere priming effects. *J. Plant Nutr. Soil Sci.* **2002**, *165*, 382–396. [CrossRef]
44. Yang, Y.; Liang, C.; Wang, Y.Q.; Cheng, H.; An, S.S.; Chang, S.X. Soil extracellular enzyme stoichiometry reflects the shift from P- to N-limitation of microorganisms with grassland restoration. *Soil Biol. Biochem.* **2020**, *149*, 107928. [CrossRef]
45. Chaparro, J.M.; Badri, D.V.; Vivanco, J.M. Rhizosphere microbiome assemblage is affected by plant development. *ISME J.* **2014**, *8*, 790–803. [CrossRef]
46. Craine, J.M.; Dybzinski, R. Mechanisms of plant competition for nutrients, water and light. *Funct. Ecol.* **2013**, *27*, 833–840. [CrossRef]
47. Jia, W.J.; Wang, S.; He, X.H.; Zhao, X.Y. Different factors drive the assembly of pine and Panax notoginseng-associated microbiomes in *Panax notoginseng*-pine agroforestry systems. *Front. Microbiol.* **2022**, *13*, 1018989. [CrossRef] [PubMed]
48. Jose, S.; Gillespie, A.R. Allelopathy in black walnut (*Juglans nigra* L.) alley cropping. I. Spatio-temporal variation in soil juglone in a black walnut-corn (*Zea mays* L.) alley cropping system in the midwestern USA. *Plant Soil* **1998**, *203*, 191–197. [CrossRef]
49. Barbeito, I.; Dassot, M.; Bayer, D.; Collet, C.; Drössler, L.; Löf, M.; Río, M.; Ruiz-Peinado, R.; Forrester, D.I.; Bravo-Oviedo, A.; et al. Terrestrial laser scanning reveals differences in crown structure of Fagus sylvatica in mixed vs. Pure European forests. *Forest Ecol. Manag.* **2017**, *405*, 381–390. [CrossRef]
50. Wu, L.F.; Zou, B.Z.; Wang, S.R.; Zhou, L.H.; Zheng, Y.; Huang, Z.Q.; He, J.Z. Effects of multispecies restoration on soil extracellular enzyme activity stoichiometry in Pinus massoniana plantations of subtropical China. *Soil Biol. Biochem.* **2023**, *178*, 108967. [CrossRef]
51. Hei, J.Y.; Wang, S.; He, X.H. Effects of exogenous organic acids on the growth, edaphic factors, soil extracellular enzymes, and microbiomes predict continuous cropping obstacles of *Panax notoginseng* from the forest understorey. *Plant Soil* **2023**, *503*, 105–122. [CrossRef]
52. Hei, J.Y.; Li, Y.; Wang, Q.; Wang, S.; He, X.H. Effects of Exogenous Organic Acids on the Soil Metabolites and Microbial Communities of *Panax notoginseng* from the Forest Understory. *Agronomy* **2024**, *14*, 601. [CrossRef]
53. Li, Y.; Hei, J.Y.; Wang, B.; Wang, S.; He, X.H. Unraveling the molecular mechanisms of flavonoid biosynthesis in *Panax notoginseng* flowers across planting patterns and developmental stages using integrated metabolomics and transcriptomics analyses. *Sci. Hortic.* **2024**, *335*, 113362. [CrossRef]
54. Zhang, J.Q.; Zhou, T.; Xiao, C.H.; Jiang, W.K.; Guo, L.P.; Kang, C.Z.; Liao, X.K.; Huang, Y.P.; Wang, X.; Lu, H. Technical evaluation and principle analysis of simulative habitat cultivation of Dendrobium nobile. *China J. Chin. Mater. Medica* **2020**, *45*, 2042–2045.
55. Chen, S.Y.; Rui, R.; Wang, S.; He, X.H. Comparative analysis of the floral fragrance compounds of *Panax notoginseng* flowers under the *Panax notoginseng-pinus* agroforestry system using SPME-GC-MS. *Molecules* **2022**, *27*, 3565. [CrossRef] [PubMed]
56. Moorhead, D.L.; Sinsabaugh, R.L.; Hill, B.H.; Weintraub, M.N. Vector analysis of ecoenzyme activities reveal constraints on coupled C, N and P dynamics. *Soil Biol. Biochem.* **2016**, *93*, 1–7. [CrossRef]
57. Sinsabaugh, R.L.; Hill, B.H.; Shah, J.J.F. Ecoenzymatic stoichiometry of microbial organic nutrient acquisition in soil and sediment. *Nature* **2009**, *462*, 795–798. [CrossRef]
58. Guan, P.T.; Yang, J.J.; Yang, Y.R.; Wang, W.; Wu, D. Land conversion from cropland to grassland alleviates climate warming effects on nutrient limitation: Evidence from soil enzymatic activity and stoichiometry. *Glob. Ecol. Conserv.* **2020**, *24*, 01328. [CrossRef]
59. Yang, Y.; Li, T.; Wang, Y.Q.; Dou, Y.X.; Cheng, H.; Liu, L.X.; An, S.S. Linkage between soil ectoenzyme stoichiometry ratios and microbial diversity following the conversion of cropland into grassland. *Agric. Ecosyst. Environ.* **2021**, *314*, 107418. [CrossRef]
60. Zhao, Y.Z.; Liang, C.F.; Shao, S.; Chen, J.H.; Qin, H.; Xu, Q.F. Linkages of litter and soil C: N: P stoichiometry with soil microbial resource limitation and community structure in a subtropical broadleaf forest invaded by Moso bamboo. *Plant Soil* **2021**, *465*, 473–490. [CrossRef]
61. Zhao, J.; Zheng, L.J.; Zhang, X.D.; Wang, L.C.; Wan, Q.L.; Wang, Z.Y. Study on changes of soil nutrient, enzyme activity and structure of microbial community before and after cropping *Panax notoginseng*. *Guangdong Agric. Sci.* **2019**, *46*, 66–73. [CrossRef]
62. Song, X.J.; Li, J.Y.; Liu, X.T.; Liang, G.P.; Li, S.P.; Zhang, M.N.; Wu, H.J. Altered microbial resource limitation regulates soil organic carbon sequestration based on ecoenzyme stoichiometry under long-term tillage systems. *Land Degrad. Dev.* **2022**, *33*, 2795–2808. [CrossRef]

63. Menge, D.N.; Batterman, S.A.; Hedin, L.O.; Liao, W.; Pacala, S.W.; Taylor, B.N. Why are nitrogen-fixing trees rare at higher compared to lower latitudes? *Ecology* **2017**, *98*, 3127–3140. [CrossRef]
64. Luo, X.Z.; Hou, E.Q.; Chen, J.Q.; Li, J.; Zhang, L.L.; Zang, X.W.; Wen, D.Z. Dynamics of carbon, nitrogen, and phosphorus stocks and stoichiometry resulting from conversion of primary broadleaf forest to plantation and secondary forest in subtropical China. *Catena* **2020**, *193*, 104606. [CrossRef]
65. Yang, S.M.; Cai, Q.K.; Liu, W.F.; Wu, J.P. Response of soil N mineralization to long-term N addition and season in *Cunninghamia lanceolata* plantation. *J. Zhejiang A&F Univ.* **2021**, *38*, 945–951.
66. Shao, X.; Yang, W.; Wu, M. Seasonal dynamics of soil labile organic carbon and enzyme activities in relation to vegetation types in Hangzhou Bay tidal flat wetland. *PLoS ONE* **2015**, *10*, 0142677. [CrossRef]
67. Fu, L.C.; Ding, Z.J.; Tang, M.; Zeng, H.; Zhu, B. Seasonal dynamics of activities, temperature sensitivities and Vector characteristics of extracellular enzymes in rhizosphere and bulk soils of two temperate forests in Mt. Dongling, Beijing. *Acta Sci. Nat. Univ. Pekin.* **2022**, *58*, 503–516. [CrossRef]
68. Toberman, H.; Evans, C.D.; Freeman, C.; Fenner, N.; White, M.; Emmett, B.A.; Artz, R.R.E. Summer drought effects upon soil and litter extracellular phenol oxidase activity and soluble carbon release in an upland *Calluna* heathland. *Soil Biol. Biochem.* **2008**, *40*, 1519–1532. [CrossRef]
69. Curtin, D.; Beare, M.H.; Qiu, W.; Chantigny, M.H. Temperature dependence of organic matter solubility: Influence of biodegradation during soil-water extraction. *Soil Biol. Biochem.* **2015**, *79*, 858–863. [CrossRef]
70. Trasar-Cepeda, C.; Gil-Sotres, F.; Leiros, M.C. Thermodynamic parameters of enzymes in grassland soils from Galicia, NW Spain. *Soil Biol. Biochem.* **2007**, *39*, 311–319. [CrossRef]
71. Sinsabaugh, R.L.; Shah, J.J.F. Integrating resource utilization and temperature in metabolic scaling of riverine bacterial production. *Ecology* **2010**, *91*, 1455–1465. [CrossRef] [PubMed]
72. Marklein, A.R.; Houlton, B.Z. Nitrogen inputs accelerate phosphorus cycling rates across a wide variety of terrestrial ecosystems. *New Phytol.* **2012**, *193*, 696–704. [CrossRef]

forests

MDPI

Article

Enzyme Activity Stoichiometry Suggests That Fertilization, Especially Nitrogen Fertilization, Alleviates Nutrient Limitation of Soil Microorganisms in Moso Bamboo Forests

Haoyu Chu [1,†], Wenhui Su [1,†], Yaqi Zhou [2], Ziye Wang [1], Yongmei Long [1], Yutong Sun [1] and Shaohui Fan [1,*]

1 International Center for Bamboo and Rattan, Key Laboratory of National Forestry and Grassland Administration, Beijing 100102, China; chuhaoyu@icbr.ac.cn (H.C.); swh612@icbr.ac.cn (W.S.); wangziye@icbr.ac.cn (Z.W.)
2 School of Karst Science, Guizhou Normal University, Guiyang 550001, China
* Correspondence: fansh@icbr.ac.cn
† These authors contributed equally to this work.

Abstract: Rational application of N fertilizer is essential for maintaining the long-term productivity of Moso bamboo forests. Microbial activity is a crucial indicator of soil quality. Changes in soil nutrient resources due to N addition can lead to microbial nutrient limitations, thereby impeding the maintenance of soil quality. Currently, there is limited research on the effects of N application on microbial nutrient limitations in Moso bamboo forest soils. To examine the changes in extracellular enzyme activity and microbial nutrient limitations in Moso bamboo forest soils following N application, we conducted an N application experiment in northern Guizhou. The findings revealed that the N3 treatment (726 kg·N·hm^{-2}·yr^{-1}) significantly reduced β-glucosidase (BG) activity by 27.61% compared to the control group (no fertilization). The N1 (242 kg·N·hm^{-2}·yr^{-1}), N2 (484 kg·N·hm^{-2}·yr^{-1}), and N3 treatments notably increased the activities of leucine aminopeptidase (LAP) and N-acetyl-β-D-glucosidase (NAG) by 11.45% to 15.79%. Acid phosphatase (ACP) activity remained unaffected by fertilization. N application treatments significantly decreased the C:Ne and C:Pe ratios, while the N:Pe ratio was less influenced by N fertilizer application. Scatter plots and vector characteristics of enzyme activity stoichiometry suggested that microorganisms in the study area were limited by C and N, and N fertilizer application reduced the vector length and increased the vector angle, indicating that N application alleviated the C and N limitation of microorganisms in Moso bamboo forests. Redundancy Analysis (RDA) demonstrated that microbial biomass phosphorus (MBP) was the most critical factor affecting extracellular enzyme activity and stoichiometry. Furthermore, Random Forest Regression analysis identified MBP and the N:Pm ratio as the most significant factors influencing microbial C and N limitation, respectively. The study demonstrated that N application modulates the microbial nutrient acquisition strategy by altering soil nutrient resources in Moso bamboo forests. Formulating fertilizer application strategies based on microbial nutrient requirements is more beneficial for maintaining soil quality and sustainably managing Moso bamboo forests. Additionally, our study offers a theoretical reference for understanding carbon cycling in bamboo forest ecosystems in the context of substantial N inputs.

Keywords: enzyme activity stoichiometry; nitrogen application; metabolic limitation; enzyme activity vector characterization

1. Introduction

China boasts a rich history of bamboo cultivation and utilization, positioning it among nations with the most abundant bamboo resources globally [1]. Bamboo has many applications in food, furniture, papermaking, and biomass energy, and the C sequestration potential of bamboo forest ecosystems is considerable [2]. According to the latest data, the area of Moso bamboo forests in China is 527.76 ha as of 2021, accounting for 69.78% of the total area of bamboo forests, with a carbon stock of 23.1 billion tons, accounting for 2.3% of the country's total carbon stock in forest vegetation [3,4]. Moso bamboo (*Phyllostachys edulis* (Carrière) J. Houz)—the most prevalent species in China—is distinguished by its rapid growth rate and robust regeneration capability [5,6]. N is the element with the highest demand and uptake during the growth and development of Moso bamboo [7,8]. The application of N fertilizer guarantees high yields of Moso bamboo, and rationally applied will not adversely influence soil quality [9]. In bamboo forest management, operators apply large amounts of N fertilizers to increase yield. However, this substantial N input can lead to soil organic carbon loss and soil acidification, potentially resulting in decreased microbial activity [10–12]. Microbial activity serves as a crucial indicator for assessing soil quality, as microorganisms contribute to the decomposition of soil organic matter by secreting extracellular enzymes [13,14]. This process not only supports their own metabolic needs but also generates additional nutrients that are readily accessible for plant growth [15]. Consequently, investigating the response of soil microbial communities to N application in Moso bamboo forests is an insightful endeavor.

N application may change the content and relative concentration of soil nutrient elements. N application significantly reduced soil organic carbon (SOC)and total N (TN) contents, had no significant effect on the total phosphorus (TP) content, but increased the soil C:P ratio [16]. N application can lead to changes in soil nutrient resources. Such changes may lead to an imbalance between soil resources and microbial nutrient requirements, resulting in microbial nutrient limitations [17]. Microorganisms regulate the balance between their own nutrient requirements and soil nutrients by secreting extracellular enzymes, and enzyme activity stoichiometry is considered to be an effective indicator of microbial nutrient limitations [18–20]. Enzymes such as β-1,4-glucosidase (BG), β-1,4-N-acetylglucosaminidase (NAG), leucine aminopeptidase (LAP), and acid phosphatase (ACP)—which are vital components of the C, N, and P cycles—have been extensively employed in investigations into the stoichiometry of extracellular enzyme activities [21–24]. Sinsabaugh et al. reported that the ecological enzyme stoichiometry of soils and sediments in geoecosystems is approximately 1:1:1 [25]. However, Luo et al. observed differences in enzyme C:N:P ratios in peatlands on the Ruoergai Plateau compared to the expected 1:1:1 ratio, with ratios closer to soil C:N:P ratios, indicating phosphorus (P) limitation in the soils of this region [26]. Moorhead et al. suggested using carrier properties of soil enzyme activities to characterize energy and nutrient limitations of soil microorganisms, and a large number of studies have been conducted to demonstrate that vector characteristics can be used to reflect the nutrient limitation of microorganisms [27–30]. Zhang et al. employed extracellular enzyme activity stoichiometry and vector features to characterize the nutrient limitations of soil microorganisms in Moso bamboo forests under varying management intensities [31]. The results indicated that intensive management (fertilization with N, P, and K) decreased BG, NAG, and ACP activity and alleviated the C, N limitation of soil microorganisms in Moso bamboo forests. Zeng et al., in studying the stoichiometry of enzyme activities in Moso bamboo forest soils after N application, also found that NAG activity decreased and ACP activity increased and that N application exacerbated the C and P limitation of Moso bamboo forest microorganisms [32]. These findings suggest that soil enzyme activities and enzyme stoichiometric characteristics can be used to characterize the nutrient acquisition strategies of soil microorganisms under N addition conditions.

Moso bamboo, with its extensive distribution in southern China, may exhibit varying soil microbial nutrient limitations across different regions. Despite this, there is a dearth of research investigating the impact of fertilization, particularly N fertilization, on the microbial nutrient limitation of Moso bamboo forest soils. In this study, we conducted a fertilization experiment (0, 242, 484, 726 $kg{\cdot}N{\cdot}hm^{-2}{\cdot}yr^{-1}$) in a pure Moso bamboo forest in northern Guizhou, China. We measured soil nutrients, microbial biomass, and extracellular enzyme activities post-fertilization. The enzyme activity stoichiometry and vector characterization were used to identify microbial nutrient limitations. Our objectives were twofold: firstly, to examine the changes in extracellular enzyme activities and microbial nutrient limitations in Moso bamboo forest soils following large-scale N fertilizer application, and secondly, to identify the key environmental factors influencing extracellular enzyme activities, enzyme activity stoichiometry, and vectorial characterization of Moso bamboo forest soils after such fertilization. To accomplish our objectives, we formulated the following hypotheses: (1) N application will modify extracellular enzyme activities and stoichiometry; (2) soil microorganisms in Moso bamboo forests within the study area are limited by C and N, and N application will either exacerbate or alleviate this nutrient limitation without altering the limiting nutrient elements; (3) the primary factors contributing to microbial nutrient limitation are alterations in both the biomass and stoichiometry of microorganisms.

2. Materials and Methods

2.1. Study Site

The research site was situated in the Hu Shi Forest Farm (28°23′~105°54′), Chishui City, Guizhou Province, China. This region is characterized by its medium to low mountains and is subject to a subtropical humid monsoon climate. The annual average temperature is 18.1 °C, with an average precipitation of 1195.7 mm. The forest farm was inaugurated in 1999 and encompasses 330 ha of pure Moso bamboo forest. Notably, there are no pests or diseases within the forest area, exhibiting a distinct size–year relationship. The average diameter at breast height for Moso bamboo is 9.86 cm, with an average height of 12.07 m and a density of 3866 plants/ha. The understory vegetation primarily comprises *Rubus buergeri* Miq, *Tetrastigma hemsleyanum* Diels et Gilg, *Curculigo capitulata* (Lour.) Kuntze, *Nothapodytes pittosporoides* (Oliv.) Sleum, and *Machilus nanmu* (Oliv.) Hemsl. According to the classification from the Chinese Soil Database, the soil in the study area is categorized as purple–yellow sandy loam. The average total carbon content, organic carbon content, and pH value (measured in distilled water) in the 0–30 cm soil layer are 19.8 g/kg, 16.94 g/kg, and 4.75, respectively [12]. The geographical location of the study area is shown in Figure 1.

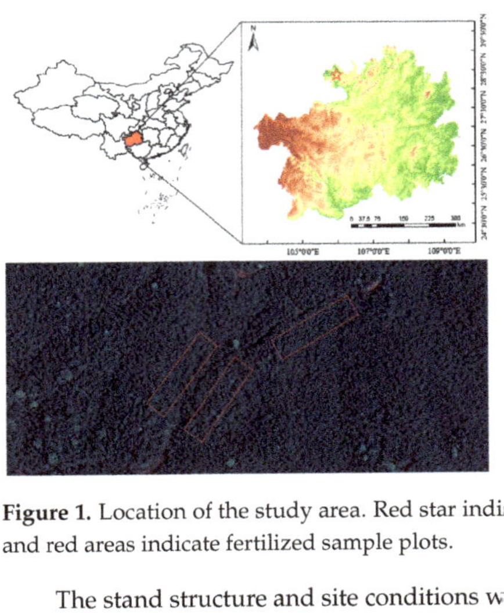

Figure 1. Location of the study area. Red star indicates the geographical location of the study area and red areas indicate fertilized sample plots.

The stand structure and site conditions were consistently applied, with a plot selected consisting of a pest-free pure bamboo forest. The N fertilizer level in this investigation was ascertained based on the research team's prior studies concerning Moso bamboo nutrient utilization and optimal scientific fertilization quantities [33]. Four distinct fertilization treatments were established in the experiment. Each plot received a consistent amount of phosphorus and potassium fertilizer as a base, while varying concentrations of N fertilizer were applied: N0 (0 kg·N·hm^{-2}·yr^{-1}), N1 (242 kg·N·hm^{-2}·yr^{-1}), N2 (484 kg·N·hm^{-2}·yr^{-1}), and N3 (726 kg·N·hm^{-2}·yr^{-1}). A no-fertilization treatment served as the control (CK). The experiment employed a randomized block design, with each plot measuring 15 m × 15 m. Each treatment was replicated three times, resulting in a total of 12 plots. A 5 m isolation zone was established between adjacent plots.

The N application experiment was executed in early October 2021 during the pre-shooting stage. The fertilizer used for this experiment comprised urea (46% N), superphosphate (178 kg/ha; 12% P_2O_5) as the base phosphorus fertilizer, and potassium chloride (147 kg/ha; 60% K_2O) as the potassium fertilizer. Based on the required proportions of fertilizers for different treatments, the prepared mixtures were dissolved in 50 L of distilled water and then applied through spraying.

2.2. Soil Sample Collection

Soil sampling was carried out in October 2022, utilizing the "S"-shaped sampling method to establish ten distinct points within each plot. Soil samples were collected from a depth of 0 to 10 cm using a soil corer with an inner diameter of 38 mm. Prior to collection, the litter layer at the sampling point was meticulously removed. The soil core was preserved in a portable thermal box and promptly transported to the laboratory for processing. From each plot, ten cores of soil were gathered. The cores from the same plot were then amalgamated to yield fifteen mixed soil samples, after which stones, roots, and other impurities were meticulously extracted. The soil mixture was divided using the quadrat method, and the excess soil was discarded. A portion of the fresh soil was passed through a 2 mm sieve to analyze the biomass of C, N, and P (MBC, MBN, MBP) and extracellular enzyme activity determination. Additionally, one portion of the soil underwent air-drying, grinding, and sieving to determine soil pH, soil organic carbon (SOC), total nitrogen (TN), total phosphorous (TP), and available phosphorous (AP).

2.2.1. Soil Chemical Properties

SOC and TN were determined by an elemental analyzer (elementar vario El cube, Langenselbold, Germany) [34]. TP was determined by the NaOH alkali fusion molybdenum antimony colorimetric method [35]. AP content was extracted with sodium fluoride hydrochloric acid and then determined by the molybdenum antimony colorimetric method [35]. The air-dried soil samples were saturated with distilled water and the soil pH was measured using a glass electrode method (ST2100, Ohaus, Parsippany, NJ, USA). The stoichiometric ratios of soil nutrients C:N, C:P, and N:P were calculated using values of SOC, TN, and TP and were expressed as C:Ns, C:Ps, and N:Ps, respectively.

2.2.2. Microbial Biomass

The MBC and MBN were determined by chloroform fumigation, and the C and N content of the extracts were measured with C and N analyzer (TOC-LCSH/CPH, Shimadzu, Beijing, China) [36]. The MBP was determined through the chloroform fumigation extraction method, with the P content in the extract being evaluated using the molybdenum antimony colorimetric method [36]. In brief, fresh soil was subjected to fumigation with ethanol-free chloroform at a temperature of 25 °C in darkness for a duration of 24 h. An equivalent volume of fresh soil, which had not undergone fumigation, was also utilized. Extracts from both fumigated and non-fumigated soils, prepared using 0.5 M K_2SO_4, were employed for MBC and MBN determinations. Meanwhile, extracts from both fumigated and non-fumigated soils, prepared using 0.5 mol/L $NaHCO_3$, were used for MBP determination. The stoichiometric C:N, C:P, and N:P ratios of microbial biomass were expressed as C:Nm, C:Pm, and N:Pm, respectively.

2.2.3. Soil Extracellular Enzyme Activity

The activity of enzymes β-1,4-glucosidase (BG), β-1,4-N-acetylglucosaminidase (NAG), leucine aminopeptidase (LAP), and acid phosphatase (ACP) was determined by enzyme-linked immunosorbent assay (ELISA) employing a double antibody one-step sandwich technique. Using BG as an example, 1 g of fresh soil was extracted with 9 milliliters of phosphate buffer (pH = 9.8), thoroughly mixed, centrifuged at 2500 revolutions per minute for 25 min at 4 degrees Celsius, and the supernatant was collected. The optical density (OD) value of each sample was measured using an enzyme marker to ascertain the sample activity. The methodology for determining soil enzyme activity was based on the protocol described by Xia et al. [37,38].

2.2.4. Enzyme Activity Stoichiometry

The stoichiometric characterization of the extracellular enzyme activities C:N, C:P, and N:P are denoted as C:Ne, C:Pe, and N:Pe, respectively. These ratios were calculated using the following equations [25]:

$$\mathrm{C:Ne} = \ln(\mathrm{BG}) / \ln(\mathrm{LAP} + \mathrm{NAG}) \quad (1)$$

$$\mathrm{C:Pe} = \ln(\mathrm{BG}) / \ln(\mathrm{ACP}) \quad (2)$$

$$\mathrm{N:Pe} = \ln(\mathrm{LAP} + \mathrm{NAG}) / l\,(\mathrm{ACP}) \quad (3)$$

2.2.5. Enzyme Activity Vector Characterization

Moorhead et al. proposed using extracellular enzyme activity vector features to characterize soil microbial growth limited by nutrients. Vector length (VL) is used to characterize soil microbial growth limited by C, while vector angle (VA) can be used to characterize soil microbial growth limited by N or P [30]. The equations for calculating these vector characterizations are as follows:

$$\mathrm{VL} = \sqrt{(\ln(\mathrm{BG}) / \ln(\mathrm{LAP} + \mathrm{NAG}))^2 + (\ln(\mathrm{BG}) / \ln(\mathrm{ACP}))^2} \quad (4)$$

$$\mathrm{VA} = \mathrm{Degrees}[\mathrm{ATAN2}(\ln(\mathrm{BG}) / \ln(\mathrm{ACP}), \ln(\mathrm{LAP} + \mathrm{NAG}) / \ln(\mathrm{ACP})] \quad (5)$$

The magnitude of the vector length serves as an indicator of the degree to which soil microbial growth is limited by C. Larger vector lengths denote a higher degree of C limitation, whereas smaller lengths indicate a lesser degree of C limitation. Using 45° as a reference, a vector angle exceeding 45° signifies that soil microbial growth is limited by P. Conversely, a vector angle below 45° indicates N limitation. The degree of limitation by either N or P increases as the vector angle deviates further from 45° and decreases as it approaches 45°.

2.3. Data Analysis

Variations in soil properties, extracellular enzyme activity, enzyme activity stoichiometry, and vector characteristics across different N application treatments were assessed using one-way analysis of variance (ANOVA). Post hoc comparisons were conducted using the Duncan test with a significance level set at $p < 0.05$. Relationships among soil properties, extracellular enzyme activity, enzyme activity stoichiometry, and vector characteristics were determined through Pearson's correlation analysis. Redundancy Analysis (RDA) was performed with soil extracellular enzyme activity and stoichiometry as response variables and soil properties as explanatory variables to identify the primary environmental factors influencing these enzyme activities and stoichiometries. Random Forest Regression Analysis was employed to discern the principal environmental determinants of soil microbial carbon limitation and nitrogen/phosphorus limitation. Data management was executed in Excel 2021 (Microsoft Corp., Redmond, WA, USA), one-way ANOVA in SPSS 27.0 (SPSS, Chicago, IL, USA), Pearson's correlation analysis in RStudio (version 3.4.2; R Foundation for Statistical Computing, Vienna, Austria) utilizing the "corrplot" package, RDA in Canoco5 (Canoco, Ithaca, NY, USA), and Random Forest Regression Analysis in RStudio (version 3.4.2) using the "rfPermute" package. All graphs presented in this study were generated in RStudio with the "ggplot2" package.

3. Results

3.1. Changes in Soil Properties

Table 1 shows that, compared to the control, the N0 treatment significantly increased TP content to 0.67 g/kg and reduced the C:Ps and N:Ps ratios by 32.99% and 32.24%, respectively. However, the N0 treatment did not significantly impact other soil properties ($p > 0.05$). The N1, N2, and N3 treatments all significantly increased TP content, ranging from 64.1% to 89.74%. Additionally, the N3 treatment significantly increased AP content to 14.34 g/kg. N fertilizer treatment decreased MBC and MBN contents. Specifically, the N1, N2, and N3 treatments significantly reduced MBC content, while the N1 and N2 treatments significantly decreased MBN content ($p < 0.05$). The response of MBP to fertilization is correlated with N fertilizer dosage. The N1 treatment significantly reduced MBP content to 2.64 mg/kg, while the N2 and N3 treatments significantly increased MBP content to 4.41 mg/kg and 6.04 mg/kg, respectively ($p < 0.05$).

Compared to the no N fertilizer treatment (N0), both N1 and N2 treatments significantly decreased SOC content by 29.17% and 24.7%, respectively ($p < 0.05$). The N1 treatment notably reduced AP content to 5.81 mg/kg, whereas the N3 treatment significantly increased AP content to 14.34 mg/kg ($p < 0.05$). N2 and N3 treatments significantly diminished MBC content by 60.96% and 19.88%, respectively. The intergroup differences revealed that variations in MBN content, MBP content, soil nutrient stoichiometry (C:Ns, C:Ps, and N:Ps), and microbial biomass stoichiometry (C:Nm, C:Pm, and N:Pm) between the N fertilizer treatment and the N0 treatment were consistent with those observed between the N fertilizer treatment and the control. This suggests that the alterations in these soil properties are primarily attributable to the addition of N fertilizer.

Table 1. N application effects on soil chemical properties of Moso bamboo forests.

Treatment	SOC g/kg	TN g/kg	TP g/kg	AP mg/kg	pH	MBC mg/kg	MBN mg/kg	MBP mg/kg	C:Ns	C:Ps	N:Ps	C:Nm	C:Pm	N:Pm
CK	16.94 ± 0.85 ab	1.25 ± 0.11 a	0.39 ± 0.02 b	7.63 ± 0.53 bc	4.59 ± 0.03 a	376.29 ± 39.94 a	55.05 ± 1.64 a	3.51 ± 0.23 c	13.56 ± 0.09 a	43.77 ± 0.58 a	3.27 ± 0.46 a	6.89 ± 0.92 a	108.27 ± 14.58 a	15.81 ± 1.17 a
N0	19.23 ± 0.54 a	1.44 ± 0.08 a	0.67 ± 0.05 a	8.8 ± 1.01 b	4.57 ± 0.03 a	415.3 ± 3.97 ab	61.59 ± 2.28 a	3.59 ± 0.33 c	13.4 ± 0.94 a	29.33 ± 2.75 b	2.21 ± 0.28 b	6.76 ± 0.26 a	117.84 ± 11.53 a	17.37 ± 1.2 a
N1	13.62 ± 1.58 b	1.13 ± 0.04 a	0.74 ± 0.07 a	5.81 ± 0.46 c	4.58 ± 0.03 a	117.58 ± 4.01 b	21.53 ± 0.48 b	2.64 ± 0.03 d	12.12 ± 1.63 a	18.94 ± 3.49 d	1.55 ± 0.1 b	5.47 ± 0.3 b	44.57 ± 1.86 b	8.16 ± 0.19 b
N2	14.48 ± 1.06 b	1.23 ± 0.14 a	0.71 ± 0.01 a	9.48 ± 0.83 b	4.68 ± 0.07 a	162.14 ± 10.51 c	33.98 ± 0.25 c	4.41 ± 0.06 b	11.98 ± 0.9 a	20.28 ± 1.11 cd	1.71 ± 0.16 b	4.77 ± 0.31 b	36.69 ± 1.97 b	7.7 ± 0.09 b
N3	17.68 ± 2.29 ab	1.41 ± 0.1 a	0.64 ± 0.01 a	14.34 ± 0.99 a	4.61 ± 0.09 a	332.72 ± 8.81 c	59.45 ± 1.51 a	6.04 ± 0.2 a	12.47 ± 1 a	27.55 ± 3.26 bc	2.2 ± 0.17 b	5.6 ± 0.01 b	55.12 ± 0.69 b	9.85 ± 0.14 b

Lowercase letters in each column indicate significant $p < 0.05$ differences between different N application treatments. Note: Values are means ± standard error (n = 3).

3.2. Changes in Extracellular Enzyme Activity and Stoichiometry

Figure 2A,B shows that, compared to the control, the N0 treatment had no significant effect on BG activity ($p > 0.05$) but significantly increased LAP + NAG activity by 6.11% ($p < 0.05$). Compared to N0 and CK, the N3 treatment significantly reduced BG activity to 217.72 U/L. In comparison to the control, the N1, N2, and N3 treatments significantly increased LAP + NAG activity, with increases ranging from 11.45% to 15.79%. Compared to the no-nitrogen treatment (N0), the application of N fertilizer increased LAP + NAG activity to varying degrees. Specifically, the N1 and N3 treatments significantly increased LAP + NAG activity by 6.5% and 9.12%, respectively. There were no significant differences in ACP activity among the treatments ($p > 0.05$) (Figure 2C).

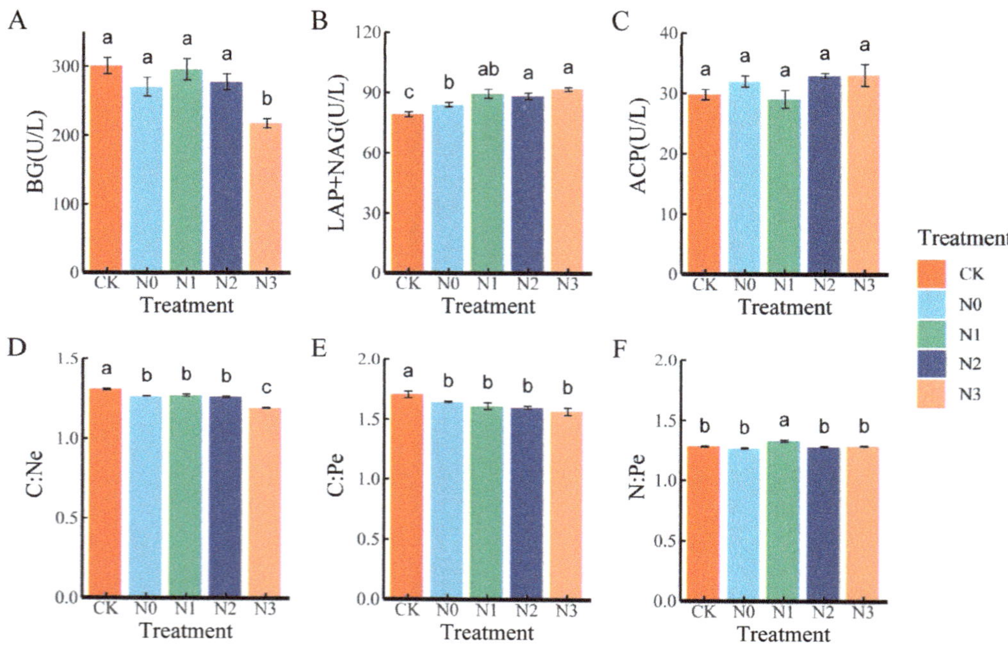

Figure 2. Characteristics of changes in soil extracellular enzyme activities (**A–C**) and enzyme activity stoichiometry (**D–F**) under different fertilization treatments. BG: β-Glucosidase; LAP: leucine aminopeptidase; NAG: N-Acetyl-β-D-glucosidase; ACP: Acid phosphatases; C:Ne: lnBG/ln(LAP + NAG); C:Pe: lnBG/lnACP; N:Pe: ln(LAP + NAG)/lnACP. Lowercase letters indicate significant ($p < 0.05$) differences between the different fertilization treatments. Values are means ± standard error (n = 3).

Figure 2D,E shows that fertilization reduced the C:Ne and C:Pe ratios to varying degrees. Compared to the control, the N0 treatment significantly reduced the C:Ne and C:Pe ratios by 3.27% and 3.5%, respectively. Among the different fertilization treatments, the C:Ne ratio in the N3 treatment was significantly lower than that in the N0, N1, and N2 treatments, with reductions ranging from 5.51% to 6.35%. There were no significant differences in the C:Pe ratio among the different fertilization treatments ($p > 0.05$). Figure 2F shows that there was no significant difference in the N:Pe ratio between the N0 treatment and the control. However, the N1 treatment significantly increased the N:Pe ratio compared to the control and other fertilization treatments, with increases ranging from 3.33% to 4.58%.

3.3. *Changes in Microbial Nutrient Limitation*

A scatter plot with the ratio of (LAP + NAG)/AP on the x-axis and the ratio of BG/ACP on the y-axis can be used to represent the nutrient limitation of microbial growth. The enzymatic stoichiometry scatter plot (Figure 3A) indicates that soil microbial growth in Moso bamboo forests is limited by both carbon and N under different treatments. Figure 3B,C depicts the characteristics of changes in vector length and angle under different treatments. Fertilization significantly reduced the vector length (VL), with decreases ranging from 3.53% to 8.3%. Among the different fertilization treatments, N input reduced the vector length to varying degrees, with the N3 treatment showing a significantly lower vector length compared to the N0 treatment. The vector angles under different treatments were all below 45°, with no significant difference between the N0 treatment and the control ($p > 0.05$). Among the different fertilization treatments, N input increased the vector angle to varying extents. The vector angles of extracellular enzyme activity in the N1 and N3 treatments were significantly higher than those in the N0 treatment ($p < 0.05$), with increases of 4.94% and 2.58%, respectively. Figure 3D illustrates the correlation between vector length and angle under different treatments, showing a significant negative correlation between these two parameters.

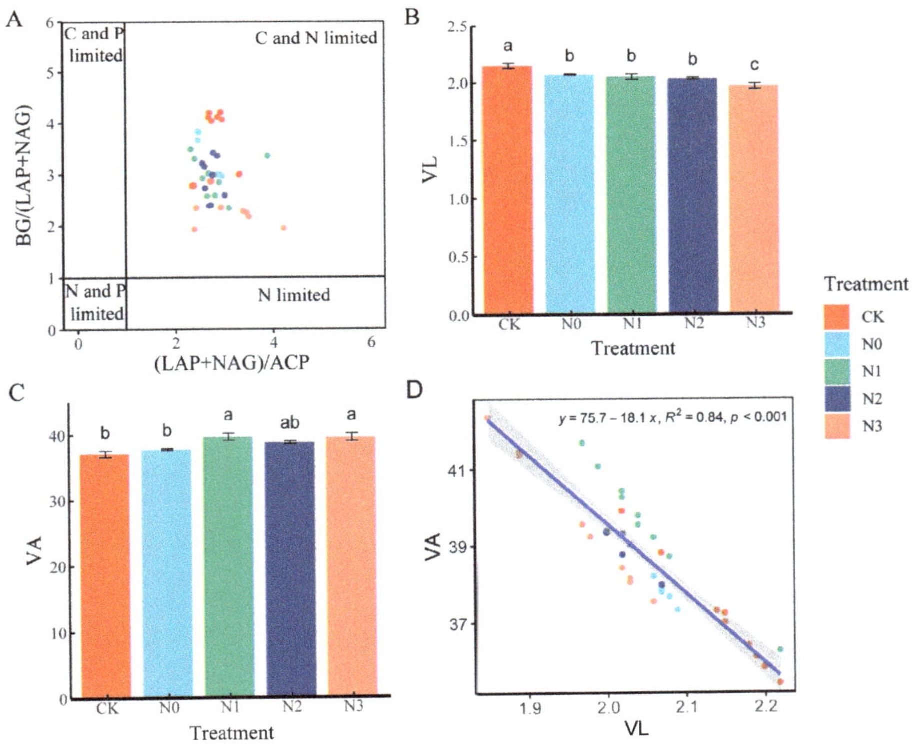

Figure 3. Chemical stoichiometric characterization of enzyme activity under different fertilization treatments (**A**), variation characteristics of vector length and vector angle (**B**,**C**), and correlation between vector length and vector angle (**D**) in soil microbial nutrient limitation. Lowercase letters indicate significant ($p < 0.05$) differences between the different fertilization treatments. Note: Values are means ± standard error (n = 3).

3.4. Environmental Factors Affecting Extracellular Enzyme Activity and Microbial Nutrient Limitation

Figure 4 illustrates the correlations between soil properties, extracellular enzyme activities, enzyme stoichiometry, and vector characteristics. BG shows a significant negative correlation with AP and MBP, while ACP shows a significant positive correlation with AP and MBP. (LAP + NAG) is significantly positively correlated with TP but significantly negatively correlated with N:Ps, MBC, and C:Nm. C:Ne exhibits a significant negative correlation with MBP. C:Pe is significantly positively correlated with C:Ps, N:Ps, C:Pm, and N:Pm but significantly negatively correlated with TP and MBP. N:Pe shows a significant negative correlation with SOC, MBC, and N:Pm. Vector length is significantly positively correlated with C:Ps, N:Ps, C:Nm, and N:Pm but significantly negatively correlated with TP, AP, and MBP. Vector angle is significantly positively correlated with TP but significantly negatively correlated with C:Ps, N:Ps, MBC, MBN, C:Nm, C:Pm, and N:Pm.

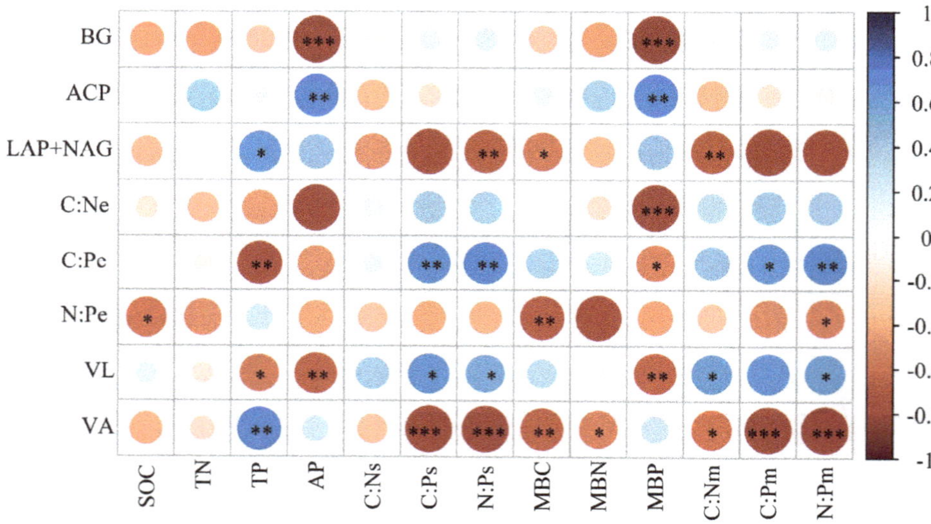

Figure 4. Correlation between extracellular enzyme activities, enzyme activity stoichiometry, and vector characteristics and soil properties under different fertilization treatments. * indicates $p < 0.05$, ** indicates $p < 0.01$, *** indicates $p < 0.001$.

RDA analysis was conducted with extracellular enzyme activity and stoichiometry as response variables and soil properties as explanatory variables to identify key environmental factors influencing enzyme activity and stoichiometry. Figure 5A shows that the first and second axes explain 93.7% and 1.61% of the variation in extracellular enzyme activity and stoichiometry under different treatments, respectively, accounting for a total of 95.31%. Figure 5B indicates that MBP, C:Nm, and TP have explanatory rates of 66.7%, 10.7%, and 10.1%, respectively, all reaching significant levels, making them the key environmental factors affecting enzyme activity and stoichiometry under different treatments. Random forest regression analysis was used to explore key environmental factors affecting vector length and vector angle. Soil property variations explain 75.95% and 65.68% of the changes in vector length and vector angle, respectively, with MBP and N:Pm identified as the most critical factors influencing vector length and vector angle (Figure 5C,D).

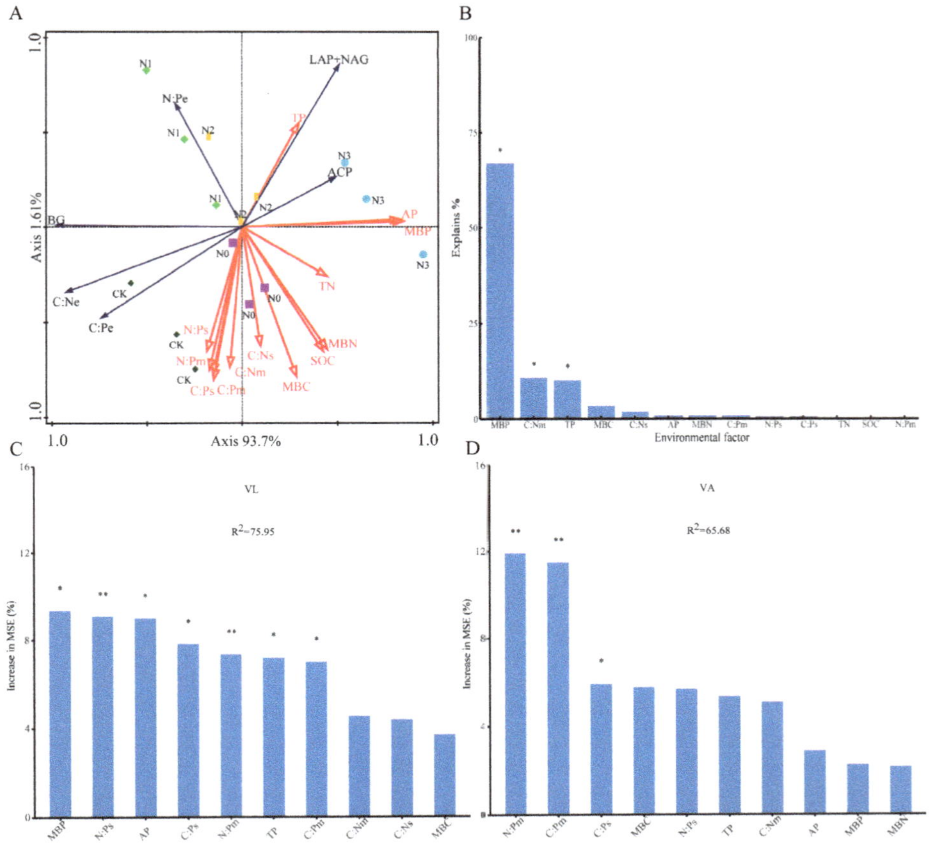

Figure 5. RDA analysis between extracellular enzyme activities and stoichiometry and soil properties under different fertilization treatments (**A**), environmental factor explanatory rate (**B**), random forest regression analysis between vector length and soil properties (**C**), and random forest regression analysis between vector angle and soil properties (**D**). In Figure 5A, the blue arrows represent response variables, and the red arrows represent explanatory variables. ** indicates $p < 0.01$ and * indicates $p < 0.05$.

4. Discussion

4.1. Extracellular Enzyme Activity and Stoichiometry in Response to N Application

Extracellular enzyme activity is related to microbial growth and metabolism. Microorganisms can adjust extracellular enzyme activity to adapt to changes in soil nutrient stoichiometry induced by N application [39]. In this study, there was no significant difference in BG activity between the N0 treatment and the control, while the N3 treatment significantly reduced BG activity (Figure 2A). This indicates that changes in BG activity are primarily influenced by N fertilizer concentration. Some meta-analyses have shown that N fertilization increases BG activity, and these changes in BG activity are closely related to microbial biomass [40,41]. As the dosage increases, the inhibitory effect of N fertilizer on BG activity becomes more pronounced. Chen et al.'s study on *Larix mastersiana* forests also observed this phenomenon (application of NH_4NO_3, 10 g m^{-2} a^{-1}), where high concentrations of N fertilizer significantly suppressed microbial activity, thereby affecting BG activity [42]. The inhibition of BG activity by N application may be related to changes in bacterial community structure [43,44]. Correlation analysis reveals a significant negative correlation between the decrease in BG activity and alterations in AP and MBP content (Figure 4). This suggests that the reduction in BG activity is associated with changes in soil nutrients and microbial biomass. The N0 treatment resulted in an increase in the activity of (LAP + NAG), indicating that the application of P and K fertilizers influences the acquisition of soil microbial N. This, in turn, promotes the activity of N acquisition enzymes [45,46]. Among the various fertilization treatments, an increase in LAP + NAG activity was observed with a rise in N fertilizer concentration (Figure 2B). This suggests that N fertilizer significantly enhances LAP + NAG activity, a finding that aligns with previous studies [47–49]. Conversely, we also found that N application had no significant effect on ACP activity, whereas Tu et al. found that N application (application of NH_4NO_3) increased ACP activity in hybrid bamboo (*Bambusa pervariabilis* McClure × *Bambusa grandis* (Q. H. Dai & X. L. Tao) Ohrnb) stands [50]. When there is a change in soil nutrient stoichiometry, microorganisms adjust their strategy of producing extracellular enzymes to obtain more of the missing nutrients to alleviate the deficient nutrient limitations [47,51,52].

The stoichiometry of extracellular enzyme activity can serve as an indicator of the dynamic equilibrium between nutrient availability [53,54]. In a meta-analysis conducted by Sinsabaugh, it was discovered that the stoichiometry of C, N, and P acquisition enzyme activities at a global scale, as represented by ln(BG):ln(LAP + NAG):ln(ACP) ratios, was approximately 1:1:1 [25]. Peng et al. found this pattern in tropical ecosystems [55]. However, this ratio may fluctuate depending on the ecosystem type and regional environmental conditions [26,56]. In this context, the ratios of C:Ne, C:Pe, and N:Pe were all greater than 1. Consequently, the ratio of lnBG:ln(LAP + NAG): lnACP also diverged from the expected 1:1:1 ratio. This observation indicates that fertilization, especially N application, induced changes in soil resource availability within our study area, prompting microbes to adapt their production strategies for C, N, and P-acquiring enzymes [57,58].

4.2. Effects of Soil Microbial Nutrient Limitation in Fertilized Moso Bamboo Forests

We employed enzyme activity stoichiometric scatter plots and enzyme activity vector features to jointly characterize the nutrient limitations of soil microbial growth in Moso bamboo forests under various treatments [59]. The findings revealed that the growth of soil microorganisms in these forests was primarily constrained by C and N. The vector angle results (Figure 3C) indicated that the vector angle remained below 45° across all treatments, suggesting that N limitation influenced the soil microorganisms of the Moso bamboo forests [22,60]. Scatter plots of enzyme activity stoichiometry further demonstrated (Figure 3A) that C and N limited the growth of soil microorganisms in Moso bamboo forests under all treatments. Additionally, we observed that the application of N fertilizer reduced the vector length and increased the vector angle, implying that N fertilization could mitigate the C and N limitations faced by soil microorganisms in the study area. Sinsabaugh et al. reported global mean values of C:Ne and N:Pe ratios at 0.62 and 0.44, respectively [17]. Our study's C:Ne and N:Pe ratios were higher than these global averages, indicating enhanced activities of BG and LAP + NAG. In accordance with the "resource allocation theory" and the "optimal allocation principle", nutrient addition alters soil enzyme activities, prompting microorganisms to secrete extracellular enzymes to address resource scarcities, thereby alleviating their own nutrient limitations [61,62]. Therefore, higher activities of BG and LAP + NAG indicate a greater microbial demand for C and N.

Zeng et al. employed enzyme stoichiometric vector characterization to examine soil microbial nutrient limitation subsequent to N application (specifically NH_4NO_3 application) in Moso bamboo forests located in Fujian, China. Their findings revealed that both C and P limitations were experienced by microbes in Moso bamboo forests, with the limitations exacerbated by the application of N [32]. Zhang et al. found that soil microorganisms in intensively managed (N, P, and K fertilization) bamboo forests in Zhejiang, China, and in control (non-managed) bamboo forests were both C and N limited [31]. A comparison of the above studies reveals that nutrient addition did not change the soil microbial nutrient limitation pattern compared to the control but rather exacerbated or reduced the limitation of a nutrient element on the basis of the control. Regarding bamboo growing regions, the nutrient limitations affecting soil microorganisms in Moso bamboo forests differ across areas. These variations can primarily be attributed to disparities in soil and climatic conditions between these regions. Jing et al. suggested that soil microbial nutrient limitation may be the result of the long-term acclimatization to soil nutrient environments and climates and that short-term effects of nutrient additions on soil stoichiometry may have a microbial nutrient limitation to a lesser extent, a view that could provide a plausible explanation for our findings [63]. Our study found the nutrient limitation status of soil microorganisms in Moso bamboo forests and their response to N application in northern Guizhou. Our results have some limitations in extrapolation due to soil and climatic conditions but provide some theoretical references for the study of microbial nutrient limitation in Moso bamboo forests in different regions.

Random forest regression analysis was used to explore the key environmental factors influencing microbial nutrient limitation. The results showed that MBP is the most critical factor affecting microbial carbon limitation, while N:Ps, AP, C:Ps, N:Pm, TP, and C:Pm also play significant roles in regulating carbon limitation. N:Pm is identified as the most crucial factor influencing microbial N limitation, with C:Pm and C:Ps also playing important roles. This indicates that changes in soil nutrient stoichiometry and microbial biomass stoichiometry caused by fertilization, especially N fertilization, are the main reasons for alleviating microbial carbon and N limitation.

Ecological stoichiometry theory posits that microbial metabolism necessitates the preservation of a balanced C, N, and P stoichiometry within an organism [64,65]. Disruptions in this stoichiometric equilibrium between microbes and their soil environment can prompt microbial responses [66,67]. These responses may include adjustments to the efficiency of elemental utilization of C, N, and P, alterations in the secretion of extracellular enzymes, or modifications in their elemental composition through shifts in community structure [68,69]. Changes in microbial biomass stoichiometry can serve as indicators of changes in microbial nutrient acquisition strategies when there is an imbalance between soil nutrient resources and the microbes' own nutrient requirements [70,71].

4.3. Shortcomings of the Study and Future Prospects

Our findings illustrate the impact of N input on nutrient limitations for soil microbes in Moso bamboo forests. However, these results were obtained with the addition of phosphorus (P) and potassium (K) fertilizers. Although the no-nitrogen treatment showed minimal differences in soil properties compared to the control, the input of P and K might obscure the effects of N on microbial activity and extracellular enzyme secretion. This makes it challenging to precisely investigate the impact of N fertilization on microbial nutrient limitations in Moso bamboo forest soils.

In routine practices, N fertilizer is typically applied alongside P and K fertilizers. Therefore, precisely exploring the effects of fertilization on microbial activity is crucial for understanding soil carbon turnover and maintaining soil quality. Further research is needed to investigate the mechanisms by which different nutrient elements, either independently or in combination, influence soil microbial metabolism.

5. Conclusions

The results indicate that N fertilization decreased BG activity, increased LAP + NAG activity, and had no significant effect on ACP activity. It also reduced enzyme stoichiometry (C:Ne, C:Pe, and N:Pe). In terms of microbial nutrient limitation, microbial growth was constrained by both carbon and N, and the addition of N alleviated these nutrient limitations. Changes in soil properties induced by fertilization were the drivers of variations in extracellular enzyme activity, enzyme stoichiometry, and vector characteristics. MBP was identified as the most critical driving factor affecting extracellular enzyme activity, enzyme stoichiometry, and microbial carbon limitation, while N:Pm was the most crucial factor influencing microbial N limitation. Nutrient addition, especially N fertilization, altered the nutrient acquisition strategies of soil microbes in Moso bamboo forests. This may change the microbial decomposition process of organic carbon, affecting the sequestration of organic carbon and the maintenance of soil quality in these forests. Our study contributes to understanding the nutrient requirements of microbes in bamboo forest ecosystems, providing a theoretical basis for the efficient use of N fertilizer and sustainable management of Moso bamboo forests. It also offers theoretical insights into the microbial mechanisms of carbon cycling in bamboo forest ecosystems under conditions of high N input.

Author Contributions: Conceptualization, S.F. and W.S.; Formal analysis, Z.W.; Methodology, Y.Z. and S.F.; Data curation, H.C., Y.L. and Y.S.; Writing—original draft, H.C.; Writing—review and editing, H.C. and W.S.; Visualization, H.C. and Y.Z. Funding acquisition, Y.Z. All authors have read and agreed to the published version of the manuscript.

Funding: National Key R&D Program of China of the 14th Five-Year Plan (grant Number 2023YFD2201202); New Agricultural Science Research and Reform Practice Project (grant Number 0322005).

Data Availability Statement: The data presented in this study are available on request from the corresponding author. The data are not publicly available due to privacy or ethical restrictions.

Conflicts of Interest: The authors declare no conflicts of interest.

References

1. Jiang, Z. *Bamboo and Rattan in the World*; China Forest Publishing House: Beijng, China, 2002.
2. Song, X.; Peng, C.; Ciais, P.; Li, Q.; Xiang, W.; Xiao, W.; Zhou, G.; Deng, L. Nitrogen Addition Increased CO_2 Uptake More than Non-CO_2 Greenhouse Gases Emissions in a Moso Bamboo Forest. *Sci. Adv.* **2020**, *6*, eaaw5790. [CrossRef] [PubMed]
3. Feng, P.; Li, Y. China's Bamboo Resources in 2021. *World Bamboo Ratt.* **2023**, *21*, 100–103.
4. Fu, L.; Su, J. Calculation of carbon sink of bamboo forest in China and its potential prediction. *China For. Econ.* **2023**, *3*, 96–102. [CrossRef]
5. Chen, X.; Zhang, X.; Zhang, Y.; Booth, T.; He, X. Changes of Carbon Stocks in Bamboo Stands in China during 100 Years. *For. Ecol. Manag.* **2009**, *258*, 1489–1496. [CrossRef]
6. Zheng, Y.; Fan, S.; Guan, F.; Xia, W.; Wang, S.; Xiao, X. Strip Clearcutting Drives Vegetation Diversity and Composition in the Moso Bamboo Forests. *For. Sci.* **2022**, *68*, 27–36. [CrossRef]
7. Zhao, J.; Cai, C. Physiological Integration Improves Nitrogen Use Efficiency of Moso Bamboo: An Isotopic (15N) Assessment. *For. Ecol. Manag.* **2023**, *542*, 121073. [CrossRef]
8. Zhao, J.; Su, W.; Fan, S.; Cai, C.; Su, H.; Zeng, X. Ammonia Volatilization and Nitrogen Runoff Losses from Moso Bamboo Forests after Different Fertilization Practices. *Can. J. For. Res.* **2019**, *49*, 213–220. [CrossRef]
9. Wu, P.; Wu, Q.; Huang, H.; Xie, L.; An, H.; Zhao, X.; Wang, F.; Gao, Z.; Zhang, R.; Bangura, K.; et al. Global Meta-Analysis and Three-Year Field Experiment Shows That Deep Placement of Fertilizer Can Enhance Crop Productivity and Decrease Gaseous Nitrogen Losses. *Field Crops Res.* **2024**, *307*, 109263. [CrossRef]
10. Zhang, X.; Zeng, Q.; Yuan, X.; Zhang, Q.; Zhou, J.; Xu, M.; Sun, H.; Chen, L.; Gao, X.; Chen, Y. Soil Dissolved Organic Matter Quality and Bacterial Community Composition Regulate the Substrate-Binding Affinity of Hydrolytic Enzymes under Short-Term Nitrogen Addition. *Geoderma* **2024**, *445*, 116885. [CrossRef]
11. Gong, H.; Li, J.; Liu, Z.; Hou, R.; Zhang, Y.; Xu, Y.; Zhu, W.; Yang, L.; Ouyang, Z. Linkages of Soil and Microbial Stoichiometry to Crop Nitrogen Use Efficiency: Evidence from a Long-Term Nitrogen Addition Experiment. *CATENA* **2024**, *240*, 107961. [CrossRef]
12. Chu, H.; Ni, H.; Su, W.; Fan, S.; Long, Y.; Sun, Y. Enhanced Nitrogen Fertilizer Input Alters Soil Carbon Dynamics in Moso Bamboo Forests, Impacting Particulate Organic and Mineral-Associated Carbon Pools. *Forests* **2023**, *14*, 2460. [CrossRef]
13. Gianfreda, L.; Ruggiero, P. Enzyme Activities in Soil. In *Nucleic Acids and Proteins in Soil*; Springer: Berlin/Heidelberg, Germany, 2006; pp. 257–311.
14. Wyszkowska, J.; Borowik, A.; Kucharski, M.; Kucharski, J. Effect of Cadmium, Copper and Zinc on Plants, Soil Microorganisms and Soil Enzymes. *J. Elem.* **2013**, *18*, 769–796. [CrossRef]
15. Cui, Y.; Fang, L.; Guo, X.; Wang, X.; Zhang, Y.; Li, P.; Zhang, X. Ecoenzymatic Stoichiometry and Microbial Nutrient Limitation in Rhizosphere Soil in the Arid Area of the Northern Loess Plateau, China. *Soil Biol. Biochem.* **2018**, *116*, 11–21. [CrossRef]
16. Yang, L.; Jia, Y.; Li, Q.; Cui, H.; Lu, J.; Ma, J.; Xu, Z. Ecoenzymatic Stoichiometry in the Rhizosphere and Bulk Soil of a Larix Principis-Rupprechtii Plantation in North China. *Forests* **2023**, *14*, 1315. [CrossRef]
17. Sinsabaugh, R.L.; Hill, B.H.; Follstad Shah, J.J. Ecoenzymatic Stoichiometry of Microbial Organic Nutrient Acquisition in Soil and Sediment. *Nature* **2009**, *462*, 795–798. [CrossRef] [PubMed]
18. Ren, C.; Chen, J.; Lu, X.; Doughty, R.; Zhao, F.; Zhong, Z.; Han, X.; Yang, G.; Feng, Y.; Ren, G. Responses of Soil Total Microbial Biomass and Community Compositions to Rainfall Reductions. *Soil Biol. Biochem.* **2018**, *116*, 4–10. [CrossRef]
19. Zhou, L.; Liu, S.; Shen, H.; Zhao, M.; Xu, L.; Xing, A.; Fang, J. Soil Extracellular Enzyme Activity and Stoichiometry in China's Forests. *Funct. Ecol.* **2020**, *34*, 1461–1471. [CrossRef]

20. Yao, B.; Wang, X.; Li, Y.; Lian, J.; Li, Y.; Luo, Y.; Li, Y. Soil Extracellular Enzyme Activity Reflects the Change of Nitrogen to Phosphorus Limitation of Microorganisms during Vegetation Restoration in Semi-Arid Sandy Land of Northern China. *Front. Environ. Sci.* **2023**, *11*, 1298027. [CrossRef]
21. Sinsabaugh, R.L.; Lauber, C.L.; Weintraub, M.N.; Ahmed, B.; Allison, S.D.; Crenshaw, C.; Contosta, A.R.; Cusack, D.; Frey, S.; Gallo, M.E.; et al. Stoichiometry of Soil Enzyme Activity at Global Scale: Stoichiometry of Soil Enzyme Activity. *Ecol. Lett.* **2008**, *11*, 1252–1264. [CrossRef]
22. Xiao, L.; Liu, G.; Li, P.; Li, Q.; Xue, S. Ecoenzymatic Stoichiometry and Microbial Nutrient Limitation during Secondary Succession of Natural Grassland on the Loess Plateau, China. *Soil Tillage Res.* **2020**, *200*, 104605. [CrossRef]
23. Li, J.; Wu, J.; Yu, J.; Wang, K.; Li, J.; Cui, Y.; Shangguan, Z.; Deng, L. Soil Enzyme Activity and Stoichiometry in Response to Precipitation Changes in Terrestrial Ecosystems. *Soil Biol. Biochem.* **2024**, *191*, 109321. [CrossRef]
24. Abay, P.; Gong, L.; Luo, Y.; Zhu, H.; Ding, Z. Soil Extracellular Enzyme Stoichiometry Reveals the Nutrient Limitations in Soil Microbial Metabolism under Different Carbon Input Manipulations. *Sci. Total Environ.* **2024**, *913*, 169793. [CrossRef]
25. Sinsabaugh, R.L.; Follstad Shah, J.J. Ecoenzymatic Stoichiometry and Ecological Theory. *Annu. Rev. Ecol. Evol. Syst.* **2012**, *43*, 313–343. [CrossRef]
26. Luo, H.; Yu, J.; Li, R.; Gu, J.D.; Luo, L.; Zhang, Y.; He, Y.; Xiao, Y.; Deng, S.; Zhang, Y.; et al. Microbial Biomass C:N:P as a Better Indicator than Soil and Ecoenzymatic C:N:P for Microbial Nutrient Limitation and C Dynamics in Zoige Plateau Peatland Soils. *Int. Biodeterior.* **2022**, *175*, 105492.
27. Xu, Z.; Yu, G.; Zhang, X.; He, N.; Wang, Q.; Wang, S.; Wang, R.; Zhao, N.; Jia, Y.; Wang, C. Soil Enzyme Activity and Stoichiometry in Forest Ecosystems along the North-South Transect in Eastern China (NSTEC). *Soil Biol. Biochem.* **2017**, *104*, 152–163. [CrossRef]
28. Zhang, W.; Xu, Y.; Gao, D.; Wang, X.; Liu, W.; Deng, J.; Han, X.; Yang, G.; Feng, Y.; Ren, G. Ecoenzymatic Stoichiometry and Nutrient Dynamics along a Revegetation Chronosequence in the Soils of Abandoned Land and Robinia Pseudoacacia Plantation on the Loess Plateau, China. *Soil Biol. Biochem.* **2019**, *134*, 1–14. [CrossRef]
29. Yang, F.; Ding, L.; Zhang, Y.; Wang, P.; Liu, J.; Zou, C.; Chen, C. Discussion on the Conversion Method of Soil Enzyme Stoichiometry from Different Vegetation Types in Guizhou Grassland, Southwest China. *J. Phys. Conf. Ser.* **2020**, *1549*, 022035. [CrossRef]
30. Moorhead, D.L.; Sinsabaugh, R.L.; Hill, B.H.; Weintraub, M.N. Vector Analysis of Ecoenzyme Activities Reveal Constraints on Coupled C, N and P Dynamics. *Soil Biol. Biochem.* **2016**, *93*, 1–7. [CrossRef]
31. Zhang, X.; Li, Q.; Zhong, Z.; Huang, Z.; Wen, X.; Bian, F.; Yang, C. Determining Changes in Microbial Nutrient Limitations in Bamboo Soils under Different Management Practices via Enzyme Stoichiometry. *CATENA* **2023**, *223*, 106939. [CrossRef]
32. Zeng, Q.; Zhang, Q.; Lin, K.; Zhou, J. Enzyme stoichiometry evidence revealed that five years nitrogen addition exacerbated the carbon and phosphorus limitation of soil microorganisms in a Phyllostachys pubescens. *Chin. J. Appl. Ecol.* **2021**, *32*, 521–528. [CrossRef]
33. Su, W. Fertilization Theory and Practice for Phyllostachys edulis Stand Based on Growth and Nutrient Accumulation Rules. Ph.D. Thesis, Chinese Academy of Forestry, Beijing, China, 2012.
34. Su, W.; Fan, S.; Zhao, J.; Cai, C. Effects of Various Fertilization Placements on the Fate of Urea-15N in Moso Bamboo Forests. *For. Ecol. Manag.* **2019**, *453*, 117632. [CrossRef]
35. Zhang, M.; Zhang, W.; Bai, S.H.; Niu, Y.; Hu, D.; Ji, H.; Xu, Z. Minor Increases in Phyllostachys Edulis (Moso Bamboo) Biomass despite Evident Alterations of Soil Bacterial Community Structure after Phosphorus Fertilization Alone: Based on Field Studies at Different Altitudes. *For. Ecol. Manag.* **2019**, *451*, 117561. [CrossRef]
36. Tietz, A.; Kirschner, A.; Langergraber, G.; Sleytr, K.; Haberl, R. Characterisation of Microbial Biocoenosis in Vertical Subsurface Flow Constructed Wetlands. *Sci. Total Environ.* **2007**, *380*, 163–172. [CrossRef] [PubMed]
37. Xia, H.; Riaz, M.; Ming, C.; Li, Y.; Wang, X.; Jiang, C. Assessing the Difference of Biochar and Aged Biochar to Improve Soil Fertility and Cabbage (*Brassica oleracea* Var. *Capitata*) Productivity. *J. Soils Sediments* **2023**, *23*, 606–618. [CrossRef]
38. Xia, H.; Riaz, M.; Tang, X.; Yan, L.; El-Desouki, Z.; Li, Y.; Wang, X.; Cuncang, J. Insight into Mechanisms of Biochar-Fertilizer Induced of Microbial Community and Microbiology of Nitrogen Cycle in Acidic Soil. *J. Environ. Manag.* **2023**, *336*, 117602. [CrossRef] [PubMed]
39. Marklein, A.R.; Houlton, B.Z. Nitrogen Inputs Accelerate Phosphorus Cycling Rates across a Wide Variety of Terrestrial Ecosystems. *New Phytol.* **2012**, *193*, 696–704. [CrossRef] [PubMed]
40. Jian, S.; Li, J.; Chen, J.; Wang, G.; Mayes, M.A.; Dzantor, K.E.; Hui, D.; Luo, Y. Soil Extracellular Enzyme Activities, Soil Carbon and Nitrogen Storage under Nitrogen Fertilization: A Meta-Analysis. *Soil Biol. Biochem.* **2016**, *101*, 32–43. [CrossRef]
41. Chen, J.; Luo, Y.; Li, J.; Zhou, X.; Cao, J.; Wang, R.; Wang, Y.; Shelton, S.; Jin, Z.; Walker, L.M.; et al. Costimulation of Soil Glycosidase Activity and Soil Respiration by Nitrogen Addition. *Glob. Chang. Biol.* **2017**, *23*, 1328–1337. [CrossRef]
42. Chen, Q.; Wang, Z.; Liu, Y.; Zheng, H.; Li, H. Response of soil enzyme activity and stoichiometric ratio to simulated nitrogen deposition in subalpine coniferous forests of western Sichuan. *Chin. J. Appl. Environ. Biol.* **2019**, *25*, 791–800. [CrossRef]
43. Kou, Z.; Li, C.; Chang, S.; Miao, Y.; Zhang, W.; Li, Q.; Dang, T.; Wang, Y. Effects of Nitrogen and Phosphorus Additions on Soil Microbial Community Structure and Ecological Processes in the Farmland of Chinese Loess Plateau. *J. Arid. Land* **2023**, *15*, 960–974. [CrossRef]
44. Carrara, J.E.; Walter, C.A.; Hawkins, J.S.; Peterjohn, W.T.; Averill, C.; Brzostek, E.R. Interactions among Plants, Bacteria, and Fungi Reduce Extracellular Enzyme Activities under Long-term N Fertilization. *Glob. Chang. Biol.* **2018**, *24*, 2721–2734. [CrossRef]

45. Fan, Z.; Wang, X.; Wang, C.; Bai, E. Effect of nitrogen and phosphorus addition on soil enzyme activities: A meta-analysis. *Chin. J. Appl. Ecol.* **2018**, *29*, 1266–1272. [CrossRef] [PubMed]
46. Sui, B. Effect of Potassium Fertilizer Application on Microbial Community in Tobacco-Growing Soil. Master's Thesis, Yunnan Agricultural University, Kunming, China, 2023.
47. Jia, X.; Zhong, Y.; Liu, J.; Zhu, G.; Shangguan, Z.; Yan, W. Effects of Nitrogen Enrichment on Soil Microbial Characteristics: From Biomass to Enzyme Activities. *Geoderma* **2020**, *366*, 114256. [CrossRef]
48. Chen, J.; Luo, Y.; van Groenigen, K.J.; Hungate, B.A.; Cao, J.; Zhou, X.; Wang, R. A Keystone Microbial Enzyme for Nitrogen Control of Soil Carbon Storage. *Sci. Adv.* **2018**, *4*, eaaq1689. [CrossRef] [PubMed]
49. Chen, H.; Li, D.; Zhao, J.; Zhang, W.; Xiao, K.; Wang, K. Nitrogen Addition Aggravates Microbial Carbon Limitation: Evidence from Ecoenzymatic Stoichiometry. *Geoderma* **2018**, *329*, 61–64. [CrossRef]
50. Tu, L.; Chen, G.; Peng, Y.; Hu, H.; Hu, T.; Zhang, J.; Li, X.; Liu, L.; Tang, Y. Soil Biochemical Responses to Nitrogen Addition in a Bamboo Forest. *PLoS ONE* **2014**, *9*, e102315. [CrossRef]
51. Merchant, S.S.; Helmann, J.D. Elemental Economy: Microbial Strategies for Optimizing Growth in the Face of Nutrient Limitation. *Adv. Microb. Physiol.* **2012**, *60*, 91–210. [PubMed]
52. Sinsabaugh, R.L.; Follstad Shah, J.J. Ecoenzymatic Stoichiometry of Recalcitrant Organic Matter Decomposition: The Growth Rate Hypothesis in Reverse. *Biogeochemistry* **2011**, *102*, 31–43. [CrossRef]
53. Deng, Q.; Hui, D.; Dennis, S.; Reddy, K.C. Responses of Terrestrial Ecosystem Phosphorus Cycling to Nitrogen Addition: A Meta-analysis. *Glob. Ecol. Biogeogr.* **2017**, *26*, 713–728. [CrossRef]
54. Averill, C.; Waring, B. Nitrogen Limitation of Decomposition and Decay: How Can It Occur? *Glob. Chang. Biol.* **2018**, *24*, 1417–1427. [CrossRef]
55. Peng, X.; Wang, W. Stoichiometry of Soil Extracellular Enzyme Activity along a Climatic Transect in Temperate Grasslands of Northern China. *Soil Biol. Biochem.* **2016**, *98*, 74–84. [CrossRef]
56. Yang, Y.; Liang, C.; Wang, Y.; Cheng, H.; An, S.; Chang, S.X. Soil Extracellular Enzyme Stoichiometry Reflects the Shift from P- to N-Limitation of Microorganisms with Grassland Restoration. *Soil Biol. Biochem.* **2020**, *149*, 107928. [CrossRef]
57. Gai, X.; Zhong, Z.; Zhang, X.; Bian, F.; Yang, C. Effects of Chicken Farming on Soil Organic Carbon Fractions and Fungal Communities in a Lei Bamboo (*Phyllostachys praecox*) Forest in Subtropical China. *For. Ecol. Manag.* **2021**, *479*, 118603. [CrossRef]
58. Wu, Y.; Chen, W.; Li, Q.; Guo, Z.; Li, Y.; Zhao, Z.; Zhai, J.; Liu, G.; Xue, S. Ecoenzymatic Stoichiometry and Nutrient Limitation under a Natural Secondary Succession of Vegetation on the Loess Plateau, China. *Land Degrad. Dev.* **2021**, *32*, 399–409. [CrossRef]
59. Wu, Z.; Chen, X.; Lu, X.; Zhu, Y.; Han, X.; Yan, J.; Yan, L.; Zou, W. Impact of Combined Organic Amendments and Chemical Fertilizers on Soil Microbial Limitations, Soil Quality, and Soybean Yield. *Plant Soil* **2024**, 1–18. [CrossRef]
60. Cui, Y.; Zhang, Y.; Duan, C.; Wang, X.; Zhang, X.; Ju, W.; Chen, H.; Yue, S.; Wang, Y.; Li, S.; et al. Ecoenzymatic Stoichiometry Reveals Microbial Phosphorus Limitation Decreases the Nitrogen Cycling Potential of Soils in Semi-Arid Agricultural Ecosystems. *Soil Tillage Res.* **2020**, *197*, 104463. [CrossRef]
61. Liu, C.; Wang, B.; Zhu, Y.; Qu, T.; Xue, Z.; Li, X.; Zhou, Z.; An, S. Eco-Enzymatic Stoichiometry and Microbial Non-Homeostatic Regulation Depend on Relative Resource Availability during Litter Decomposition. *Ecol. Indic.* **2022**, *145*, 109729. [CrossRef]
62. Song, S.; Wang, X.; He, C.; Chi, Y. Effects of Utilization Methods on C, N, P Rate and Enzyme Activity of Artificial Grassland in Karst Desertification Area. *Agronomy* **2023**, *13*, 1368. [CrossRef]
63. Jing, X.; Chen, X.; Xiao, W.; Lin, L.; Wang, C.; He, J.-S.; Zhu, B. Soil Enzymatic Responses to Multiple Environmental Drivers in the Tibetan Grasslands: Insights from Two Manipulative Field Experiments and a Meta-Analysis. *Pedobiologia* **2018**, *71*, 50–58. [CrossRef]
64. Li, J.; Sang, C.; Yang, J.; Qu, L.; Xia, Z.; Sun, H.; Jiang, P.; Wang, X.; He, H.; Wang, C. Stoichiometric Imbalance and Microbial Community Regulate Microbial Elements Use Efficiencies under Nitrogen Addition. *Soil Biol. Biochem.* **2021**, *156*, 108207. [CrossRef]
65. Wang, F.; Liu, Q.; Hu, X.; Fang, X.-M.; Wang, S.; Chen, F.-S. Divergent Responses of Soil Microbial Community to Long-Term Nitrogen and Phosphorus Additions in a Subtropical Chinese Fir Plantation. *CATENA* **2024**, *242*, 108132. [CrossRef]
66. Cui, Y.; Bing, H.; Moorhead, D.L.; Delgado-Baquerizo, M.; Ye, L.; Yu, J.; Zhang, S.; Wang, X.; Peng, S.; Guo, X.; et al. Ecoenzymatic Stoichiometry Reveals Widespread Soil Phosphorus Limitation to Microbial Metabolism across Chinese Forests. *Commun. Earth Environ.* **2022**, *3*, 184. [CrossRef]
67. Cui, Y.; Moorhead, D.L.; Peng, S.; Sinsabaugh, R.L.; Peñuelas, J. Predicting Microbial Nutrient Limitations from a Stoichiometry-Based Threshold Framework. *Innov. Geosci.* **2024**, *2*, 100048. [CrossRef]
68. Cui, Y.; Bing, H.; Fang, L.; Jiang, M.; Shen, G.; Yu, J.; Wang, X.; Zhu, H.; Wu, Y.; Zhang, X. Extracellular Enzyme Stoichiometry Reveals the Carbon and Phosphorus Limitations of Microbial Metabolisms in the Rhizosphere and Bulk Soils in Alpine Ecosystems. *Plant Soil* **2021**, *458*, 7–20. [CrossRef]
69. Mooshammer, M.; Wanek, W.; Hämmerle, I.; Fuchslueger, L.; Hofhansl, F.; Knoltsch, A.; Schnecker, J.; Takriti, M.; Watzka, M.; Wild, B.; et al. Adjustment of Microbial Nitrogen Use Efficiency to Carbon: Nitrogen Imbalances Regulates Soil Nitrogen Cycling. *Nat. Commun.* **2014**, *5*, 3694. [CrossRef] [PubMed]

70. Mao, X.; Sun, T.; Zhu, L.; Wanek, W.; Cheng, Q.; Wang, X.; Zhou, J.; Liu, X.; Ma, Q.; Wu, L.; et al. Microbial Adaption to Stoichiometric Imbalances Regulated the Size of Soil Mineral-Associated Organic Carbon Pool under Continuous Organic Amendments. *Geoderma* **2024**, *445*, 116883. [CrossRef]
71. Cai, M.; Zhang, Y.; Zhao, G.; Zhao, B.; Cong, N.; Zhu, J.; Zheng, Z.; Wu, W.; Duan, X. Excessive Climate Warming Exacerbates Nitrogen Limitation on Microbial Metabolism in an Alpine Meadow of the Tibetan Plateau: Evidence from Soil Ecoenzymatic Stoichiometry. *Sci. Total Environ.* **2024**, *930*, 172731. [CrossRef]

forests

MDPI

Review

Impacts of Harvesting and Prescribed Burning on Forest Soil Carbon Dynamics: A Global Meta-Analysis

Moeka Ono * and Asko Noormets

Department of Ecology and Conservation Biology, Texas A&M University, College Station, TX 77843, USA; noormets@tamu.edu
* Correspondence: mono@tamu.edu

Abstract: Forest management practices such as clearcutting, thinning, and prescribed burning are widely implemented to achieve various ecological and silvicultural objectives, yet their effects on soil carbon dynamics and belowground processes remain uncertain. We conducted a global meta-analysis of 414 observations from 110 studies to quantify the impacts of these practices on total soil respiration (SR), its autotrophic (Ra) and heterotrophic (Rh) components, and associated biophysical and soil variables. Clearcutting and prescribed burning both reduced SR by an average of 11%, driven largely by Ra declines following reductions in live biomass, forest floor inputs, and microbial biomass. Thinning caused no significant change in SR, likely due to the limited belowground disturbance and residual vegetation compensatory growth, although impacts intensified when combined with post-treatments (e.g., residue removal or site-preparation burns), resembling those of clearcutting or repeated burns. In contrast, post-burn treatments following clearcutting did not substantially alter biological factors or SR components. Across practices, soil temperature increased due to the opening of the canopy, middle- and understory vegetation, and forest floor disturbance, but this warming showed no consistent relationship with Rh or SR. Instead, responses were primarily governed by substrate availability, highlighting its central role in soil carbon fluxes under management disturbances.

Keywords: forest management; clearcut; thinning; prescribed fire; soil respiration; autotrophic respiration; heterotrophic respiration; soil carbon dynamics

1. Introduction

Forests cover approximately 30% of the Earth's land surface and store nearly 40% of terrestrial carbon [1,2], making them a critical component of the global carbon sink. Globally, ~7% of forests are managed plantations [3], and ~74% comprise secondary forests recovering from past disturbances [4]. Recent analyses have documented an increasing trend in soil heterotrophic respiration [5] and a concurrent decline in global forest carbon sink capacity [6,7], raising concerns about the resilience of forest carbon storage. Forest management practices are often designed to enhance tree growth, health, risk mitigation, or habitat quality [8], but they also represent disturbances that affect soil carbon fluxes and dynamics [9,10]. Quantifying their effects is therefore essential for improving predictions of terrestrial carbon feedback under global change [11].

Forest management activities, such as harvesting and prescribed burning, can alter key structural and environmental drivers of forest carbon cycling, such as light availability, microclimate, and the substrates and nutrients supply [10,12]. In response, trees may shift carbon allocation strategies to balance growth, maintenance, and stress tolerance, for

https://doi.org/10.3390/f16101555

example, by modifying biomass investment between above- and belowground components [13]. Changes in detrital input and rhizosphere substrates can constrain microbial activity, whereas shifts in microclimates, such as soil warming, may stimulate microbial decomposition and root growth [14]. Physical soil disturbance can also expose previously protected organic matter, thus accelerating mineralization [15]. These interacting factors contribute to the high variability in soil carbon responses across forest types, sites, and management regimes.

Recent meta-analyses have assessed the impact of disturbances on soil respiration in global forest ecosystems [16], and under specific management disturbances, such as forest fires [17–19], thinning [20–22], and harvesting [23]. These studies suggest that soil respiration (SR) tends to increase under light harvesting, driven by fine root biomass growth and increased soil nutrients, while both autotrophic (Ra) and heterotrophic (Rh) respiration generally decline with greater harvest intensity, like clearcutting, because of substantial loss in detritus inputs. In contrast, different studies have arrived at contrasting conclusions about the effect of fires on both Ra and Rh [17,18]. Fire effects can also depend on burn intensity, with prescribed burns typically being low-intensity compared to wildfires. As a result, syntheses focused on prescribed fire often reflect conditions that are not directly comparable to high-intensity fires.

Despite these advances, several information gaps remain. With accruing data, it is increasingly possible to resolve "harvesting" effects to those caused by thinning and clearcutting, and "fire" effects to prescribed burns and wildfires. At the same time, management practices are often applied in sequence or combination, such as clearcutting or thinning, followed by site-preparation burning [24–26], herbicide application [27], fertilization [28], or the use of repeated prescribed burning for fuel reduction [29]. However, the limited availability of such datasets has not allowed a thorough assessment of the modifying effects of these secondary treatments, and their long-term consequences for belowground carbon cycling remain poorly resolved [8,30]. Building on previous meta-analyses that primarily examined single management practices, this study will review their combined impacts on soil respiration components and associated soil and biophysical variables. A better quantitative understanding of these interactions will help balance climate mitigation, economic, and soil health considerations in land management decisions [11,31].

In this meta-analysis, we synthesized peer-reviewed studies on three major forest management practices—clearcutting, thinning, and prescribed burning—to quantify their effects on SR components (Ra and Rh) and associated biophysical and soil environmental variables. The objectives of this study were to (1) assess the effects of major forest management practices (clearcutting, thinning, and prescribed burn) on SR, its components, and soil carbon stocks, and (2) explore how soil carbon dynamics respond to variation in management strategies, climate, forest types, and recovery stages. We hypothesized that (i) clearcutting would substantially reduce Ra by increasing root mortality and limiting photosynthate supply from aboveground biomass and would concurrently reduce Rh by limiting both the photosynthate supply and the inputs of aboveground detritus; (ii) prescribed burning would reduce Ra through damage to fine root biomass and concurrently reduce Rh due to the subsequent decrease in root exudates, though the overall reduction would be smaller than with clearcutting; (iii) thinning effect would be minimal on Ra and Rh; and (iv) sequence or combined management effects would amplify the effects on Ra and Rh, resulting in greater cumulative impacts than individual practices alone.

2. Materials and Methods

2.1. Data Collection and Compilation

Peer-reviewed journal articles published before August 2024 were obtained by searching the ISI Web of Science database, using the search terms "(soil CO2 OR soil carbon dioxide OR soil carbon efflux OR soil carbon emission OR root respiration OR autotrophic respiration OR microb* respiration OR belowground respiration OR heterotrophic respiration) AND (harvest* OR thinn* OR log* OR understory OR litter* OR manag* OR clear cut* OR clearcut* OR burn* OR slash* OR fire) AND (boreal OR temperate OR tropical OR mediterranean) AND (forest)". In addition to these search results, publications from previous meta-analyses on soil respiration [16–18] and the Soil Respiration Database (SRDB) Version 5 [32] that met the selection criteria were also included.

The retrieved articles were screened based on the following criteria: (1) field-based studies involving forest management practices; (2) inclusion of at least one soil respiration component (total, autotrophic, or heterotrophic respiration); (3) a study duration of more than one growing season; and (4) inclusion of both control and treatment groups. Regarding criterion (1), we excluded incubation studies, as this method does not fully account for the essential biophysical link between live plants and soil (e.g., photosynthate supply to rhizosphere), which may yield divergent results from field settings [33]. For chronosequence studies, the longest undisturbed stand was treated as the control. In total, 414 observations from 110 articles published between 1987 and 2025 met the selection criteria and were included in the meta-analysis (Figure 1, Table S2). The dataset includes 189 observations for clearcutting, 175 observations for thinning, and 50 observations for prescribed burning.

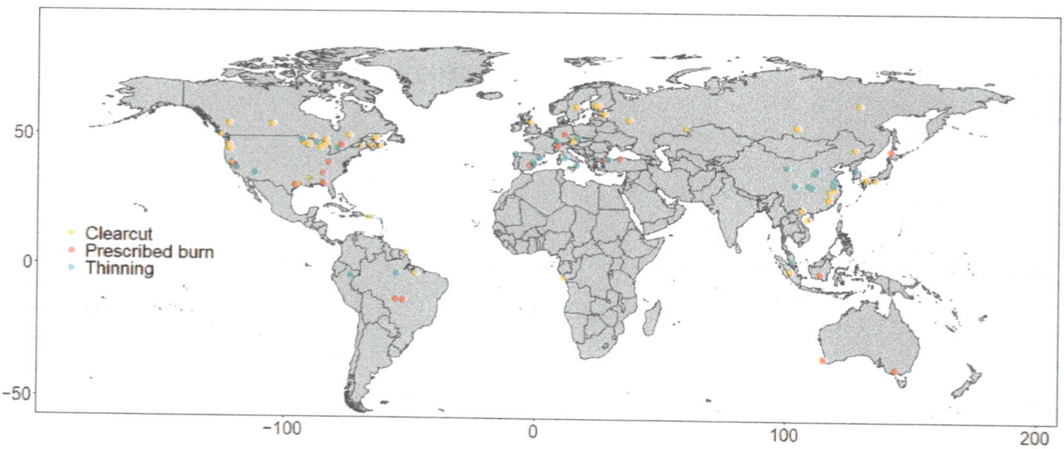

Figure 1. The global distribution of studies included in the meta-analysis, with management types indicated by color.

2.2. Data Extraction

Following the initial screening, relevant data were directly obtained from texts and tables, or extracted figures using PlotDigitizer (https://plotdigitizer.com/; accessed on 14 July 2025). Specifically, the following variables were extracted: (1) geographical and climate information, including latitude (°), longitude (°), climate zone (tropical, subtropical, temperate, Mediterranean, or boreal), mean annual temperature (MAT, °C), and mean annual precipitation (MAP, mm); (2) forest stand characteristics, including forest type (needleleaf, broadleaf, or mixed) and leaf habit (evergreen, deciduous, or mixed); (3) information on management practices, including the type of management (clearcutting, prescribed burn, or thinning), disturbance intensity, the time since application (years), whether harvest residues were retained, whether any additional practices were applied (e.g., site-preparation burn, thinning), and whether the practice was repeated; (4) soil respiration components, including total soil respiration (SR), autotrophic respiration (Ra), and heterotrophic respiration; (5) soil environmental and chemical properties of mineral soils, including soil temperature (Ts, °C), soil moisture (SM), bulk density (BD), pH, soil organic carbon and nitrogen (SOC and SON), total soil carbon and nitrogen (TC and TN; including both organic and inorganic), soil C:N ratio (CN), and (6) biological variables, including fine and coarse root biomass (FRB and CRB), microbial biomass carbon (MBC), stand leaf are index (LAI), forest floor mass (FF), aboveground litterfall (LF), and aboveground biomass (ABG). Dissolved organic carbon (DOC) and nitrogen (DON), and microbial biomass nitrogen (MBN) were also collected, but given the low number of studies that reported them, they could not be used in the meta-analysis.

Disturbance intensity and recovery time were categorized to evaluate the effects of forest management on soil carbon dynamics. Clearcutting intensity included stem-only harvesting, whole-tree harvesting, and whole-tree harvesting with either soil surface removal or soil compaction. Thinning intensity was classified as low (<25% tree or <20% basal area removal), medium (up to 50% trees or 20–35% basal area), and high (>50% trees or >35% basal area). Prescribed burn intensity was based on understory vegetation and damages to overstory trees: low (burned understory vegetation with unburned canopy), medium (partial canopy damage with 20–90% scorched), and high (complete canopy damage with >90% scorched). Recovery stages were grouped as early (<2 years), medium (2–5 years), and late (>5 years). The listed groupings were selected as the most common categories used in the literature.

For studies that reported multiple disturbance levels or site conditions (e.g., variations in tree density or species composition), each level was treated as a separate observation. Similarly, for multi-year studies with distinct annual measurement records, each year was considered as an independent observation. Data were extracted directly from the original publications whenever possible. When specific variables were unavailable, the data were sourced from studies conducted at the same location.

The compiled dataset had the greatest representation of studies from temperate forests (46%), followed by boreal (23%) and Mediterranean forests (17%), especially those located in Western Europe and North America. The geographic distribution of study sites ranged from 37.5° S to 62.3° N (Figure 1), with MAT ranging from −10.0 °C to 27.0 °C and MAP ranging from 146 mm to 3500 mm (Table S1). The time since the management disturbance ranged from 0 to 100 years (Table S1).

2.3. Data Analysis

We quantified the effect of forest management on soil respiration (SR), heterotrophic respiration (Rh), autotrophic respiration (Ra), and related variables using the natural logarithm of the response ratio (*lnRR*) as the effect size. The *lnRR* was calculated as the natural log of the ratio of the mean in the treatment group ($\overline{x_t}$) to that in the control group ($\overline{x_c}$), following the formula:

$$\ln RR = \ln\left(\frac{\overline{x_t}}{\overline{x_c}}\right) = \ln\overline{x_t} - \ln\overline{x_c} \quad (1)$$

Since standard errors or standard deviations were not consistently reported across studies, we used sample sizes to compute the weighting factors, following previous meta-analyses [16,23,34,35]. The weight W_r for each observation was calculated as:

$$W_r = \frac{n_c \times n_t}{n_c + n_t} \quad (2)$$

where n_c and n_t are the number of replicates in the control and treatment groups, respectively.

To evaluate the overall effects of management practices on soil respiration components, environmental conditions, and biophysical factors, we used weighted mixed-effects models with study ID included as a random effect. The model can be expressed as:

$$Y_{ij} = \beta_0 + u_j + \varepsilon_{ij} \quad (3)$$

where Y_{ij} is the response variable for observation i in study j, β_0 is the overall intercept, and u_j is the random effect of study j, with weight W_r applied to each observation. Models were fitted using the "lmer" function in the "lme4" package [36]. To minimize overinterpretation, only variables reported in more than three independent studies were included. The grand mean of *lnRR* ($lnRR_{++}$) and its 95% confidence interval (CI) were estimated using a bootstrap approach implemented in the "boot" package [37]. The treatment effect was considered statistically significant if the 95% CI did not overlap with zero. For interpretability, *lnRR* estimates and CIs were back-transformed to represent the percentage change relative to the control group, calculated as:

$$\text{Effect Size}(\%) = \left(e^{lnRR_{++}} - 1\right) \times 100\% \quad (4)$$

We also examined correlations between responses of soil respiration components and key soil or biological factors, as well as their variation across disturbance severity, season of implementation, forest type, climate zone, and the time since application for each management type. Differences among categorical variables were assessed using one-way analysis of variance (ANOVA), followed by Tukey's post hoc test when the ANOVA results were significant. Results were considered statistically significant at $p < 0.05$. To identify influential moderators of soil respiration, we used the "glmulti" package, which generates all possible model combinations and ranks variable importance based on corrected Akaike information criterion (AICc) [38]. Relative importance was calculated as the sum of Akaike weights across all models. Variables with importance values ≥ 0.8 were considered essential. Due to inconsistent data coverages on responses of soil respiration components as well as key soil or biological factors, the moderator analysis was restricted to geographical and climatic variables (i.e., latitude, MAT, MAP, climate zone, and forest type) and management-related variables (i.e., time since treatment, severity level [for clearcutting and thinning], repetition [only prescribed burn], and the presence of post-harvest residues [for clearcutting and thinning], post-harvest management [for clearcutting and thinning]). The numerical values were standardized using z-score normalization, calculated as:

$$z = (x -)/\sigma \quad (5)$$

where x is the original value, μ is the mean of the variable, and σ is the standard deviation.

Publication bias was assessed using a funnel plot and Egger's regression for asymmetry [39]. No significant asymmetry was detected, indicating minimal publication bias in the dataset (Figure S1). All the data analyses were performed in R (version 4.3.3) [40] and implemented in RStudio (version 2023.12.1) [41].

3. Results

3.1. Management Effects on Soil Properties and Respiration Components

Forest management practices—clearcutting, thinning, and prescribed burning—altered key biological and environmental factors, which resulted in soil respiration dynamics and soil pools (Figure 2). These responses may also reflect post-disturbance vegetation recoveries, including understory vegetation, shrubs, and root sprouts, as well as post-treatment practices (e.g., residue removal, herbicide application, bedding). Among the 130 clearcutting observations reporting post-disturbance vegetation recovery, 47% documented natural regeneration of understory and woody vegetation, whereas 40% involved plantation establishment. These indicate that ecosystem responses (i.e., soil environmental conditions, soil respiratory factors) to management practices are shaped not only by the immediate effects of disturbance but also by the subsequent recovery of both overstory and understory vegetation.

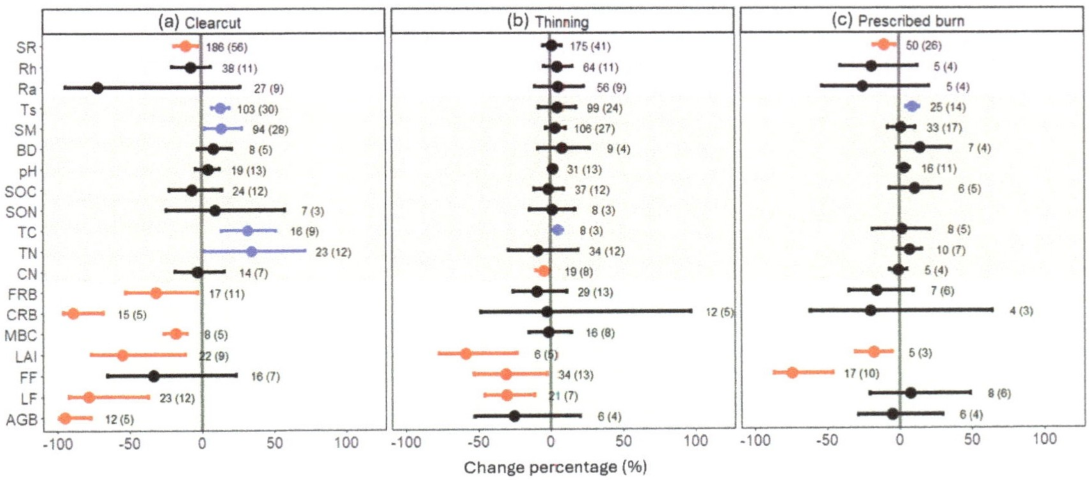

Figure 2. Effects of (**a**) clearcutting, (**b**) thinning, and (**c**) prescribed burn on soil respiration components, soil environmental conditions, and biophysical factors. Variables include (1) soil respiration (pink): total soil respiration (SR), heterotrophic respiration (Rh), autotrophic respiration (Ra); (2) soil environmental and chemical properties (yellow): soil temperature (Ts), soil moisture (SM), soil bulk density (BD), soil pH (pH), soil organic carbon (SOC) and nitrogen (SON), total carbon (TC) and nitrogen (TN), soil C/N ratio (CN); (3) biological variables (green): fine and coarse root biomass (FRB and CRB), microbial biomass carbon (MBC), canopy leaf area index (LAI), forest floor mass (FF), aboveground litterfall (LF), and aboveground biomass (ABG). The number of observations and studies (in parentheses) is shown for each variable. Error bars represent 95% confidence intervals. Effects are considered significant when the confidence intervals do not overlap zero, with blue indicating significantly positive effects, red indicating significantly negative effects, and black indicating non-significant effects. Data was only displayed when more than two studies were available for each variable.

Clearcutting led to the most pronounced shifts in forest structure and soil microclimate. Aboveground biomass (AGB) declined by 95% (95% CI: −99 to −77%), accompanied by reductions in leaf area index (LAI; −55%, 95% CI: −77 to −12%), coarse root biomass (CRB; −90%, 95% CI: −96 to −69%), and fine root biomass (FRB; −32%, 95% CI: −53 to −2.9%). These structural changes were coupled with significant increases in soil temperature (Ts; +13%, 95% CI: +7.0 to +19%) and soil moisture (SM; +13%, 95% CI: +2.1 to +28%). Thinning induced similar but largely non-significant trends, including ABG (−26%, 95% CI: −53 to

+21%), and FRB (−9.5%, 95% CI: −26 to +11%), and a significant reduction in LAI (−59%, 95% CI: −77 to −23%). Associated changes in Ts (+4.5%, 95% CI: −8.2 to +18%) and SM (+2.9%, 95% CI: −3.6 to +10%) were not statistically significant. Prescribed burning caused comparatively modest shifts, with LAI reduced by 18% (95% CI: −30 to −5.5%) and a non-significant increase in AGB (+5.2%; 95% CI: −28 to +29%). Ts increased significantly (+9.2%, 95% CI: +4.7 to +13%), while SM increased more often than it decreased (+1.1%, 95% CI: −7.8 to +11%).

Aboveground litterfall (LF) significantly declined under clearcutting (−79%; 95% CI: −92 to −37%) and thinning (−31%; 95% CI: −46 to −12%), but exhibited a non-significant tendency toward increase under prescribed burning (+7.5%, 95% CI: −20 to +49%). Microbial biomass carbon (MBC) declined across treatments, with the greatest reduction observed under clearcutting (−18%, 95% CI: −26 to −10%), and to a smaller, non-significant decrease under thinning (−1.7%, 95% CI: −15 to +14%). For prescribed burning treatments, the number of available MBC observations was insufficient to support a quantitative analysis.

Soil organic carbon (SOC) and soil organic nitrogen (SON) exhibited no significant changes under any management treatment. SOC showed a slight decline under clearcutting (−6.5%; 95% CI: −23 to +14%) and thinning (−1.7%; 95% CI: −11 to +8.9%), but an increasing trend under prescribed burning (+11%; 95% CI: −6.6 to +28%). SON had increasing trends under clearcutting (+9.3%; 95% CI: −25 to +57%) and thinning (+0.7%; 95% CI: −14 to +17%), whereas data for prescribed burning were limited (n = 1).

Clearcutting, however, resulted in significant increases in both total carbon (TC; +31%, 95% CI: +13% to +51%) and total nitrogen (TN; +34%, 95% CI: +2.1% to +71%). Thinning induced a smaller, but significant increase in TC (+4.8%, 95% CI: +1.1% to +8.5%), without a significant change in TN (−9.1%, 95% CI: −29 to +19%). The soil C:N ratio (CN) significantly decreased by 5.2% following thinning (95% CI: −10 to −0.4%). In contrast, prescribed burning did not significantly alter TC (+1.9%; 95% CI: −18 to +21%) or TN (+6.8%; 95% CI: −1.3 to +15%).

These shifts in biological and soil microclimate data correlated with changes in soil respiration dynamics. Total soil respiration (SR) significantly decreased following clearcutting (−11%; 95% CI: −19 to −2.2%) and prescribed burning (−11%; 95% CI: −17 to −2.8%), while thinning had little effect (+1.2%, 95% CI: −5.0 to +7.7%). Though not statistically significant, decreases in both autotrophic (Ra) and heterotrophic (Rh) respiration appeared to contribute to the overall SR decline under clearcutting (Ra: −72%; 95% CI: −94 to +27% and Rh: −7.1%; 95% CI: −20 to +6.5%) and burning (Ra: −25%; 95% CI: −53 to +21% and Rh: −19%; 95% CI: −41 to +12%). In contrast, Ra and Rh increased slightly under thinning (Ra: +5.4%; 95% CI: −11 to + 23% and Rh: +4.5%; 95% CI: −4.5 to + 15%).

3.2. *Responses of Soil Pools and Carbon Fluxes to Management Strategies, Climate, and Recovery Stage*

3.2.1. Clearcutting

The decrease in soil respiration (SR) was pronounced immediately following clearcutting (−19%), but gradually recovered and shifted to a slight increase (+6.0%) at the late stage (>5 years). Among climate zones, a significant increase was only detected in Mediterranean forests (+18%), although this may be due to a selection bias (Figure 3a). A stepwise decline in SR was observed with increasing harvest intensity; however, this was not statistically significant due to limited sample size, particularly under more intensive treatments.

Autotrophic respiration (Ra) declined consistently across all treatment categories (Figure 3c). While Ra tended to recover over time, the changes were not statistically significant, again likely constrained by limited sample size. Heterotrophic respiration (Rh) also showed a general decreasing trend, though the magnitude of change was less pronounced than for Ra (Figure 3b). No consistent recovery trajectory was detected for Rh across stages. Although residue removal and fuel treatments (e.g., site-preparation burning [30 observations] or thinning [1 observation]) following clearcutting were generally associated with reductions in SR and Rh, these effects were not statistically significant.

Soil organic carbon (SOC) declined more substantially in boreal forests (−30%), whereas it increased by 14% in temperate forests (Figure 3d). Significant SOC decreases were also observed following harvest residue removal (−14%) and during the early recovery stage (−15%; <2 years). In contrast, total carbon (TC) generally increased across climate zones, forest types, and recovery stages, with significant gains detected in boreal forests (+63%). This increase in TC was also detected during both the early (+42%) and late (+37%) recovery stages (Figure 3e). Total nitrogen (TN) exhibited a similar pattern to TC, showing overall increases across the examined categories (Figure 3f).

FRB significantly increased in boreal forests (+30%), but decreased in temperate forests (−38%), although this latter pattern may be an artifact of the limited number of studies, particularly from boreal regions (n = 2). Across recovery stages, mean FRB was slightly positive in the late stage (+4.5%), but did not differ significantly from the early stage (−37%; <2 years). Litterfall (LF) exhibited a sharp decline during the early recovery stage (−94%), which partially recovered in the late stage (−49%; >5 years) (Figure 3i). Responses of the forest floor were variable, but overall, fuel reduction management did not produce significant changes (Figure 3h).

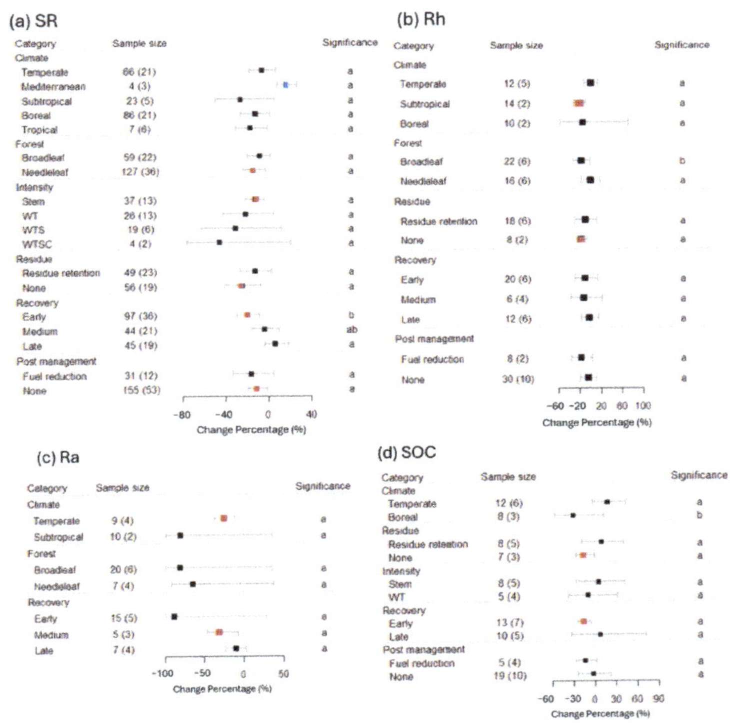

Figure 3. *Cont.*

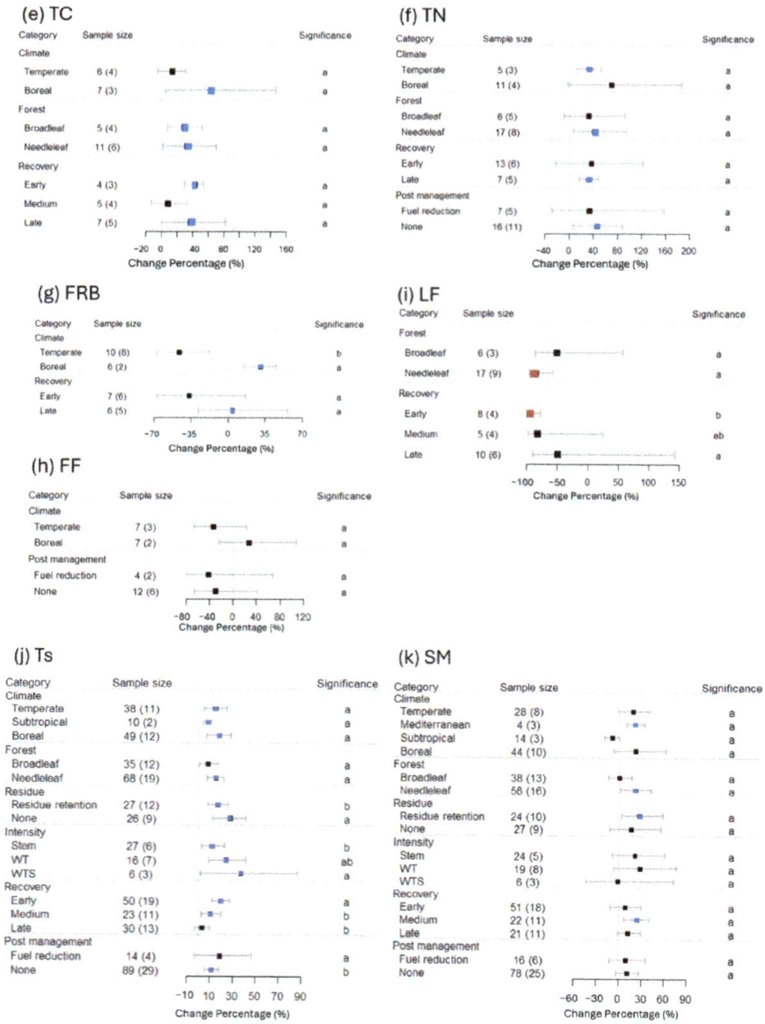

Figure 3. Clearcutting effect on soil respiration components, carbon and nitrogen pools, and environmental variables in response to forest management practices across different categorical variables. Panels represent (**a**) total soil respiration (SR), (**b**) heterotrophic respiration (Rh), (**c**) autotrophic respiration (Ra), (**d**) soil organic carbon (SOC), (**e**) total carbon (TC), (**f**) total nitrogen (TN), (**g**) fine root biomass (FRB), (**h**) forest floor mass (FF), (**i**) litterfall (LF), (**j**) soil temperature (Ts), and (**k**) soil moisture (SM). Categories include climate zone (boreal, Mediterranean, temperate, subtropical, tropical), forest type (broadleaf, needleleaf), residue management (residue retention, none), clearcutting intensity (stem-only [stem], whole tree [WT], whole tree with soil surface removal [WTS], whole tree with soil compaction [WTSC]), recovery stage (early [<2 years], medium [2–5 years], late [>5 years]), and post management (fuel reduction, none). Error bars represent 95% confidence intervals. Effects are considered significant when the confidence intervals do not overlap zero, with blue indicating significantly positive effects, red indicating significantly negative effects, and black indicating non-significant effects. Letters denote groupings based on post hoc significance tests ($p < 0.05$); categories sharing the same letter are not significantly different. Sample sizes and number of studies used for each category are listed in parentheses. Data was only displayed when more than three observations were available for each variable.

Among environmental variables, soil temperature (Ts) increased consistently across all treatment categories (Figure 3j). However, retention of harvest residues, implementation of less intensive harvesting (e.g., stem-only removal), and the absence of post-harvest interventions mitigated the extent of temperature increases. Furthermore, temperature effects began to diminish during the medium recovery stage (2 to 5 years). Soil moisture (SM) exhibited a general increasing trend, although the magnitude of change was smaller compared to Ts, and no statistically significant differences were detected among groups (Figure 3k).

3.2.2. Thinning

Thinning resulted in moderate changes in SR and its components. SR significantly increased in subtropical forests (+37%), while changes in other regions were not significant (Figure 4a). When thinning was followed by a prescribed burn, SR decreased more sharply (−18%) compared to thinning alone (+3.8%). No significant differences in SR were observed across forest types, thinning severity, residue treatments, or recovery stages. Rh exhibited a general increasing trend across thinning severity levels and recovery stages, though these changes were not statistically significant (Figure 4b). Ra significantly increased under moderate severity thinning (+14%), which was greater than the change observed under low severity thinning (−3.9%) (Figure 4c).

Soil organic carbon (SOC) did not change significantly across any categories, but was significantly higher under moderate thinning severity (+15%) compared with high severity (−10%) (Figure 4d). Although a stepwise decline in SOC was observed with increasing time since thinning, this trend was not statistically supported. Total nitrogen (TN) significantly declined in subtropical forests (−9.8%) (Figure 4e).

Fine root biomass (FRB) varied significantly across climate zones and thinning intensities, with higher values in temperate forests (−1.1%) compared to Mediterranean forests (−31%), and under high-intensity thinning (+1.6%) compared to low-intensity thinning (−27%) (Figure 4f). FRB declined more sharply when thinning was followed by site-preparation burning (−40%) than when no post-treatment was applied (−3.7%). The mean FRB during the early recovery stage was negative (−24%), but shifted to positive values at the medium (2–5 years; +5.2%) and late (>5 years; +6.5%) recovery stage, although these differences were not statistically significant. Forest floor mass (FF) generally declined, with the pronounced reductions in the early recovery stage (< 2 years; −45%) (Figure 4g). Similarly to FRB, although no significant differences were detected across recovery stages, mean FF during the medium stage showed a slight positive response (+1.6%). Site preparation burning following thinning effectively reduced FF by 64%, compared to no post-treatment (−4.3%). Litterfall (LF) generally declined following thinning, with a significant reduction under high-intensity thinning (−35%), but no significant difference compared with moderate thinning (−33%). In contrast to FRB and FF, LF did not show any signs of recovery, even in the late recovery stage (−32%; >5 years) (Figure 4h).

Soil warming following a thinning declined over time, with the early (+11%), the medium (+5.1%), and the late stage (−7.3%) (Figure 4i). Ts rose more in subtropical forests (+30%) than in temperate forests (−2.8%). SM showed an overall increasing trend across most categories, although the changes were not statistically significant across climate zones, forest types, thinning severity, or recovery stages (Figure 4j). Notably, residue retention after thinning mitigated the SM increase (+6.2%) compared to residue removal (−5.8%).

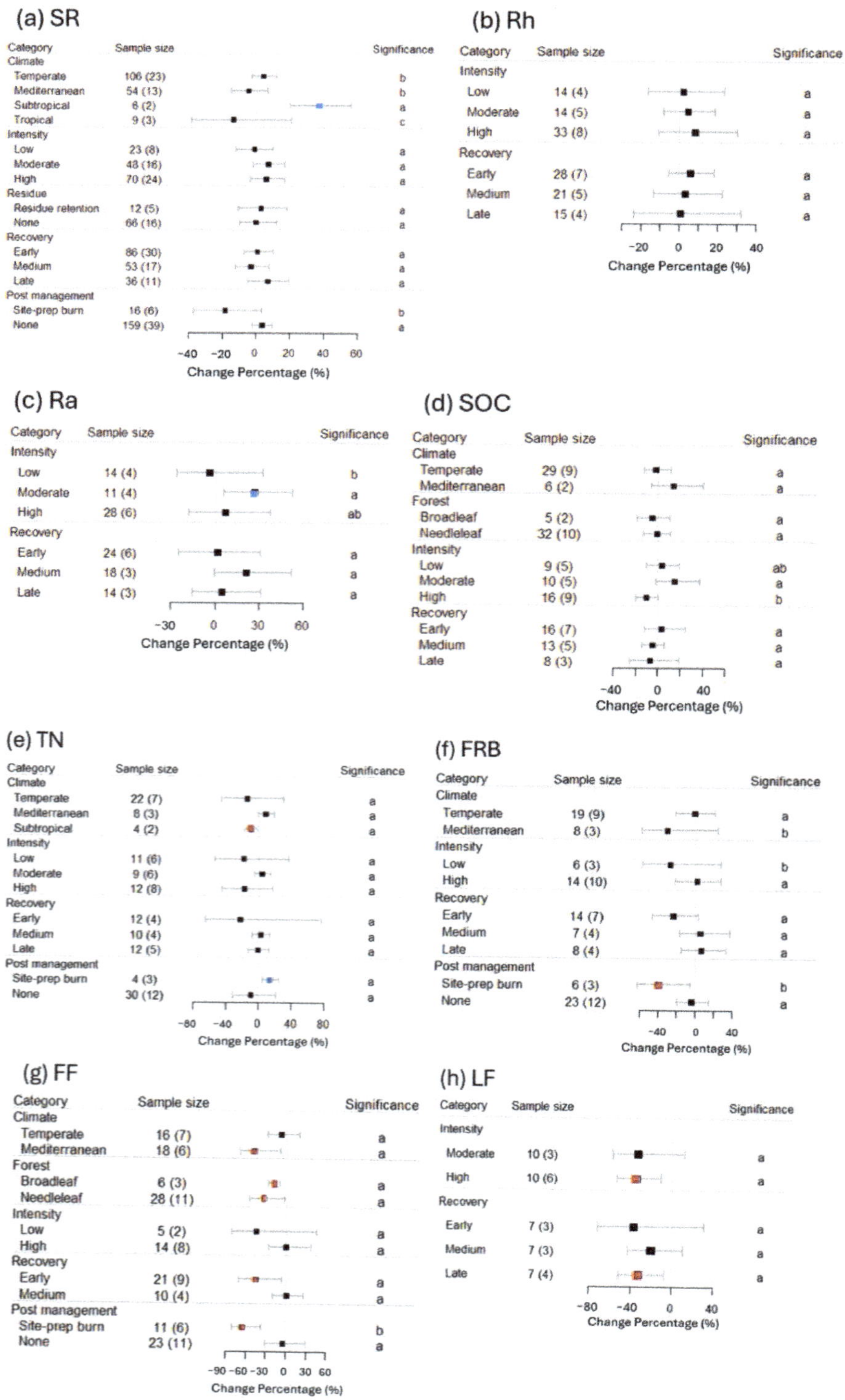

Figure 4. *Cont.*

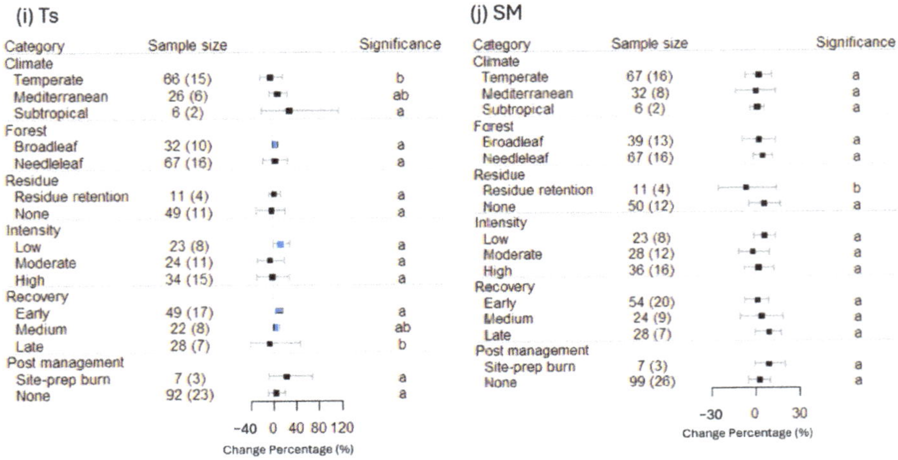

Figure 4. Thinning effect on soil respiration components, carbon and nitrogen pools, and environmental variables in response to forest management practices across different categorical variables. Panels represent (**a**) total soil respiration (SR), (**b**) heterotrophic respiration (Rh), (**c**) autotrophic respiration (Ra), (**d**) soil organic carbon (SOC), (**e**) total nitrogen (TN), (**f**) fine root biomass (FRB), (**g**) forest floor mass (FF), (**h**) litterfall (LF), (**i**) soil temperature (Ts), and (**j**) soil moisture (SM). Categories include climate zone (boreal, Mediterranean, temperate, subtropical, tropical), forest type (broadleaf, needleleaf), residue management (residue, no residue), thinning intensity (low, moderate, high), recovery stage (early [<2 years], medium [2–5 years], late [>5 years]), and post management (prescribed burn, none). Error bars represent 95% confidence intervals. Effects are considered significant when the confidence intervals do not overlap zero, with blue indicating significantly positive effects, red indicating significantly negative effects, and black indicating non-significant effects. Letters denote groupings based on post hoc significance tests ($p < 0.05$); categories sharing the same letter are not significantly different. Sample sizes and number of studies used for each category are listed in parentheses. Data was only displayed when more than three observations were available for each variable.

3.2.3. Prescribed Burning

Among the 29 observations that reported burn intensity, 27 were classified as low-intensity, commonly used for fuel reduction. SR declined significantly under repeated burning (−17%), a reduction that was substantially greater than that observed under single burns (−6.4%) (Figure 5a). No significant differences in SR were detected across climate zones, forest types, or recovery stages. Only 10% of prescribed-burn observations reported Rh and Ra, which limited further categorical analysis.

TN significantly increased by 12% in needleleaf forests, although the wide confidence intervals and limited sample sizes constrained the ability to detect a significant difference from broadleaf forests (Figure 5c). In contrast, TC did not show any significant changes or differences in category (Figure 5b). Forest floor mass (FF) significantly decreased, and the reduction in FF was greater in single burn (−87%) than repeated burns (−31%; Figure 5d). Litterfall (LF) showed insignificant increase in both single and repeated burns with no significant difference (Figure 5e).

Ts consistently increased across forest types and fire frequencies, but no significant differences were detected within categories (Figure 5f). In contrast, SM responses were more variable. SM significantly increased under repeated burns (+13%), in contrast to a slight decrease observed after single burns (−5.6%) (Figure 5g).

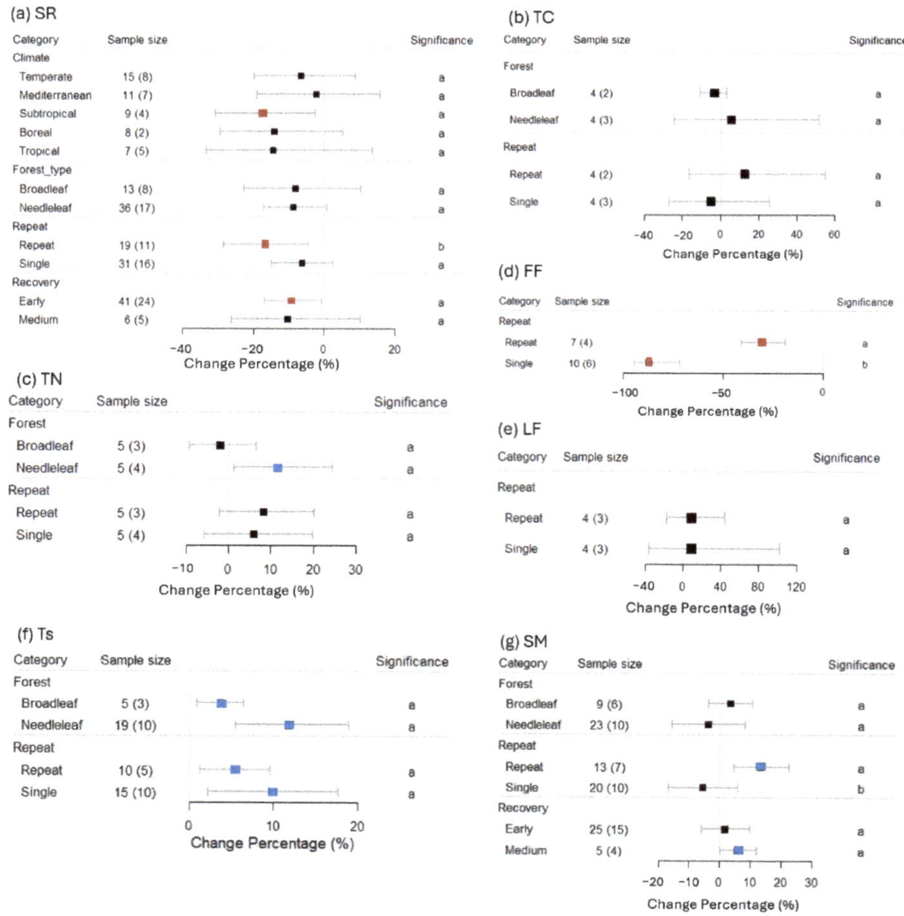

Figure 5. Prescribed burn effect on soil respiration components, carbon and nitrogen pools, and environmental variables in response to forest management practices across different categorical variables. Panels represent (**a**) total soil respiration (SR), (**b**) total carbon (TC), (**c**) total nitrogen (TN), (**d**) forest floor mass (FF), (**e**) litterfall (LF), (**f**) soil temperature (Ts), and (**g**) soil moisture (SM). Categories include climate zone (boreal, Mediterranean, temperate, subtropical, tropical), forest type (broadleaf, needleleaf), burn frequency (repeat, single), and recovery stage (early [<2 years], medium [2–5 years], late [>5 years]). Error bars represent 95% confidence intervals. Effects are considered significant when the confidence intervals do not overlap zero, with blue indicating significantly positive effects, red indicating significantly negative effects, and black indicating non-significant effects. Letters denote groupings based on post hoc significance tests ($p < 0.05$); categories sharing the same letter are not significantly different. Sample sizes and number of studies used for each category are listed in parentheses. Data was only displayed when more than three observations were available for each variable.

3.3. Relationship Between Soil Respiration Components and Biophysical Variables

Under clearcutting, SR declined with decreases in both Ts ($R^2 = 0.08$, $p = 0.004$; Figure 6a) and soil moisture ($R^2 = 0.13$, $p < 0.001$; Figure 6b). In contrast, change in microbial biomass carbon (MBC) was positively associated with the change in SR, with a relatively strong correlation ($R^2 = 0.65$, $p = 0.016$; Figure 6c). In thinning treatments, changes in soil respiration and its components were positively correlated with biological variables. SR increased with increasing forest floor mass (R^2 = 0.22, p = 0.005; Figure 6d), and Rh response was positively correlated with that in MBC (R^2 = 0.42, p = 0.041; Figure 6e) but negatively with that in litterfall (LF) (R^2 = 0.53, p = 0.011; Figure 6e). Change in Ra was positively related to that in fine root biomass (FRB) (R^2 = 0.79, p < 0.001; Figure 6g). No other variables under clearcutting or thinning, and none under prescribed burns, showed significant correlations with soil respiration components.

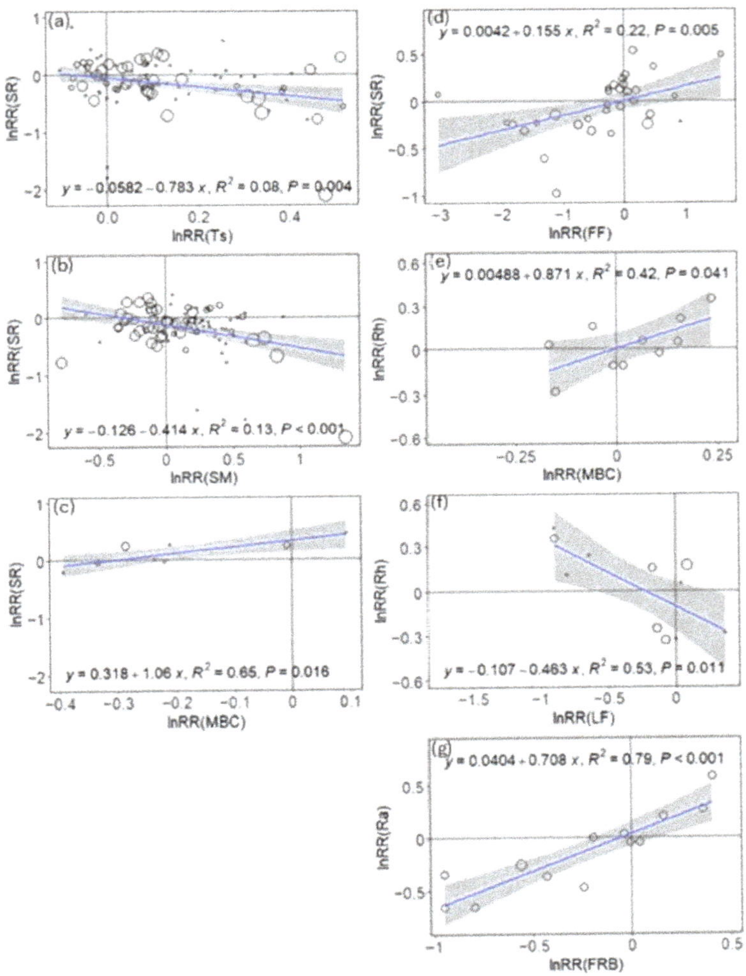

Figure 6. Relationships between log response ratios (lnRR) of soil respiration components and environmental or biological variables following (**a**–**c**) clearcutting and (**d**–**g**) thinning. Panels show: (**a**) total soil respiration (SR) vs. soil temperature (Ts), (**b**) SR vs. soil moisture (SM), (**c**) SR vs. microbial biomass carbon (MBC), (**d**) SR vs. forest floor mass (FF), (**e**) heterotrophic respiration (Rh) vs. MBC, (**f**) Rh vs. aboveground litterfall (LF), and (**g**) autotrophic respiration (Ra) vs. fine root biomass (FRB). Circle sizes indicate the relative weight of each data point based on the number of replicates per study. Solid lines represent fitted regression models, and shaded areas denote 95% confidence intervals.

The key moderators of harvest and prescribed burning effects on SR were determined with model selection based on the sum of Akaike weights (Figure 7). For clearcutting, the retention of logging residue and the intensity of the disturbance were the most important and significant moderators of SR change (Figure 7a). For thinning, important moderators included thinning intensity, climate zone, post-treatment (i.e., prescribed burning), and mean annual temperature (MAT) (Figure 7b). For prescribed burning, MAT was the primary moderator of SR response (Figure 7c).

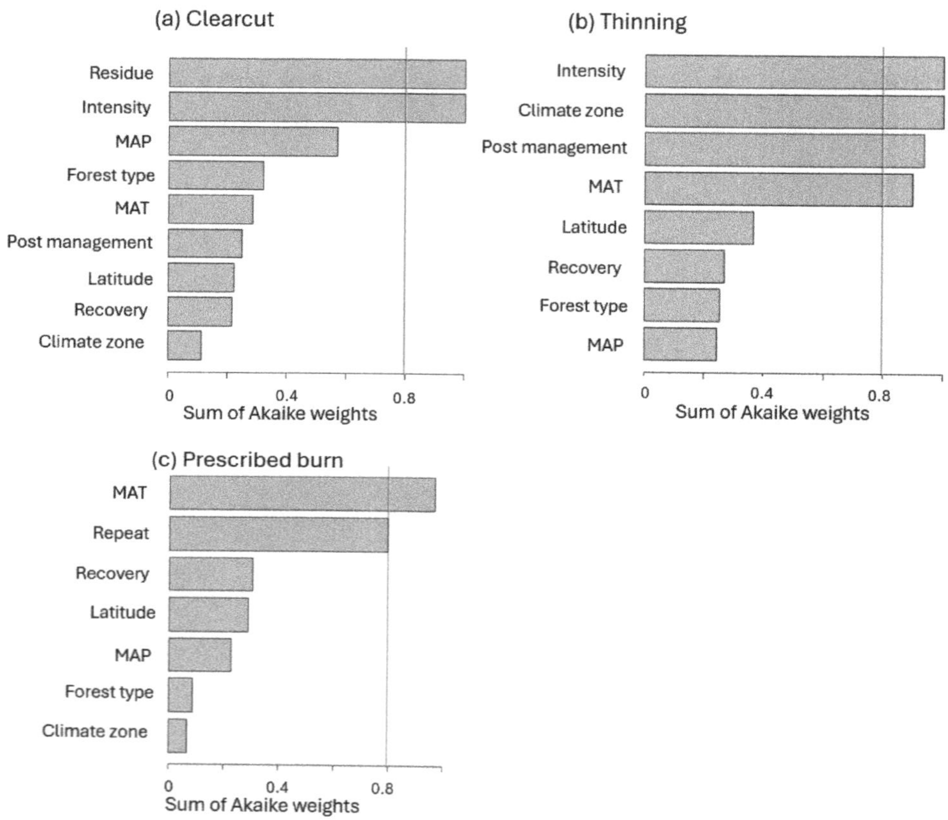

Figure 7. Model-averaged variable importance for predicting the effects of (**a**) clearcutting, (**b**) thinning, and (**c**) prescribed burn, on the natural logarithm of the response ratio of soil respiration. Importance values represent the sum of Akaike weights from model selection using the corrected Akaike Information Criterion (AICc). A threshold of 0.8 (red line) was used to identify essential predictor variables.

4. Discussion

4.1. Effects of Forest Management on Above- and Belowground Biomass and Soil Respiration Components

Clearcutting significantly reduced both fine and coarse root biomass (Figure 2a), whereas thinning and prescribed burning resulted in non-significant reductions (Figure 2b,c). Correspondingly, both Ra and Rh declined under clearcutting and prescribed burning, with the reduction being more pronounced in Ra than in Rh (Figure 2b,c). The stronger decline in Ra can be attributed to losses of live plant biomass, which directly suppresses substrate supply and root activity, and thus Ra. The reduction in Ra was clearly stronger in clearcutting (−72%), as the damage to both aboveground biomass and root systems was greater. Prescribed burning also reduced Ra (−25%), likely due to the mortality of fine roots in the surface soils, consistent with earlier reviews [17,18]. Such effects can arise even under low-intensity burns, especially when fires are applied repeatedly [42–44] or occur under unfavorable seasonal conditions [45,46]. Ra showed a recovery trend following a clearcut over time (Figure 3c), likely with the regrowth of understory vegetation and the regeneration of overstory species [47,48]. In contrast, the change in Rh was small, likely reflecting the offsetting effects of reduced root exudates but increased detritus both above- and belowground [49–51]. We should note that while the predominant effect of declining Rh reported here and in Akande et al. [16], the meta-analysis by Yang et al. [23] reported an overall increase. It is likely that their conclusion resulted from combining clearcutting and thinning into a general "harvesting" category. Furthermore, their finding of a negative relationship between Rh response ratio and harvest intensity is consistent with the current observation of the greatest Rh decline following clearcutting.

Unlike clearcutting and burning, thinning tended to maintain or even enhance SR components (Figure 2b). Non-significant increases in SR, Rh, and Ra were observed under thinning, consistent with previous meta-analyses reporting either significant or non-significant increases in SR or Ra following thinning [20,21,52]. Importantly, Ra increased significantly by 27% under moderate thinning intensity (Figure 4c), consistent with Zhao et al. [47], whereas SR and Rh did not respond significantly to thinning of any intensity (Figure 4a,b). The increase in Ra likely reflects enhanced fine root growth under moderate thinning (Figure 4f), supported by positive correlations between Ra (and SR) responses and fine root biomass (Figure 6g and Figure S2). Improved light availability and reduced competition for water and nutrients under moderate to high thinning intensity stimulate the growth of residual trees [22,53], promoting their root [54,55] and mycorrhizal development [56] and contributing to elevated Ra. Furthermore, earlier studies suggest that moderate thinning can also enhance understory vegetation growth, further increasing root biomass and Ra [55].

Harvesting, both thinning and clearcutting, reduced tree density and consequently decreased litterfall inputs, and this reduction persisted into the late recovery stage (>5 years; Figures 3i and 4h), consistent with long-term declines in litter production observed under thinning [20]. Under clearcutting, MBC also declined significantly (Figure 2a), consistent with previous syntheses that reported a reduction in microbial biomass by 19% following harvesting [57]. Nevertheless, Rh did not exhibit a significant decline under either harvesting treatment; rather, under thinning, it even showed a slight, though non-significant, increase (Figures 2b and 6b). As with clearcutting, the opposing effects of increased detritus production and reduced exudates may balance one another out, or one may temporarily dominate. For instance, Wang et al. [55] reported that thinning stimulated both fine root production and turnover in overstory and understory vegetation of a Chinese fir plantation, with the effect being more pronounced under intense thinning. López et al. [54] also reported that fine root mortality increased by 32% two years after a heavy thinning in the Mediterranean oak forest. In our analysis, signs of fine root biomass recovery were evident in both thinning and clearcutting treatments (Figures 3g and 4f), likely contributing to the stability of Rh. Enhanced aboveground productivity following thinning [22] may further increase belowground carbon inputs; for example, Zhao et al. [58] observed increased root exudates during the growing season following a heavy thinning in a Chinese fir plantation. In addition to altering substrate inputs, thinning may also affect microbial community composition and functionality [52], thus affecting Rh. However, data on these effects are sparse in our dataset and were therefore not included in the current meta-analysis.

4.2. Effects of Forest Management on Soil Environment and Properties

Harvesting and prescribed burns reduced the forest floor and increased light availability by opening the canopy and reducing mid- and understory vegetation. These changes elevated soil temperature, particularly during the early recovery stage. Contrary to our hypothesis, however, higher temperature in response to all of these treatments, as well as increased soil moisture under clearcutting, did not enhance Rh or SR. The negative correlation of SR with Ts and SM under clearcutting (Figure 6a,b) aligns with earlier studies [16,59] and may be attributed to the restoration of carbohydrate supply and fine root biomass with understory regrowth and overstory regeneration over time that override correlations with temperature [48]. Furthermore, the loss of obligate symbiont microbes after clearcutting may also restrict soil C processing [51,60].

Our meta-analysis showed that overall effects on mineral soils were limited. The main exception was clearcutting, where TC and TN significantly increased, a pattern that persisted even during late recovery (>5 years), likely due to delayed mortality of belowground detritus [61,62]. In contrast, SOC declined under certain conditions, particularly when residue removal or site-preparation burning was applied, as these practices can extend depletion into deeper soil layers [63,64]. The magnitude and direction of these responses may also be shaped by climate and vegetation type [65,66]. Although the residue retention was similar in temperate (40%) and boreal forests (49%), SOC tended to increase more often in the temperate forests and decline in boreal forests. This pattern aligns with Mayer et al. [67], who reported the strongest SOC losses in cold-climate forests with large pre-harvest SOM pools. Nevertheless, SOC recovery is expected during early to mid-succession in boreal forests, supported by plant and mycorrhizal hyphal production, and the mycorrhizal suppression of saprotrophs (i.e., Gadgil effect) [68].

By contrast, prescribed burning led to slight, non-significant increases in SOC, TC, and TN, a pattern consistent with earlier reviews [30,69]. These modest changes may result from the incorporation of unburned or partially charred slash into the mineral soil, or from incomplete combustion of organic matter under low fire temperatures [70]. On the other hand, more severe fires can cause greater combustion of the forest floor and mineral soils, often resulting in lower soil C and N pools [71].

4.3. Effects of Repeated and Post-Treatment Burns on Soil Dynamics and Respiration

Repeated low-intensity prescribed burns are widely applied to manage forest structure, reduce wildfire risk, and promote ecosystem resilience [29]. In our dataset, 38% of observations involved repeated burns. These burns reduced SR by 17%, a decline greater than that observed for single burns (Figure 5a). Although low-intensity fires are often assumed to have negligible effects on soil processes [30], our results show that repeated applications generate cumulative impacts on both forest structure and belowground dynamics, thereby reducing SR. This meta-analysis detected that repeated burns, compared to single burns, led to smaller reductions in forest floor mass (Figure 5d), likely due to progressive fuel depletion, but greater increases in soil moisture (Figure 5g), potentially from reduced fine root density or reduced transpiration and increased throughfall following long-term canopy loss [70]. Yet, despite these shifts, SR declined more strongly under repeated burns, suggesting that limitations in substrate availability, and root and microbial activity were dominant controls. Interpretation of the full magnitude and temporal trajectory of this decline, however, remains constrained by the limited, short-term nature of many prescribed burning observations. Consistent evidence from prior work points to reductions in root biomass [44,72], losses of surface soil C and N [73,74], lowered litter quality [75], and altered microbial and mycorrhizal communities [42,43,76]. In a different study, we hypothesized that long-term frequent burning may have reduced fungal colonization of roots [44], amplifying declines in the autotrophic component of SR. Repeated fires may introduce pyrogenic C inputs [19,45,77], which can increase soil recalcitrance to decomposition [78,79], but its production under low-intensity fires is variable and less consistent than that from wildfires [30]. Collectively, these cumulative effects of organic matter loss, reduced root inputs, and altered microbial–plant interactions explain the stronger SR declines observed under repeated burns than single burns.

Harvesting combined with site-preparation burning is also widely implemented to suppress competing vegetation, reduce wildfire risk, and facilitate plantation establishment (e.g., [24,25]). Following clearcutting, logging residues can buffer soils against compaction and erosion, enhance moisture and nutrient retention, and provide organic inputs that support plant regrowth [80,81]. However, when residues are removed or combusted through site-preparation burning, long-term negative impacts may arise, including mineral soil carbon losses [82,83], shifts in understory plant compositions [84], and, in some cases, elevated mortality of remaining trees [8,85]. The effects of post-harvest burning were more pronounced under thinning than under clearcutting. After thinning, reductions in forest floor, fine root biomass, and SR were significantly greater with post-burn than without (Figure 4a,f,g), whereas post-burn following clearcutting did not substantially alter biological factors or SR components, although Ts increased (Figure 3j). Compared with clearcutting, thinning leaves more standing trees and residual root systems, which continue to supply carbon to the soil. Site-preparation burn, therefore, may disproportionately disrupt root and mycorrhizal activity and microbial processes by damaging residual fine roots, forest floor, and understory vegetation, cumulatively leading to reductions in SR, analogous to the effects of repeated prescribed burns. Additionally, among selected thinning studies, only 5 of the 67 observations (7%) explicitly reported that the forest floor was left intact; thus, it is possible that a site-preparation burning may directly affect soil surface roots and microbes. By contrast, clearcutting removes the majority of overstory trees, so subsequent fuel reduction fires did not significantly affect forest floor mass (Figure 3h). While SR and Rh showed trends of decrease when residues were removed or post-burn applied, these differences were not statistically significant (Figure 3a,b). As discussed earlier, recovery of SR and Ra appears more dependent on regrowth of understory vegetation and overstory tree regeneration, such that additional burning does not markedly intensify reductions in SR or Rh.

5. Conclusions

This study evaluated the effects of three major forest management practices—clearcutting, thinning, and prescribed burning—on soil respiration (SR), its component fluxes (Ra and Rh), and associated biophysical and soil environmental variables. Our results showed that clearcutting and repeated prescribed burning significantly reduced SR, particularly Ra, although for potentially different reasons. While after clearcutting the decline is clearly attributable to the removal of live crowns and thus the supply of carbon, prescribed burning may, on the one hand, combust detritus and, on the other hand, kill roots and microbes in the surface soil. While increased sink strength following prescribed burning may also be possible, the data evaluated here did not indicate this. In contrast, thinning caused non-significant changes in SR components, likely because the damage to the belowground compartment was minor and compensatory growth of remaining trees and understory vegetation offset long-term reductions in aboveground litterfall. However, this non-significant pattern may shift toward negative effects when thinning is combined with post-treatments, such as site-preparation burn or residue removal, which can further impair root systems and microbial communities, resembling impacts of clearcutting or repeated fire. Crucially, despite observed increases in soil temperature associated with canopy opening and/or reduced understory cover, SR and its components were not consistently correlated with warming, reinforcing the role of substrate availability rather than temperature as the dominant driver of soil carbon flux responses to management. We recommend that future research prioritize long-term field experiments, particularly under prescribed burn treatments, to better understand long-term soil carbon stabilization. Overall, the meta-analysis demonstrates that forest management decisions regulate not only aboveground biomass and stand structure but also fundamental belowground carbon processes, with important implications for soil health and climate mitigation.

Supplementary Materials: The following supporting information can be downloaded at: https://www.mdpi.com/article/10.3390/f16101555/s1, Table S1: Site characteristics and management practices across forest disturbance types; Table S2: List of publications used for the meta-analysis; Figure S1: Funnel plots of management effects on soil respiration response (lnSR); Figure S2: Relationship between log response ratios (lnRR) of soil respiration (SR) and fine root biomass (FRB) under thinning.

Author Contributions: M.O.: Writing—original draft, Visualization, Methodology, Investigation, Software, Formal analysis, Data curation, Conceptualization. A.N.: Writing—review and editing, Supervision, Validation, Project administration, Investigation, Funding acquisition, Conceptualization. All authors have read and agreed to the published version of the manuscript.

Funding: M.O. was supported by the ITO Foundation and the McMillan-Ward Fellowship.

Data Availability Statement: The list of articles used in the meta-analysis is presented as Table S2. All the data and code files for the analyses in this manuscript were archived in Zenodo: https://doi.org/10.5281/zenodo.17290818.

Conflicts of Interest: The authors declare no conflicts of interest.

References

1. Bonan, G.B. Forests and Climate Change: Forcings, Feedbacks, and the Climate Benefits of Forests. *Science* **2008**, *320*, 1444–1449. [CrossRef]
2. Pan, Y.; Birdsey, R.A.; Fang, J.; Houghton, R.; Kauppi, P.E.; Kurz, W.A.; Phillips, O.L.; Shvidenko, A.; Lewis, S.L.; Canadell, J.G.; et al. A Large and Persistent Carbon Sink in the World's Forests. *Science* **2011**, *333*, 988–993. [CrossRef]
3. Xu, H.; He, B.; Guo, L.; Yan, X.; Zeng, Y.; Yuan, W.; Zhong, Z.; Tang, R.; Yang, Y.; Liu, H.; et al. Global Forest Plantations Mapping and Biomass Carbon Estimation. *J. Geophys. Res. Biogeosci.* **2024**, *129*, e2023JG007441. [CrossRef]
4. FAO. *Global Forest Resources Assessment 2015: How Are the World's Forests Changing?* FAO: Rome, Italy, 2016.
5. Bond-Lamberty, B.; Bailey, V.L.; Chen, M.; Gough, C.M.; Vargas, R. Globally rising soil heterotrophic respiration over recent decades. *Nature* **2018**, *560*, 80–83. [CrossRef]
6. Gibbs, D.A.; Rose, M.; Grassi, G.; Melo, J.; Rossi, S.; Heinrich, V.; Harris, N.L. Revised and updated geospatial monitoring of 21st century forest carbon fluxes. *Earth Syst. Sci. Data* **2025**, *17*, 1217–1243. [CrossRef]
7. Pan, Y.; Birdsey, R.A.; Phillips, O.L.; Houghton, R.A.; Fang, J.; Kauppi, P.E.; Keith, H.; Kurz, W.A.; Ito, A.; Lewis, S.L.; et al. The enduring world forest carbon sink. *Nature* **2024**, *631*, 563–569. [CrossRef] [PubMed]
8. Kalies, E.L.; Kent, L.L.Y. Tamm Review: Are fuel treatments effective at achieving ecological and social objectives? A systematic review. *For. Ecol. Manag.* **2016**, *375*, 84–95. [CrossRef]
9. Mayer, M.; Prescott, C.E.; Abaker, W.E.A.; Augusto, L.; Cécillon, L.; Ferreira, G.W.D.; James, J.; Jandl, R.; Katzensteiner, K.; Laclau, J.-P.; et al. Tamm Review: Influence of forest management activities on soil organic carbon stocks: A knowledge synthesis. *For. Ecol. Manag.* **2020**, *466*, 118127. [CrossRef]
10. Ameray, A.; Bergeron, Y.; Valeria, O.; Montoro Girona, M.; Cavard, X. Forest Carbon Management: A Review of Silvicultural Practices and Management Strategies Across Boreal, Temperate and Tropical Forests. *Curr. For. Rep.* **2021**, *7*, 245–266. [CrossRef]
11. Kaarakka, L.; Cornett, M.; Domke, G.; Ontl, T.; Dee, L.E. Improved forest management as a natural climate solution: A review. *Ecol. Solut. Evid.* **2021**, *2*, e12090. [CrossRef]
12. Jandl, R.; Lindner, M.; Vesterdal, L.; Bauwens, B.; Baritz, R.; Hagedorn, F.; Johnson, D.W.; Minkkinen, K.; Byrne, K.A. How strongly can forest management influence soil carbon sequestration? *Geoderma* **2007**, *137*, 253–268. [CrossRef]
13. Noormets, A.; Epron, D.; Domec, J.C.; McNulty, S.G.; Fox, T.; Sun, G.; King, J.S. Effects of forest management on productivity and carbon sequestration: A review and hypothesis. *For. Ecol. Manag.* **2015**, *355*, 124–140. [CrossRef]

14. Liu, Y.; Liu, S.; Wan, S.; Wang, J.; Wang, H.; Liu, K. Effects of experimental throughfall reduction and soil warming on fine root biomass and its decomposition in a warm temperate oak forest. *Sci. Total Environ.* **2017**, *574*, 1448–1455. [CrossRef] [PubMed]
15. Zummo, L.M.; Friedland, A.J. Soil carbon release along a gradient of physical disturbance in a harvested northern hardwood forest. *For. Ecol. Manag.* **2011**, *261*, 1016–1026. [CrossRef]
16. Akande, O.J.; Ma, Z.L.; Huang, C.Y.; He, F.L.; Chang, S.X. Meta-analysis shows forest soil CO_2 effluxes are dependent on the disturbance regime and biome type. *Ecol. Lett.* **2023**, *26*, 765–777. [CrossRef]
17. Gui, H.; Wang, J.; Hu, M.; Zhou, Z.; Wan, S. Impacts of fire on soil respiration and its components: A global meta-analysis. *Agric. For. Meteorol.* **2023**, *336*, 109496. [CrossRef]
18. Zhou, L.; Liu, S.; Gu, Y.; Wu, L.; Hu, H.-W.; He, J.-Z. Fire decreases soil respiration and its components in terrestrial ecosystems. *Funct. Ecol.* **2023**, *37*, 3124–3135. [CrossRef]
19. Cheng, Y.; Luo, P.; Yang, H.; Li, H.; Luo, C.; Jia, H.; Huang, Y. Fire effects on soil carbon cycling pools in forest ecosystems: A global meta-analysis. *Sci. Total Environ.* **2023**, *895*, 165001. [CrossRef]
20. Zhang, X.; Guan, D.; Li, W.; Sun, D.; Jin, C.; Yuan, F.; Wang, A.; Wu, J. The effects of forest thinning on soil carbon stocks and dynamics: A meta-analysis. *For. Ecol. Manag.* **2018**, *429*, 36–43. [CrossRef]
21. Yang, L.; Niu, S.; Tian, D.; Zhang, C.; Liu, W.; Yu, Z.; Yan, T.; Yang, W.; Zhao, X.; Wang, J. A global synthesis reveals increases in soil greenhouse gas emissions under forest thinning. *Sci. Total Environ.* **2022**, *804*, 150225. [CrossRef]
22. Zhang, H.; Liu, S.; Yu, J.; Li, J.; Shangguan, Z.; Deng, L. Thinning increases forest ecosystem carbon stocks. *For. Ecol. Manag.* **2024**, *555*, 121702. [CrossRef]
23. Yang, L.; Zhang, H.; Qin, J.; Liu, X.; Mayer, M. A global meta-analysis of forest harvesting effects on soil respiration, its components, and temperature sensitivity. *Agric. For. Meteorol.* **2024**, *358*, 110259. [CrossRef]
24. Piqué, M.; Domènech, R. Effectiveness of mechanical thinning and prescribed burning on fire behavior in *Pinus nigra* forests in NE Spain. *Sci. Total Environ.* **2018**, *618*, 1539–1546. [CrossRef] [PubMed]
25. Shive, K.L.; Coppoletta, M.; Wayman, R.B.; Paulson, A.K.; Wilson, K.N.; Abatzaglou, J.T.; Saberi, S.J.; Estes, B.; Safford, H.D. Thinning with follow-up burning treatments have increased effectiveness at reducing severity in California's largest wildfire. *For. Ecol. Manag.* **2024**, *572*, 122171. [CrossRef]
26. Vilà-Vilardell, L.; De Cáceres, M.; Piqué, M.; Casals, P. Prescribed fire after thinning increased resistance of sub-Mediterranean pine forests to drought events and wildfires. *For. Ecol. Manag.* **2023**, *527*, 120602. [CrossRef]
27. Wagner, R.G.; Newton, M.; Cole, E.C.; Miller, J.H.; Shiver, B.D. The role of herbicides for enhancing forest productivity and conserving land for biodiversity in North America. *Wildl. Soc. Bull.* **2004**, *32*, 1028–1041. [CrossRef]
28. Fox, T.R.; Jokela, E.J.; Allen, H.L. The Development of Pine Plantation Silviculture in the Southern United States. *J. For.* **2007**, *105*, 337–347. [CrossRef]
29. Davis, K.T.; Peeler, J.; Fargione, J.; Haugo, R.D.; Metlen, K.L.; Robles, M.D.; Woolley, T. Tamm review: A meta-analysis of thinning, prescribed fire, and wildfire effects on subsequent wildfire severity in conifer dominated forests of the Western US. *For. Ecol. Manag.* **2024**, *561*, 121885. [CrossRef]
30. Fontúrbel, T.; Carrera, N.; Vega, J.A.; Fernández, C. The Effect of Repeated Prescribed Burning on Soil Properties: A Review. *Forests* **2021**, *12*, 767. [CrossRef]
31. Vargas, R.; Paz, F.; de Jong, B. Quantification of forest degradation and belowground carbon dynamics: Ongoing challenges for monitoring, reporting and verification activities for REDD+. *Carbon Manag.* **2013**, *4*, 579–582. [CrossRef]
32. Jian, J.; Vargas, R.; Anderson-Teixeira, K.; Stell, E.; Herrmann, V.; Horn, M.; Kholod, N.; Manzon, J.; Marchesi, R.; Paredes, D.; et al. A restructured and updated global soil respiration database (SRDB-V5). *Earth Syst. Sci. Data* **2021**, *13*, 255–267. [CrossRef]
33. Rinkes, Z.L.; Sinsabaugh, R.L.; Moorhead, D.L.; Grandy, A.S.; Weintraub, M.N. Field and lab conditions alter microbial enzyme and biomass dynamics driving decomposition of the same leaf litter. *Front. Microbiol.* **2013**, *4*, 260. [CrossRef]
34. Liu, Y.; Men, M.; Peng, Z.; Chen, H.Y.H.; Yang, Y.; Peng, Y. Spatially explicit estimate of nitrogen effects on soil respiration across the globe. *Glob. Change Biol.* **2023**, *29*, 3591–3600. [CrossRef]
35. Ma, Z.; Chen, H.Y. Effects of species diversity on fine root productivity in diverse ecosystems: A global meta-analysis. *Glob. Change Biol.* **2016**, *25*, 1387–1396. [CrossRef]
36. Bates, D.; Mächler, M.; Bolker, B.; Walker, S. Fitting Linear Mixed-Effects Models Using lme4. *J. Stat. Softw.* **2015**, *67*, 1–48. [CrossRef]
37. Canty, A.; Ripley, B. *boot: Bootstrap R (S-Plus) Functions*, R package Version 1.3-29. 2024. Available online: https://cran.r-project.org/web/packages/boot/boot.pdf (accessed on 14 July 2025).
38. Calcagno, V. *glmulti: Model Selection and Multimodel Inference Made Easy*, R package Version 1.0.8. 2020. Available online: https://CRAN.R-project.org/package=glmulti (accessed on 14 July 2025).
39. Egger, M.; Smith, G.D.; Schneider, M.; Minder, C. Bias in meta-analysis detected by a simple, graphical test. *BMJ* **1997**, *315*, 629. [CrossRef] [PubMed]

40. R Core Team. *R: A Language and Environment for Statistical Computing*; R Foundation for Statistical Computing: Vienna, Austria, 2024.
41. Posit team. *RStudio: Integrated Development Environment for R*; Posit Software, PBC: Boston, MA, USA, 2024.
42. Fox, S.; Taylor, M.K.; Callaham, M.; Jumpponen, A. Fire-excluded and frequently burned longleaf pine forests have contrasting soil microbial communities. *For. Ecol. Manag.* **2024**, *551*, 121519. [CrossRef]
43. Hart, S.C.; Classen, A.T.; Wright, R.J. Long-term interval burning alters fine root and mycorrhizal dynamics in a ponderosa pine forest. *J. Appl. Ecol.* **2005**, *42*, 752–761. [CrossRef]
44. Ono, M.; Noormets, A.; Mitchell, S. The effect of the frequency of prescribed burning on annual soil carbon balance in a loblolly-shortleaf pine forest in East Texas. *Front. For. Glob. Change* **2025**, *8*, 1602557. [CrossRef]
45. Matosziuk, L.M.; Alleau, Y.; Kerns, B.K.; Bailey, J.; Johnson, M.G.; Hatten, J.A. Effects of season and interval of prescribed burns on pyrogenic carbon in ponderosa pine stands in the southern Blue Mountains, Oregon, USA. *Geoderma* **2019**, *348*, 1–11. [CrossRef]
46. Smith, J.E.; McKay, D.; Niwa, C.G.; Thies, W.G.; Brenner, G.; Spatafora, J.W. Short-term effects of seasonal prescribed burning on the ectomycorrhizal fungal community and fine root biomass in ponderosa pine stands in the Blue Mountains of Oregon. *Can. J. For. Res.* **2004**, *34*, 2477–2491. [CrossRef]
47. Zhao, B.; Cao, J.; Geng, Y.; Zhao, X.; von Gadow, K. Inconsistent responses of soil respiration and its components to thinning intensity in a Pinus tabuliformis plantation in northern China. *Agric. For. Meteorol.* **2019**, *265*, 370–380. [CrossRef]
48. Yang, L.; Qin, J.; Geng, Y.; Zhang, C.; Pan, J.; Niu, S.; Tian, D.; Zhao, X.; Wang, J. Long-term effects of forest thinning on soil respiration and its components in a pine plantation. *For. Ecol. Manag.* **2022**, *513*, 120189. [CrossRef]
49. Mayer, M.; Matthews, B.; Rosinger, C.; Sanden, H.; Godbold, D.L.; Katzensteiner, K. Tree regeneration retards decomposition in a temperate mountain soil after forest gap disturbance. *Soil Biol. Biochem.* **2017**, *115*, 490–498. [CrossRef]
50. Mayer, M.; Sandén, H.; Rewald, B.; Godbold, D.L.; Katzensteiner, K. Increase in heterotrophic soil respiration by temperature drives decline in soil organic carbon stocks after forest windthrow in a mountainous ecosystem. *Funct. Ecol.* **2017**, *31*, 1163–1172. [CrossRef]
51. Kohout, P.; Charvátová, M.; Štursová, M.; Mašínová, T.; Tomšovský, M.; Baldrian, P. Clearcutting alters decomposition processes and initiates complex restructuring of fungal communities in soil and tree roots. *ISME J.* **2018**, *12*, 692–703. [CrossRef]
52. Zhou, T.; Wang, C.; Zhou, Z. Impacts of forest thinning on soil microbial community structure and extracellular enzyme activities: A global meta-analysis. *Soil Biol. Biochem.* **2020**, *149*, 107915. [CrossRef]
53. Zhou, D.; Zhao, S.Q.; Liu, S.; Oeding, J. A meta-analysis on the impacts of partial cutting on forest structure and carbon storage. *Biogeosciences* **2013**, *10*, 3691–3703. [CrossRef]
54. López, B.C.; Sabate, S.; Gracia, C.A. Thinning effects on carbon allocation to fine roots in a Quercus ilex forest. *Tree Physiol.* **2003**, *23*, 1217–1224. [CrossRef]
55. Wang, D.; Olatunji, O.A.; Xiao, J. Thinning increased fine root production, biomass, turnover rate and understory vegetation yield in a Chinese fir plantation. *For. Ecol. Manag.* **2019**, *440*, 92–100. [CrossRef]
56. Caihong, Z.; Nier, S.; Hao, W.; Honglin, X.; Hailong, S.; Ling, Y. Effects of thinning on soil nutrient availability and fungal community composition in a plantation medium-aged pure forest of *Picea koraiensis*. *Sci. Rep.* **2023**, *13*, 2492. [CrossRef]
57. Holden, S.R.; Treseder, K.K. A meta-analysis of soil microbial biomass responses to forest disturbances. *Front. Microbiol.* **2013**, *4*, 163. [CrossRef]
58. Zhao, J.; Ye, Y.; Sun, X.; Shi, L.; Chen, X.; Guan, Q. Root exudation patterns of Chinese fir after thinning relating to root characteristics and soil conditions. *For. Ecol. Manag.* **2023**, *541*, 121068. [CrossRef]
59. Euskirchen, E.S.; Chen, J.; Gustafson, E.J.; Ma, S. Soil Respiration at Dominant Patch Types within a Managed Northern Wisconsin Landscape. *Ecosystems* **2003**, *6*, 595–607. [CrossRef]
60. Mushinski, R.M.; Gentry, T.J.; Boutton, T.W. Organic matter removal associated with forest harvest leads to decade scale alterations in soil fungal communities and functional guilds. *Soil Biol. Biochem.* **2018**, *127*, 127–136. [CrossRef]
61. Noormets, A.; McNulty, S.G.; Domec, J.-C.; Gavazzi, M.; Sun, G.; King, J.S. The role of harvest residue in rotation cycle carbon balance in loblolly pine plantations. Respiration partitioning approach. *Glob. Change Biol.* **2012**, *18*, 3186–3201. [CrossRef] [PubMed]
62. Avera, B.N.; Rhoades, C.C.; Calderón, F.; Cotrufo, M.F. Soil C storage following salvage logging and residue management in bark beetle-infested lodgepole pine forests. *For. Ecol. Manag.* **2020**, *472*, 118251. [CrossRef]
63. Prest, D.; Kellman, L.; Lavigne, M.B. Mineral soil carbon and nitrogen still low three decades following clearcut harvesting in a typical Acadian Forest stand. *Geoderma* **2014**, *214–215*, 62–69. [CrossRef]
64. Achat, D.L.; Deleuze, C.; Landmann, G.; Pousse, N.; Ranger, J.; Augusto, L. Quantifying consequences of removing harvesting residues on forest soils and tree growth—A meta-analysis. *For. Ecol. Manag.* **2015**, *348*, 124–141. [CrossRef]
65. Nave, L.E.; DeLyser, K.; Domke, G.M.; Holub, S.M.; Janowiak, M.K.; Kittler, B.; Ontl, T.A.; Sprague, E.; Sucre, E.B.; Walters, B.F.; et al. Disturbance and management effects on forest soil organic carbon stocks in the Pacific Northwest. *Ecol. Appl.* **2022**, *32*, e2611. [CrossRef]

66. Nave, L.E.; DeLyser, K.; Domke, G.M.; Holub, S.M.; Janowiak, M.K.; Ontl, T.A.; Sprague, E.; Viau, N.R.; Walters, B.F.; Swanston, C.W. Soil carbon in the South Atlantic United States: Land use change, forest management, and physiographic context. *For. Ecol. Manag.* **2022**, *520*, 120410. [CrossRef]
67. Mayer, M.; Baltensweiler, A.; James, J.; Rigling, A.; Hagedorn, F. A global synthesis and conceptualization of the magnitude and duration of soil carbon losses in response to forest disturbances. *Glob. Ecol. Biogeogr.* **2024**, *33*, 141–150. [CrossRef] [PubMed]
68. Gundale, M.J.; Axelsson, E.P.; Buness, V.; Callebaut, T.; DeLuca, T.H.; Hupperts, S.F.; Ibáñez, T.S.; Metcalfe, D.B.; Nilsson, M.-C.; Peichl, M.; et al. The biological controls of soil carbon accumulation following wildfire and harvest in boreal forests: A review. *Glob. Change Biol.* **2024**, *30*, e17276. [CrossRef] [PubMed]
69. Alcañiz, M.; Outeiro, L.; Francos, M.; Úbeda, X. Effects of prescribed fires on soil properties: A review. *Sci. Total Environ.* **2018**, *613–614*, 944–957. [CrossRef] [PubMed]
70. Scharenbroch, B.C.; Nix, B.; Jacobs, K.A.; Bowles, M.L. Two decades of low-severity prescribed fire increases soil nutrient availability in a Midwestern, USA oak (Quercus) forest. *Geoderma* **2012**, *183–184*, 80–91. [CrossRef]
71. Nave, L.E.; Vance, E.D.; Swanston, C.W.; Curtis, P.S. Fire effects on temperate forest soil C and N storage. *Ecol. Appl.* **2011**, *21*, 1189–1201. [CrossRef]
72. Rocha, W.; Metcalfe, D.B.; Doughty, C.E.; Brando, P.; Silverio, D.; Halladay, K.; Nepstad, D.C.; Balch, J.K.; Malhi, Y. Ecosystem productivity and carbon cycling in intact and annually burnt forest at the dry southern limit of the Amazon rainforest (Mato Grosso, Brazil). *Plant Ecol. Divers.* **2014**, *7*, 25–40. [CrossRef]
73. Butler, O.M.; Lewis, T.; Rashti, M.R.; Chen, C.R. Energetic efficiency and temperature sensitivity of soil heterotrophic respiration vary with decadal-scale fire history in a wet sclerophyll forest. *Soil Biol. Biochem.* **2019**, *134*, 62–71. [CrossRef]
74. Pellegrini, A.F.A.; Ahlström, A.; Hobbie, S.E.; Reich, P.B.; Nieradzik, L.P.; Staver, A.C.; Scharenbroch, B.C.; Jumpponen, A.; Anderegg, W.R.L.; Randerson, J.T.; et al. Fire frequency drives decadal changes in soil carbon and nitrogen and ecosystem productivity. *Nature* **2018**, *553*, 194–198. [CrossRef]
75. Coates, T.A.; Chow, A.T.; Hagan, D.L.; Wang, G.G.; Bridges, W.C., Jr.; Dozier, J.H. Frequent Prescribed Burning as a Long-term Practice in Longleaf Pine Forests Does Not Affect Detrital Chemical Composition. *J. Environ. Qual.* **2017**, *46*, 1020–1027. [CrossRef]
76. Oliver, A.K.; Callaham, M.A.; Jumpponen, A. Soil fungal communities respond compositionally to recurring frequent prescribed burning in a managed southeastern US forest ecosystem. *For. Ecol. Manag.* **2015**, *345*, 1–9. [CrossRef]
77. Pellegrini, A.F.A.; Caprio, A.C.; Georgiou, K.; Finnegan, C.; Hobbie, S.E.; Hatten, J.A.; Jackson, R.B. Low-intensity frequent fires in coniferous forests transform soil organic matter in ways that may offset ecosystem carbon losses. *Glob. Change Biol.* **2021**, *27*, 3810–3823. [CrossRef]
78. González-Pérez, J.A.; González-Vila, F.J.; Almendros, G.; Knicker, H. The effect of fire on soil organic matter—A review. *Environ. Int.* **2004**, *30*, 855–870. [CrossRef]
79. Knicker, H. How does fire affect the nature and stability of soil organic nitrogen and carbon? A review. *Biogeochemistry* **2007**, *85*, 91–118. [CrossRef]
80. Udali, A.; Chung, W.; Talbot, B.; Grigolato, S. Managing harvesting residues: A systematic review of management treatments around the world. *For. Int. J. For. Res.* **2025**, *98*, 117–135. [CrossRef]
81. Eisenbies, M.H.; Vance, E.D.; Aust, W.M.; Seiler, J.R. Intensive Utilization of Harvest Residues in Southern Pine Plantations: Quantities Available and Implications for Nutrient Budgets and Sustainable Site Productivity. *BioEnergy Res.* **2009**, *2*, 90–98. [CrossRef]
82. James, J.; Page-Dumroese, D.; Busse, M.; Palik, B.; Zhang, J.; Eaton, B.; Slesak, R.; Tirocke, J.; Kwon, H. Effects of forest harvesting and biomass removal on soil carbon and nitrogen: Two complementary meta-analyses. *For. Ecol. Manag.* **2021**, *485*, 118935. [CrossRef]
83. Hume, A.M.; Chen, H.Y.H.; Taylor, A.R. Intensive forest harvesting increases susceptibility of northern forest soils to carbon, nitrogen and phosphorus loss. *J. Appl. Ecol.* **2018**, *55*, 246–255. [CrossRef]
84. Haskins, K.E.; Gehring, C.A. Long-term effects of burning slash on plant communities and arbuscular mycorrhizae in a semi-arid woodland. *J. Appl. Ecol.* **2004**, *41*, 379–388. [CrossRef]
85. Gibbons, P.; Lindenmayer, D.B.; Barry, S.C.; Tanton, M.T. The effects of slash burning on the mortality and collapse of trees retained on logged sites in south-eastern Australia. *For. Ecol. Manag.* **2000**, *139*, 51–61. [CrossRef]

forests

MDPI

Review

Forest Soil Microbiomes: A Review of Key Research from 2003 to 2023

Aurelia Onet [1], Paola Grenni [2,3,*], Cristian Onet [1,*], Vlad Stoian [4] and Vlad Crisan [5]

1 Faculty of Environmental Protection, University of Oradea, 26 General Magheru Street, 410048 Oradea, Romania; onetaurelia@gmail.com
2 Water Research Institute, National Research Council (IRSA-CNR), via Salaria km 29,300, 00015 Rome, Italy
3 NBFC, National Biodiversity Future Center, Università degli Studi di Palermo, Piazza Marina 61, 90133 Palermo, Italy
4 Department of Microbiology, Faculty of Agriculture, University of Agricultural Sciences and Veterinary Medicine of Cluj-Napoca, 3-5 Calea Mănăştur, 400372 Cluj-Napoca, Romania; vlad.stoian@usamvcluj.ro
5 National Institute for Research and Development in Forestry "Marin Drăcea", 13 Cloşca Street, 500040 Braşov, Romania; vlad_crsn@yahoo.com
* Correspondence: paola.grenni@cnr.it (P.G.); cristyonet@yahoo.com (C.O.)

Abstract: Forests have a key role in mitigating both non-biological and biological ecological disturbances. However, major disturbances (soil pollution, shift from native forest species to exoticones, forested watersheds and climate changes) can have different impacts on a forest's soil microbiome. Because the soil microbial community of forests has a key role in a variety of ecosystem services that promote the forest's health, this review tries to answer the following questions: (i) Which are the main ecological disturbances that drive the responses of the forest soil microbiome? (ii) How can we measure these changes? For this aim, the review summarizes details on the tree vegetation type, the microbial communities in forest ecosystems, and the mutual influence between plants, soil, and microbiomes. Microbial communities are shaped by factors such as soil type and composition, plant and vegetation types, nutrient levels and soil fertility, disturbance patterns, symbiotic associations, biotic interactions, and the progression of forest succession. Anthropogenic activities produce a rapid response in the microbial communities, leading to both short- and long-term alterations. Harvesting processes reduce drastically the microbiome diversity, forcing a shift from specialized to more generalist microorganisms. Restoration scenarios indicate a re-establishment of microbial communities to a level similar to the native forest, but with a high percentage of replaced native microorganisms. This review emphasizes that the forest soil microbiome is shaped by a range of environmental, ecological, and biotic factors. The primary drivers of the soil microbiome in forest ecosystems discussed in this review include soil composition and nutrient availability, plant community structure, microbial interactions within the soil, disturbances, succession, and temporal dynamics. When considered together, these factors interact in complex ways, influencing the diversity, function, and resilience of the soil microbiome in forest ecosystems.

Keywords: holobiont; ecological disturbances of forest; climate change; anthropogenic forest conversion; microbial community profile; diversity and abundance of forest soil biota; disturbance responses of soil bacterial and fungal communities

1. Introduction

The biosphere's health relies mainly on forest ecosystems that regulate essential nutrient and energy patterns, which, in turn, can significantly influence the climate system [1].

https://doi.org/10.3390/f16010148

Undisturbed forest landscapes, unaffected by human pressures, serve as significant carbon sinks, mitigating greenhouse gas emissions. On the other hand, tropical deforested regions and poorly managed forests can exacerbate climate change [2–5]. Forest ecosystems are important not only for regulating the climate but also because they include a vast biodiversity that holds crucial ecological and economic value. For this reason, the United Nations General Assembly proclaimed 21 March as the International Day of All Types of Forests in 2012, and the specific 2023 theme was "Forests and Health".

The aim is to emphasize the role of healthy forests for healthy people, in the framework also of the "One Health" concept (UN, 2023, https://www.un.org/en/un-chronicle/healthy-forests-are-crucial-human-health-and-sustainable-development, accessed on 1 June 2024) and in line with the sustainable development goals (SDG 15, https://www.un.org/en/chronicle/article/goal-15-seeing-forest-trees-making-most-synergies-achieve-sdgs-constrained-environment#:~:text=A%20case%20in%20point%20is, degradation%20and%20halt%20biodiversity%20loss%E2%80%9D, accessed on 1 June 2024). Current human management practices like deforestation, agriculture on a large scale, small-scale farming, and the development of infrastructures are unfortunately leading to rapid changes in forest biodiversity. Moreover, as forests increasingly become a source of renewable materials and fuels, their economic significance has been growing, forcing their harvesting in several parts of the planet. It is thus imperative to adopt conservation and sustainable management strategies to preserve the functionality of forests, protect global biodiversity and, at the same time, match the socio-economic needs [6,7].

Belowground microbiota, which includes soil, the rhizosphere, and root-associated microbial communities, is crucial for sustaining the physiology and the growth of plants [8]. These beneficial effects depend on nutrient-absorbing processes that increase nutrient uptake efficiency (plant growth enhancement) and protect against various negative abiotic and biotic conditions [8]. Moreover, specific bacteria and fungi interconnected with perennial tree crops interface with soil invertebrate food webs, supporting several regulation mechanisms [9]. On the other hand, trees have also a great influence on soil microbiota diversity, functioning, and fitness. For example, it was recently found that tree diversity has a positive influence on soil microbial resistance to drought [10]. Nevertheless, knowledge of how microbial communities of soil react to forest management is still very scarce. It has been well known that important biomarkers of forest ecosystem status are microbial community structure, their patterns of gene expression, and the metabolic activities. In fact, these factors can increase the capability to observe the health of forest ecosystems, to assess the effects of management practices, and in some way, to identify variations in nutrient and energy flow patterns before they have permanent effects [11].

The use of community profiling approaches such as molecular, biochemical, and physiological techniques have advanced the knowledge on the role of microbial diversity in forest soils. It was recently found that forest type influences microbial community composition and the inter- and intra-ecosystem variability of communities between and within forest ecosystems and that microbial communities respond to disturbance and forest management activities. Finally, numerous works highlight the relationship between microbial community and forest soil processes [12–14].

Even though different long-term responses to forest disturbance are observed in bacteria and fungi, several studies validate definite changes in the responses of important bacterial and fungal functional groups (e.g., nitrifying bacteria and mycorrhizal fungi), and indicate that both microbial groups play important functions in the long-term modifications of biogeochemical processes detected as a result of disturbances to forests [15,16].

Understanding the factors that drive the forest soil microbiome has become a prominent research topic worldwide [17]. Soil microbial communities have pivotal functions in promoting soil health by contributing to organic matter turnover and nutrient cycling. As an indispensable part of the soil ecosystem, the soil microbiome drives processes such as soil formation, fertility, plant growth and stress tolerance, as well as the nutrient cycles and carbon storage [18]. In forests, soil microbes contribute to forest ecosystem health by decomposing organic matter, cycling carbon and nutrients, incorporating humic compounds into mineral soils, and connecting plant and ecosystem functions. Moreover, they have a crucial role in carbon storage and are responsible for plant photosynthetic carbon inputs to the soil [19]. However, soil microbial communities are highly sensitive to changes in forest land use, and the impact of such changes on these communities remains poorly understood despite their functional significance [20,21]. Recently, Baldrian et al. [5] indicated that, although significant research progress in the forest microbial ecology framework has taken place, there are still many gaps and challenges that researchers face. Among others, some key limitations and challenges in this field are the complexity of microbial communities, spatial and temporal variation, limited knowledge on microbial functions, methodological constraints, a lack of long-term data, climate change and disturbances, data integration, and collaboration [7]. However, by addressing these issues and working collaboratively, scientists can make significant strides in unraveling the intricate relationships between microorganisms and forest ecosystems, ultimately aiding in forest conservation and management efforts.

The holobiont concept fits into the better understanding of forest ecosystems and their future evolution. The terms holobiont refers to a host organism (in this case the plant) and its associated microbial communities, collectively functioning as a unit [22]. This concept have been challenged the traditional view of organisms as individuals and emphasizes the importance of symbiotic relationships between a host and its microbiome [23]. The microbiome includes bacteria, archaea, viruses, fungi, and other microorganisms that inhabit a specific environment, such as the surface of plant roots or the internal parts of the plants. The holobiont concept suggests that the host and its associated microbiome should be considered as a single ecological and evolutionary unit [24]. Thus, the interactions within the holobiont can play a crucial role in the adaptation and response of the organism to ecological pressures. The microbiome can contribute to the host organism's ability to adapt to environmental changes. For example, certain microbes in the soil can enhance a plant's resistance to pathogens or help it acquire nutrients more efficiently [25]. Moreover, symbiotic relationships within the holobiont can provide benefits to the host under different ecological pressures. For instance, gut bacteria in animals can aid in digestion, contribute to the immune system, and even play a role in protecting against harmful pathogens [26]. Based on the frameworks proposed in this concept, new interdisciplinary research areas can be developed, and new functional solutions identified.

Faure et al. [10] consider that the assemblage processes of the microbiota and the function of the host in microbiota selectivity are valuable research topics. The paper highlighted that the exudates from plants can selectively influence the microbiota, causing plants to have contrasting strategies for nutrient resources.

The proposition of interactions between plant and earthworm holobionts for enlisting microbiota that aids in organic matter recycling and mineral nutrition corresponds with the increasing acknowledgment of the intricate relationships between organisms and their microbiomes in ecological processes [12]. The plant–earthworm interaction might entail the enlistment of specific microbial communities by both entities. For instance, plants can release root exudates that attract beneficial microbes engaged in nutrient cycling and promoting plant growth. Earthworms, through their interactions with the soil and organic matter, might harbor specific microbiota in their gut or on their skin. As they traverse the soil, these earthworm-associated microbes could influence the soil microbiome. The concept of holobionts implies that plants, earthworms, and their associated microbiota operate as a cooperative entity. The microbiota linked with both the plant and earthworm could contribute to the overall well-being and functionality of the holobiont in the context of organic matter recycling and nutrient cycling.

Dessaux et al. [27] examined the mutually interdependence between plants and microbiota and their subsequent possible applicability to improve performance and healthy of plants. Holobiont functioning is another important issue when examining holobiont studies. Which functional traits of the holobiont are derive from the microbiota and the host assemblage, and what are these functional traits that support the holobiont's adaptation in a mutational environment?

The holobiont functional traits resulting from the assembly of microbiota and host can encompass various aspects, all of which contribute to the holobiont's adaptation in a changing environment [28,29]. Microbiota associated with the host can facilitate nutrient uptake and metabolism, including the acquisition of essential nutrients from the environment and the breakdown of complex organic compounds [30]. This can enhance the holobiont's ability to thrive in nutrient-limited or variable environments. Certain microbiota can confer resistance or tolerance to pathogens and pests by directly antagonizing them through the production of antimicrobial compounds or by priming the host's immune system [31]. This contributes to the holobiont's ability to withstand disease pressure. Microbiota can enhance the host's tolerance to various abiotic stresses such as drought, salinity, or extreme temperatures by modulating physiological processes or producing stress-protective compounds [32]. This helps the holobiont to survive and thrive in challenging environmental conditions. Beneficial symbiotic relationships between the host and specific microbiota can provide mutualistic benefits, such as the exchange of nutrients or growth-promoting factors. These interactions contribute to the overall fitness of the holobiont in its environment [33]. Its microbiota can influence host development and growth through various mechanisms, including hormone production, nutrient availability, and regulation of gene expression. This can lead to enhanced growth rates or altered developmental patterns that optimize the holobiont's performance in its environment. Overall, the diverse functional traits resulting from the interaction between host and microbiota contribute to the adaptability and resilience of the holobiont in the face of environmental variability and change [34,35]. Understanding these traits is essential for harnessing the potential of microbiota-mediated adaptations in agriculture, ecosystem management, and other fields.

The diversity of microbial communities within a holobiont can contribute to its resilience in the face of ecological challenges [36]. A diverse microbiome can provide functional redundancy, ensuring that essential functions are maintained even if some microbial species are affected. Understanding the holobiont concept has implications for fields such as medicine, agriculture, and ecology [37–39]. It highlights the interconnectedness of organisms and their associated microbiomes and emphasizes the importance of considering this holistic perspective when studying adaptation to ecological pressures.

To understand these relationships is fundamental for sustainable and responsible agriculture and the management of ecosystems. By appreciating the collaborative relationships between different organisms and their microbiomes, researchers and practitioners can develop strategies to enhance soil health, nutrient cycling, and organic matter decomposition in a more holistic manner. This approach aligns with the broader shift towards ecological thinking in agriculture and ecosystem science.

In this review, we explore how bacteria play a role in the complex forest ecosystem, and summarise findings from recent literature to understand the main factors that determine their abundance and diversity. We discuss how microbial populations of forest ecosystems have been responding to climate change, and the implications of this response on ecosystem processes. Finally, current knowledge on the bacteria's role in addressing ecosystem processes and their capacity to adapt to global change are also discussed.

2. Materials and Methods

This review considers articles published between 2003 and 2023 concerning the main factors that drive the responses of the forest soil microbiome based on an overview of scientific publications on the disturbance responses of forest microbial communities. The articles are original studies as well as literature reviews, including scoping reviews and systematic reviews.

The selection criteria were:

- Peer-reviewed articles;
- Type of publication (only original studies or reviews were considered).

The main directions of the systematic review consisted in reviewing the impact of tree vegetation and tree composition on soil microbial communities; the influence of the soil microbial community on the plant-associated microbiome in the context of forest anthropogenic disturbance; the effect of harvesting on the diversity and structure of soil bacterial and fungal communities; the study of the soil microbial community successional patterns during forest ecosystem restoration; the diversity and response to global change; and the screening of bacterial communities in forest ecosystems.

Exclusion criteria comprised:

- Editorials;
- Published studies other than in English.

ERIC (Education Resources Information Centre) and numerous databases, including Web of Science, Science Direct, SpringerLink and Google Scholar, were used to study key relevant research. The following keywords in the various databases were used: forest soil microbiome, disturbance responses of forest microbial communities, influence of climate change on microbial diversity, drivers of abundance and diversity of forest soil microorganisms, forest anthropogenic disturbance, harvesting, ecosystem restoration.

3. Results and Discussion

The literature review included 102 articles that met the criteria for our analysis. The relevant data extracted from the articles were categorized by continent and research topic (Table 1).

Table 1. Important data from articles distributed by continent and research topic.

Subject	Article Distribution by Continents				
General	**Europe**	**Australia**	**Asia**	**North America**	**South America**
Bacterial communities in forest ecosystems	23	2	7	6	1
Impact of tree vegetation and tree composition on soil microbial communities	4	-	2	-	-
Influence of soil microbial community on plant-associated microbiome in the context of forest anthropogenic disturbance	2	-	2	3	1

Table 1. *Cont.*

Subject	Article Distribution by Continents				
General	**Europe**	**Australia**	**Asia**	**North America**	**South America**
Influence of harvesting on diversity and structure of soil bacterial and fungal communities	5	-	3	5	3
The patterns of soil microbial community succession during the restoration of forest ecosystems	1	2	3	7	-
The impact of global change on bacterial communities within forest ecosystems	6	2	6	6	-
Total	41	6	23	27	5

3.1. *Bacterial Ecology in Forest Soils*

The ecological factors that drive the microbiome in forest soils are diverse and impact the composition, diversity, and function of microbial communities. Forest soils are dynamic environments, and the microbes that inhabit them have key roles in nutrient cycling, decomposition, and the overall health of the ecosystem.

Forest soil ecology studies have predominantly focused on fungi, although bacteria are essential portion of the microbial community in soil of forests [40]. It is important to note that the functions of bacteria and fungi in forest ecosystems do not have to be considered as separate entities, as they interact with each other in various ways (Figure 1). The fungal biomass in forest soils has numerous effects on bacteria, such as the development of specific niches in patches of soil colonized from mycorrhizal fungi (i.e., the mycorrhizosphere) and microbial mats in the soil [41,42]. Moreover, bacteria occupy various habitats within forest soils, including bulk soil, the rhizosphere, litter, and deadwood habitats [42]. Their communities are influenced by soil type and composition, vegetation and plant communities, nutrient availability and soil fertility, disturbance regimes, symbiotic relationships, biotic interactions, and forest succession. Bacteria are involved in essential soil processes such as carbon, nitrogen, and phosphorus cycling, and they contribute to the decomposition of dead plant biomass and fungal mycelia. In fact, recent studies have revealed that bacteria frequently possess genes that encode plant cell wall-degrading enzymes, indicating that they are also involved in the plant material decomposition.

They interface with the roots of plants and with the mycorrhizal fungi either by acting as commensals or helpers of mycorrhizae. Additionally, bacteria are crucial for several phases in the nitrogen cycle, including nitrogen fixation. The effects of global change, such as climate warming and increased carbon dioxide levels, can impact bacterial communities in forest soils. However, the responses are specific to each forest ecosystem, and it is currently challenging to incorporate bacteria into predictive models. While significant progress has been made in understanding bacterial ecology in forest soils, the exact extent of their contributions to forest ecosystem processes remains incomplete. A comprehensive understanding of all soil community members' activities is necessary to fully recognize the bacterial role in forest ecosystems [43].

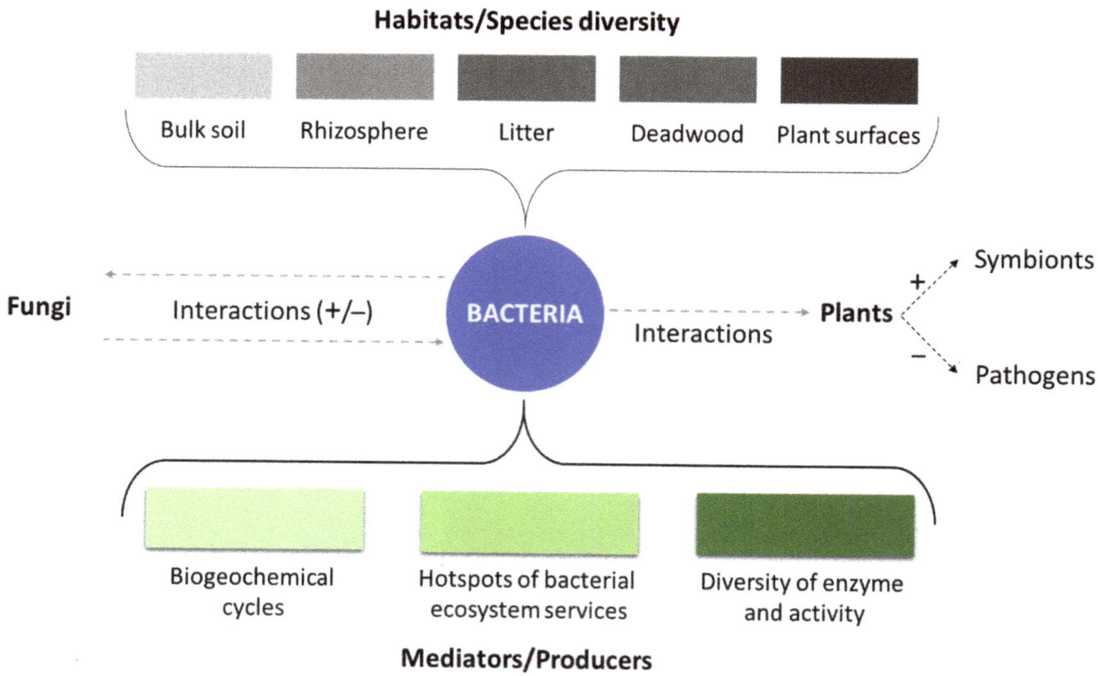

Figure 1. The role and impact of bacterial communities in forests. Grey dash lines indicate the possible interactions among bacteria, fungi and plants. The different bacterial habitats that influence species diversity are displayed at the top of the figure (in grey). At the bottom of the figure, the different functions of bacteria as mediators or producer are displayed in green.

Forest microorganisms, including bacteria and fungi, have been recognized to have crucial roles in the forest ecosystem. As decomposers, symbionts, or pathogens, they affect carbon turnover and retention, as well as the availability of other nutrients [9]. In addition, research has shown that microbial communities are essential for facilitating biogeochemical cycles, and understanding their role in ecosystem processes is crucial for predicting how forests will respond to future environmental conditions [44–46]. Fungi are among the most extensively studied microorganisms in temperate and boreal forest soils, where they form diverse communities of saprotrophic and mycorrhizal fungal taxa. They are considered the primary decomposers in forest soils due to their capability to produce several extracellular enzymes that efficiently break down the recalcitrant fraction of dead plant biomass [47,48].

In the context of plant–microbial interactions, negative interactions occur when the collective impact of soil organisms, such as pathogens, mutualists, and decomposers, leads to reduced plant performance. Conversely, positive interactions occur when soil communities provide benefits that enhance plant performance, such as increased biomass production and survival. Due to the crucial role of these interactions in shaping ecosystem properties, it is crucial to understand how soil microbe–microbe and soil microbe–plant relations react to climate change. This research topic will bring insights into important ecosystem functions, including the carbon storage in the soil and net primary productivity [16]. However, it is unclear which way microbial activity will influence carbon feedbacks among plants, soil and the atmosphere [48]. While the direct effects of climate change affecting microbial function have been thoroughly investigated, more indirect effects through changes in the interactions between plants and soil microbes and between microbes and soil microbes are less well known, even though they can influence vital processes such as plant chemistry, plant community composition, and mineralization rates. Plants and microbes can react to climate change by altering their population ranges, symbiotic partners, or timing of phonological actions. The relative abundance and function of soil communities are also affected by climatic changes, as soil community members vary in their physiology, temperature sensitivity, and growth rates. De Angelis et al. [49] demonstrated that a 5 °C warming in a temperate forest changed the relative abundances of soil bacteria and increased the bacterial-to-fungal ratio of the community.

Bacteria are found in various habitats, which represent different layers within forest ecosystems, including plant tissues and surfaces, streams, rocks, and especially the forest floor, soil, and litter. According to Lauber et al. [50], the five most abundant phyla in most soils are *Acidobacteria*, *Actinobacteria*, *Proteobacteria*, *Bacteroidetes*, and *Firmicutes*. Besides, the biogeographical patterns of the microbial community show site-uniqueness, and most of the species can be considered as "rare", due to their site-specific conditions. The pH of the soil appears to be the most significant factor that determines the composition of bacterial communities in the soil, but other factors, such as organic matter content, accessibility of nutrients, climate conditions, and biotic interactions (especially with vegetation), also play crucial roles [51–53]. Within the groups described by Lauber et al. [50], only *Acidobacteria* shows a very high relative abundance at pH between 4–5, while the other ones exhibit a high relative abundance (30%–40%) at neutral and alkaline values. Fungal taxa are 8–10 times less abundant than bacterial ones in forest litter and soil, but have a higher specificity for one or two tree species than bacteria [52]. In general, up to 80% of bacteria are associated with more than six tree species, which indicates the additional effect of litter and pH in their diversity.

Spatial variability of chemical parameters accounts for the presence of hotspots of microbial activity with higher abundance and activity in the soil, such as in and on plant litter and dead wood, or on surrounding plant roots [54–56]. Each of these niches has distinct characteristics and, therefore, a characteristic bacterial community. Numerous studies have shown that fungi dominate the litter habitat and play a critical role in litter decomposition, but recent research has also highlighted the active involvement of bacteria in litter transformation. In coniferous forest litter, *Betaproteobacteria*, *Bacteroidetes*, and *Acidobacteria* were found to incorporate relatively more cellulose-derived carbon than fungi. This trend was observed in different pine forest soils in North America as well, where *Bacteroidetes* (*Sphingobacteriales*), *Proteobacteria* (*Caulobacteriales*, *Burkholderiales*, *Xanthomonadales* and *Rhizobiales*), and subdivision 1 *Acidobacteria* were identified as the primary accumulators of cellulose-derived carbon. Studies conducted on deciduous forests have also reported that at least 10% of litter bacteria are capable of decomposing cellulose, with the most common groups being *Proteobacteria*, *Actinobacteria*, *Bacteroidetes*, and *Acidobacteria* [57]. While the acidic soils of coniferous forests are connected to mainly Proteobacteria, Acidobacteria, and Actinobacteria in temperate deciduous forests, litter bacterial communities look especially enriched with Bacteroidetes and Proteobacteria [58,59]. The changes in the chemical structure of litter and root exudates due to the tree species types influence the soil bacterial communities because of the changes in substrate chemistry [60].

Bacterial and fungal communities in the soil ecosystem have varying spatial distributions. Fungal taxa are more confined to either the litter or organic horizon of soil and have a more heterogeneous distribution throughout the ecosystem. Environmental factors are more influential than geographic dispersal limitation in determining the bacterial community structure of forest soils [61].

The impact of soil properties and vegetation types on the structure and diversity of soil bacteria is widely acknowledged. Soil pH is recognized as a crucial factor that regulates the composition of bacterial communities [62]. Additionally, other soil characteristics, such as nutrient availability and plant diversity, also affect the composition and diversity of soil bacterial communities [63,64]. Previous studies have demonstrated that the distribution of bacterial communities is influenced by soil properties, climate, and other factors [64]. However, the specific controlling factors differ across different ecosystem types and spatial scales.

Wei et al. [65] analyzed the structure, diversity, and function of soil bacterial communities in different forest types. They selected two natural and mature forest sites in China and employed high-throughput sequencing to compare bacterial community diversity and function in four different seasons. The authors found that the soil bacterial composition at the phylum level (top 10 phyla) was similar, with differences only observed in relative abundance. The bacterial communities at the two sites exhibited two significant predicted functions related to the carbon cycle and three significant predicted functions related to the nitrogen cycle. This study is among the few to investigate the relationship between bacterial community structure and function and examine bacterial diversity in two typical forest sites under natural conditions in climatically transitional regions of China.

Bacterial communities in forest ecosystems are of significant importance for several reasons, as they play a vital role in maintaining the health, functioning, and sustainability of these ecosystems. Some key points highlighting the importance of bacterial communities in forest ecosystems are nutrient cycling, nitrogen fixation, disease control, and mycorrhizal symbiosis. The forest soil microbiome is shaped by a complex interplay of abiotic (e.g., soil composition, moisture, temperature) and biotic (e.g., plant species, soil fauna) factors. The overall health and functioning of the forest ecosystem are tightly linked to microbial processes driven by these ecological factors. Understanding these interactions is key to improving forest management and conservation efforts. The study and conservation of these communities is essential for the health and sustainability of forest ecosystems, especially in the face of ongoing environmental changes.

3.2. The Impact of Tree Vegetation and Tree Composition on Soil Microbial Communities

It is well known that soil physicochemical properties are strongly correlated with microbial communities. For example, a recent study investigated the seasonal dynamics of six soil enzyme activities (protease, urease, acid phosphatase, cellulase, peroxidase, and polyphenol oxidase) and microorganisms (bacteria, fungi, actinomycetes, and total microorganisms), in different vegetation types, and their reactions to shifts under soil physicochemical properties [66]. Four types of shelter forests (*Bambusaoldhamii, Pinuselliottii, Eucalyptus robusta,* and *Casuarina equisetifolia*) in China were chosen to investigate changes in seasonal patterns of enzyme activities (urease, acid phosphatase, cellulase, and polyphenol oxidase) and microorganism presence in soil correlated with soil physicochemical properties (pH, moisture, temperature, nutrients). The results highlighted that different shelter forests had seasonal differences in the selected microbial activities. The soil bacteria, fungi, and total microbial presence in four shelter forests considerably improved in autumn for *Actinomycetes*. Soil enzymatic activities were correlated to total microbial presence and soil physicochemical properties. The research also confirmed that soil enzyme activities and total microbial presence had significant seasonal changes among the four shelter forests.

Jin et al. [67] found that bacteria are dominant populations in shelter-forest soils in the extreme arid areas, accounting for more than 80% of total soil microbes, and fungi are not more than 1% of the total soil microbes, although soil microbial community structure changes among three soil layers. Zhang et al. [68] studied bacterial surface soil community diversity and function in the shelterbelts of agricultural soils of five forest types. Dominant forest soil bacterial phyla include *Proteobacteria, Actinomycetes, Acidobacteria, Chlorobacteria,* and *Bacillus*, and soil bacteria were found to be better linked and to have more severe competition in the top layer of the soil than in the bottom layer.

On the other hand, Heo et al. [18] showed the capacity of the microbial community as a proactive marker of climate-driven vegetation change. The authors studied the soil bacterial and fungal communities in Korea (Odaesan) over a four-year interval through eDNA meta-barcoding and analyzed the differences in composition and function between forest types (Mongolian oak, *Quercusmongolica*, forest with, as well as without, Manchurian fir, *Abiesholophylla*) and sampling years. The results showed that denitrifiers were predominant in Manchurian firs, but no difference in the influence of climate change according to forest type. While the tree vegetation was stable, the microbial communities modified significantly over the course of four years. Climate change alters microbial communities significantly, although it is not sufficient to trigger a change in vegetation, so a microbial indicator can be developed to assess the disturbance of accumulated pressure on the forest ecosystem. The authors showed the impact of Manchurian fir trees and the effects of climate change on soil microbial communities in temperate forests.

As it is reasonable to understand, not only the activity, but also the microbial community structure, is greatly influenced by the type of vegetation, and in particular, the type of tree species (Figure 2). For example, Prada-Salcedo et al. [69] analyzed the bacterial and fungal diversity in relation to forest composition using amplicon sequencing. They found that microbial Shannon diversity was affected by the proportion of evergreen trees, and the fungi were influenced by forest tree species composition.

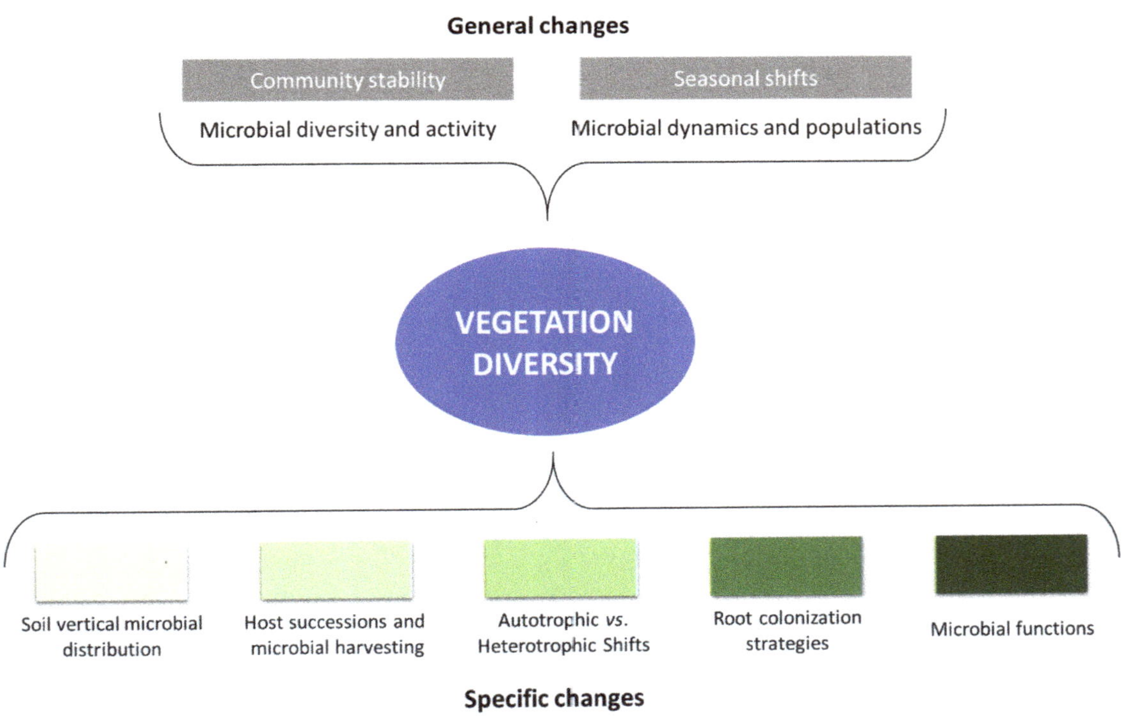

Figure 2. General (in grey, on the top of the figure) and specific (in green, on the bottom) changes in forest microbial communities due to vegetation type and diversity.

Gillespie et al. [55] analyzed the effect of forest litter and absorptive roots on wide-ranging functions of soil microbial communities under mixed tree species conditions. The work suggested influences indirectly through the effects of tree mixture on surface litter and absorptive root traits and soil parameters. Mixed forests consisting of any three tree species affected soil microbial functioning, influencing nutrient availability in the forest soil litter and resource acquisition by the roots. The mixing of trees also changed soil microbial functioning and catabolic diversity, impacting soil fertility and physicochemical characteristics. Thus, the results indicate an indirect but existing impact of tree species mixing on the microbial heterotrophic soil community activity in four different forest ecosystems, from Mediterranean to boreal forests.

Mycorrhizal communities are greatly influenced by both the composition of tree associations and their cover within the canopy. Symbiotic mycorrhizal fungi sustain the diversity within forest trees, but act with a general-to-specific functionality based on a spatial gradient from large (forest ecosystem) to low scale (host tree) [70]. Arbuscular mycorrhizas in associations with forest ecosystems vary in terms of species distribution: from generalist species with a wide distribution and associated with multiple hosts to specialized species associated with only one or a reduced number of hosts [71]. Tree status and the overall status of the forest plant community are visible in the feedback between mycorrhizas and forest future dynamics, seedling survival, and performance [72]. Mycorrhizal colonization strategy is visible in the diversity of forests [73], with differences induced by arbuscular mycorrhizas or ectomycorrhizas domination [74], both contributing at the establishment of a specific flux of nutrients toward their host, the ecosystem nutrient cycling, and host successions [75,76]. The host species and its physiological traits and status are responsible for harvesting single or multiple mycorrhizal species, either arbuscular, ectomycorrhizal, or their dual colonization [77]. This mechanism is visible in the colonization status related to the type of ecosystem and its tree diversity [78], with the complexity of interactions increasing complementarily to the increase in species composition [79].

Understanding the complex interactions between tree vegetation, composition, and soil microbial communities is essential for sustainable forest management, reforestation efforts, and the conservation of biodiversity [80]. It also has implications for ecosystem functioning and resilience, particularly in the context of changing environmental conditions. Researchers continue to study these relationships to gain a better understanding of the intricate connections between trees and the soil microbiome.

3.3. Interactions Between Soil Microbial Communities and Plant-Associated Microbes Under Forest Anthropogenic Stress

Nutrient availability/presence has a great influence on microbial communities (Figure 3). It is well known that plants release nutrients (e.g., root exudates) in the rhizosphere, which is the main area of the soil that influences the microbial community [6]. The presence of contaminants also has a great influence on the microbial community and activity [81].

In particular, Yurong et al. [82] investigated the rhizosphere's influence on soil characteristics, bacterial communities and enzyme activities in *Robinia pseudoacacia* L. throughout a gradient of heavy metal content contamination. The authors identified that soil organic matter, accessible nitrogen, and phosphorous contents were clearly higher in the rhizosphere than in the bulk soil at heavy metal contaminated sites. An increase in the activities of four soil enzymes indicative of the C-cycle (β-glucosidase), N-cycle (protease, urease), and P-cycle (alkaline phosphatase) were found in the rhizosphere soil at all study sites compared to the bulk soil. Quantitative PCR (qPCR) and restriction fragment length polymorphism (RFLP) were used to determine the relative abundance, composition, and diversity of bacteria in the bulk and rhizosphere soils, respectively. The bacterial 16S rRNA gene copy number in the bulk soil was markedly less than in the rhizosphere and had distinctly detrimental correlations with total extractable Pb concentrations. At the various study sites, *Firmicutes, Gammaproteobacteria*, and *Alphaproteobacteria* were the most prevalent bacteria. The bacterial diversity index of species richness (S) and Margalef (dMa) were both clearly higher in the rhizosphere soil than in the bulk soil, even though no differences were found with Simpson's index (D) between the bulk and rhizosphere soil. The results of the redundancy analysis revealed that soil pH, electrical conductivity, soil organic matter, and total/extractable Pb contents were the main factors influencing the relative abundance, composition, and diversity of bacteria. The native forest replacement with exotic species can alter microbial-related soil ecosystem services. This was found by Almonacid-Muñozet al. [81] in a Chilean native forest. In particular, the rhizosphere bacterial and fungal communities of the native deciduous tree *Nothofagus obliqua* (a pioneer species that is dominant in structure and composition after disturbance), grown within an anexotic *Pinus radiata* forest of a plantation growing nearby, were analysed. β-diversity analysis indicated that soil type (rhizosphere or bulk soil) and location type (native forest or *P. radiata* plantation) were responsible for most of the change between the bacterial and fungal communities. *Proteobacteria* and *Basidiomycota* were the dominant bacterial and fungal phyla in both soil types and sites. Bacteria displayed comparable and abundant taxa at the family level, regardless of soil type or site. The most abundant fungal taxa related to native forests were *Tricholomataceae* and *Cantharellales*, while *Russulaceae* and *Hyaloscyphaceae* were the dominant families in *P. radiata*. The key bacterial functional groups were chemoheterotrophic and aerobic chemoheterotrophic, with no significant variation among soil types or sites. Furthermore, the authors showed that the composition and diversity of bacterial and fungal communities typical of the *N. obliqua* native forest are affected by the adjacent forest and primarily reflect site properties, such as the available source of lignin-rich wood. In this study, the different leaf litter and root exudates between *Nothofagus oblique* and *Pinus radiate* probably were key factors that modified chemical and physical parameters of soil, influencing the soil microbial community.

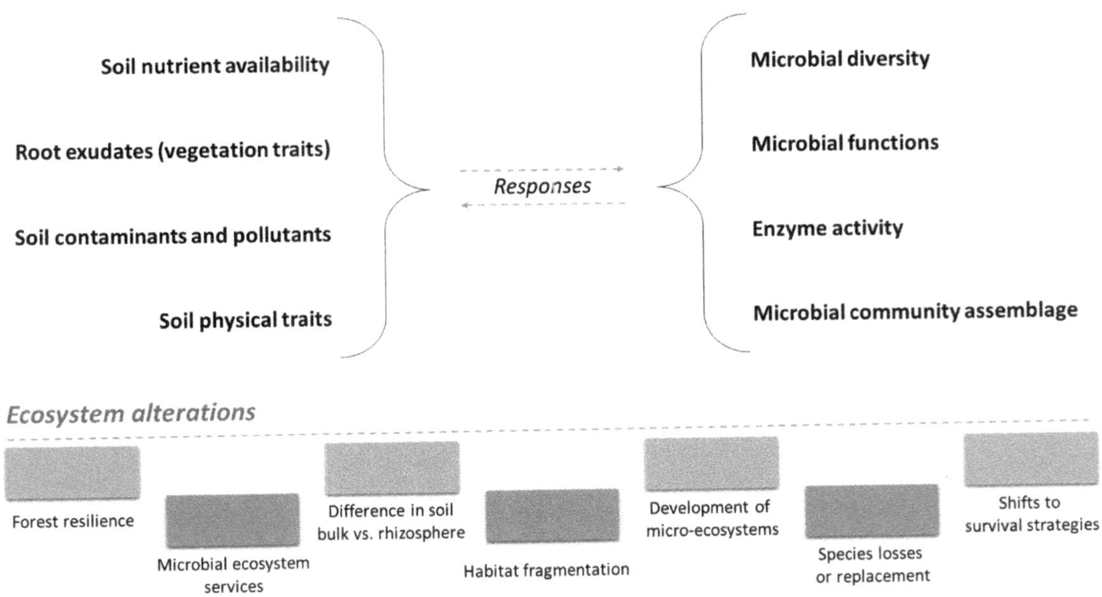

Figure 3. Effects of ecosystem alterations (due to the combined influence of anthropogenic disturbances and soil–plant conditions) on the various soil factors (on the left) and, consequently, on forest microbiome (on the right).

Forest disturbance is recognized for its enduring impact on biogeochemical cycling, particularly on microbially mediated processes such as nitrogen (N) cycling in Appalachian ecosystems. Osburn et al. [20] used sequencing of 16S and ITS to investigate bacterial (16S) and fungal (ITS) soil communities in forested watersheds with varying disturbance histories and in neighbouring reference forests at the Coweeta Hydrologic Laboratory in the Appalachian Mountains of North Carolina. The study revealed that forests that had been altered previously exhibited significant differences in the composition, including increased alpha diversity and abundance of copiotrophic bacterial phyla (e.g., *Proteobacteria*) and N-cyclophytes (e.g., *Nitrospirae*). The fungal community structure also revealed detrimental effects, particularly in mycorrhizal taxa.

Nevertheless, fungal alpha diversity was not impacted by disturbance, and disturbance effects were inconsistent at the fungal class level. Co-occurrence networks established for bacteria and fungi displayed that disturbed communities had better-linked and closely clustered networks, suggesting that disturbance impacts not only community structure but similarly potential ecological interactions between taxa. While bacteria and fungi showed various long-term reactions to forest disturbance, the study shows strong responses of relevant bacterial and fungal functional groups (e.g., nitrifying bacteria and mycorrhizal fungi), suggesting both microbial groups have a central role within the long-term modifications of observed biogeochemical pathways resulting from forest disturbance in the region.

Soil fungal communities in forest, especially mycorrhizal symbionts, are greatly affected by anthropogenic disturbances, which often lead to a habitat fragmentation and the development of micro-ecosystems with different environmental conditions, symbiotic functionalities, and health status [83]. The magnitude of anthropogenic pressure is visible in the plant mycorrhizal status, the survival of native species in the competition with invasive (alien) ones, the maintenance of a homogenous native habitat and niche, or the conversion to an artificial heterogeneous one [84]. Fragmentation of forest ecosystems changes the assemblage of mycorrhizal communities, reducing the dimension of fungal networks developed between many different hosts to a simpler and less diverse network between fewer hosts [85]. Along with changes in the mycorrhizal community due to anthropogenic pressure, a shift in the mycorrhizal type and dominance is visible in the rhizosphere of trees, an increase in the incidence of pathogens, and an overall forest decline [86–89]. The reduction in mycorrhizal diversity due to anthropogenic pollution leads to a decrease in their potential functionality, with the magnitude of the effect caused by single or combined pollutants [90,91]. Currently, the levels of nitrogen and phosphorus, as well as metal pollution, greatly alter the diversity of mycorrhizal communities emerging to a critical load that drastically reduces both fungal species and their hosts [92–94]. Human activities in forests alter the resilience of these ecosystems, forcing host species to access mycorrhizal symbionts to a greater level in the roots, changing their specific nutritional function to a larger number of functions, for both survival and protection from abiotic and biotic pressure [95,96].

Overall, the relationship between soil microbial communities and the plant-associated microbiome in the context of forest anthropogenic disturbance is intricate and multifaceted. Research in this area continues to provide insights into how human activities can impact these ecosystems and how to manage and restore them for long-term ecological health and sustainability.

3.4. Influence of Harvesting on the Diversity and Structure of Soil Bacterial and Fungal Communities

Several studies have shown that all forest harvesting systems have an impact on soils, with significant short-term and long-term effects on microbial communities, particularly in response to organic matter removal and soil compaction, which are considered major disturbances (Figure 4). Different levels of disturbance can result in distinct responses, with some effects persisting for several years after tree harvesting, as demonstrated by Hartman et al. [46]. The most sensitive microbial groups in forest systems are the symbiotic and saprobe species, which can potentially serve as indicators for monitoring recovery. It is necessary to conduct long-term monitoring over the course of a forest stand rotation to assess the resilience of the microbial community structure over time. Although the functional diversity of both bacteria and fungi is high, fungi have demonstrated a much stronger response to harvesting than bacteria.

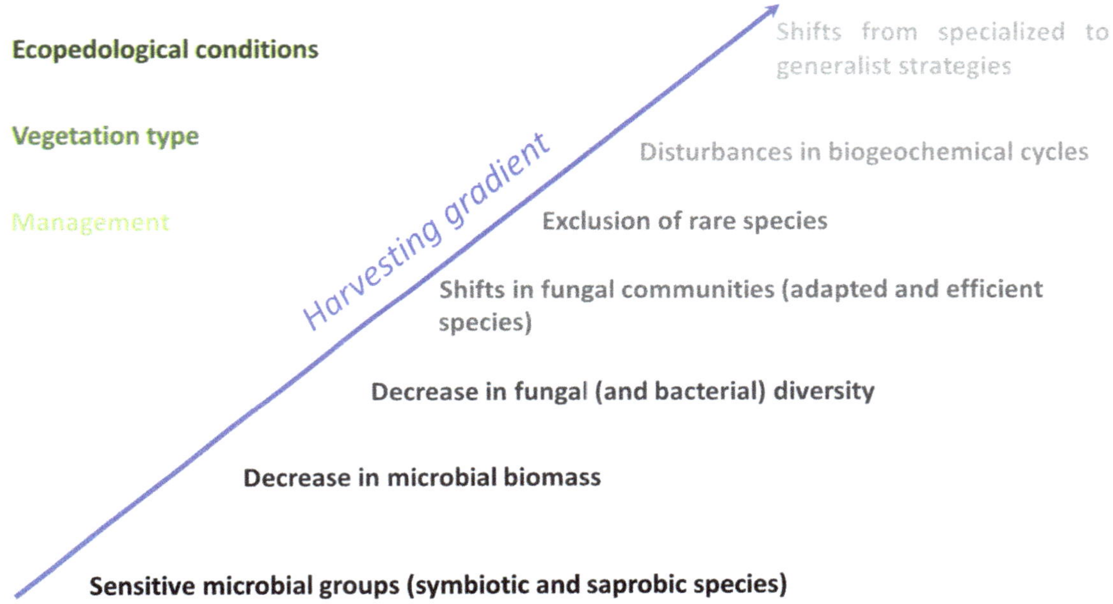

Figure 4. The impact of harvesting gradient on forest microbial communities (on the right) and on the various factors controlling forest changes (on the left). The higher color intensity indicates the greater impact.

According to Chatterjee et al. [97] the response of forest species to disturbances such as timber harvesting depends on various factors, including the magnitude of the disturbance, soil type, and environmental conditions. Additionally, the biomass of soil microbial communities can decrease after timber harvesting, as found by Smith et al. [98].

The significance of fungal and bacterial communities in forest ecosystems, as highlighted by Dighton et al. [99], means that the removal of overstory trees has a direct impact on these communities. This impact is dependent on carbon inputs, as indicated by several authors, and can also affect the structure of litter inputs through saprotrophic fungi and *Actinomycete* bacteria [100–104]. However, some studies have contradictory results, suggesting that microbial communities may not respond to perturbations in the same way as trees do [105]. For instance, some studies indicate that microbial biomass C and N content and respiration may not be affected by whole tree harvest, soil compaction [106], or forest floor removal [107]. Meanwhile, other studies have shown that these properties may increase or decrease [108].

Forest ecosystem structure and function are influenced directly or indirectly by stand density. Wang et al. [109] studied the influence of Chinese fir plantation density on the functional diversity of the soil microbial community and concluded that density changes influence the soil physicochemical characteristics, composition, and metabolic functional diversity of microbial communities in Chinese fir plantations. Soil microbial functional diversity was investigated by Biolog ECO technology. Soil pH, content of oxidizable organic carbon, available N, available P, and available K were found to be affected by stand density. Chinese fir plantation density had a considerable influence on the utilization of carbohydrates, amino acids, carboxylic acids, and phenolic acids by the soil microbial community, but did not have a significant effect on the utilization of polymers. Principal component analysis (PCA) showed carbohydrates, polymers, and phenolic acids were susceptible carbon sources, which determined changes in the metabolic functions of soil microbial communities in the studied forest.

Harvesting is a management process that produces, in both the short and long term, a decrease in diversity in fungal communities, along with a reduction in host natural life cycle [110]. Observed shifts in fungal communities are associated with the quantity of mineralizable nitrogen, organic carbon, and lead to an increase in adapted and efficient species, while producing a decrease in the mycelium biomass in soil and in less-represented (rare) species in the community which are excluded from the new formed ecological niche [111–113]. In this context, the decomposition process of deadwood and remains suffers a reduction in efficiency, due to the narrowing of fungal diversity [114]. Both management and vegetation type act coupled on soil fungi and their biological processes, with the changes produced in the litter quantity, and its quality, visible in the restriction of nutrient cycling [115,116]. The long-term effect of harvesting on fungal communities causes a rapid reduction in species diversity, with the disappearance of multiple species, followed by a slow re-establishment of fungi to occupy the empty niches [117,118]. This process ends after 30–50 years, and the re-established fungal community presents a different species assemblage and composition compared to the native one. The shift in fungal processes is related to the harvesting intensity, from a high-diversity fungal community with specialized strategies in nutrient acquisition, to a lower-diversity community with generalist strategies.

Understanding the influence of harvesting on soil bacterial and fungal communities is vital for sustainable land and forest management. Properly managed harvesting and land-use practices can help minimize disruption to these microbial communities, ensuring healthy and productive ecosystems in the long term. Research in this area continues to provide insights into the ways in which these impacts can be mitigated and managed.

3.5. *Patterns of Soil Microbial Community Succession During the Restoration of Forest Ecosystems*

The assessment of soil microbial communities is becoming progressively important for evaluating the sustainability of forest ecosystems and their response to stress and disturbances. However, scant information exists regarding the discernible patterns in microbial community structure and composition during secondary succession or ecosystem restoration. The translation of forests to managed states, such as agriculture or timber plantations, has contributed to the modification of 75% of ice-free terrestrial ecosystems globally [119]. While approximately 50% of the earth's land surface was afforested in the prehistoric era, about 40% of that forest cover has been disappeared, and a large part of the surviving forest has been subjected to different types of disturbance, particularly in the past two centuries [120]. Forest conversion still accelerating in the 21st century [121], highlighting the importance of characterising the effects of forest disturbance on both terrestrial biodiversity and ecosystem functions. Such comprehension is essential biogeochemically, as forest ecosystems play a pivotal role in the elemental cycles and deliver valuable ecosystem services, such as carbon storage, nutrient cycle regulation, and the provision of clean drinking water [122]. Soil microorganisms are key drivers of these biogeochemical processes and associated ecosystem services, performing essential functions like litter decomposition and carbon and nitrogen cycling, highlighting an urgency to investigate the effects of forest perturbation on soil microbial communities [123].

The successions observed within forest ecosystems and the mechanisms that produce them represent an interconnected phenomenon. Disturbances produced especially by anthropic activities are visible in the occurrence of different successional patterns, which act as restoration stages. The changes in terrestrial microorganisms, plant biomass, species composition, and soil physicochemical properties are effects of the forest disturbances. Moreover, soil C and N stocks, which are known as drivers of microbial community structure, are affected by the forest disturbance processes [123], and the magnitude of change is proportional with the resilience of these communities. Several studies focus on the influence on soil microbial communities of timber harvest (e.g. Kohout et al. [124]), and prescribed fire (Shen et al. [125]), which can all modify bacterial and/or fungal diversity. Zhou et al. [126] found a connection between bacterial community variations in previously disturbed forests, as well as increased relative abundance of r-selected bacterial phyla (e.g., *Proteobacteria*) with previous disturbance. Thus, the selection within the native microflora assures a microbial pool that can persist in the restoration process. Moreover, changes in forest soil microbial communities are likely to occur due to disturbances. No studies fully investigate the long-term activity of bacterial communities, as well as fungi, as a result of past disturbances such as timber harvesting, agricultural conversion and timber plantation conversion [127]. Responses to perturbations of soil microbial communities have been found to be relevant in the forests of the Appalachian region (USA), where about 70% of the territory is covered by forests and almost all forest ecosystems in the region have suffered from past disturbances due to human activities, which includes the logging of commercial timber and/or switching to agriculture [128,129]. In this region, Keiser et al. [130] assessed long-term effects of disturbance on soil microbial N-cycle functions, finding elevated nitrification rates in previously disturbed forests. These results have been accentuated by Lin et al. [131], who found an elevated abundance of nitrifying microorganisms.

Hyphal networks developed in soil by mycorrhizas plays an important role in the success of forest restoration processes [132]. These networks act as a biological transfer system for nutrients, from areas apart from plant root activity up to their host root system. The restoration period is linked with the gradual increase in diversity of mycorrhizal species and the establishment of mycorrhizal links between different tree species [133]. This biological mechanism relies on the mycorrhizas' capacity to form symbiosis with multiple hosts, but induces a series of negative interactions between native and inoculated mycoflora [134]. Multiple successional scenarios arise from these interactions: (a) Native mycorrhizal species act rapidly to colonize the new hots, completely eliminating the non-native (or inoculated) species; (b) Part of the non-native (or inoculated) species persist in the restored community, followed by a stratified successional process until a climax based on a dual origin of species in the community; (c) The non-native (or inoculated) mycoflora restrict the development of the native one and its access to plant hosts, which leads to a complete change in the long-term assemblage of the fungal community. The direction toward one successional scenario, or a combination of scenarios, is due to the mycorrhizal ability to adapt the harsh environmental conditions during the first stages of restoration [135–137]. Restoration is a process that targets forest function recovery, with the number of targeted functions positively correlated to the number of species used [138]. In this process, the feedback established between plants and their mycorrhizal partners leads to host's survival, increased growth and development, and achievement of dominance in community [139]. The inoculum (both native and/or non-native mycorrhizas) used in restoration is responsible for the success of this process, with different efficiency levels between species [140].

The influence of harvesting, particularly in the context of forestry, on the diversity and structure of soil bacterial and fungal communities has a considerable ecological importance. The impact of harvesting on these microbial communities can have both short-term and long-term effects. Understanding the patterns of soil microbial community succession during forest ecosystem restoration is crucial for planning and executing successful restoration projects. By promoting the recovery of microbial diversity and function, restoration efforts can help rebuild resilient and healthy ecosystems with improved ecological services and long-term sustainability.

3.6. Impact of Global Change on Bacterial Communities Within Forest Ecosystems

Covering over 40 million km^2 and accounting for 30% of the global land area, forests are among the most extensive and crucial ecosystems on earth [141]. Forests are present in most of the earth's biomes and host a significant proportion of global diversity. The Northern Hemisphere's temperate and boreal forests, for instance, occupy a significant portion of its land surface and house 46% of all trees on the planet—0.66 and 0.74 trillion, respectively [142]. Global change, which includes both climate change and other human-induced changes, has been identified as a primary threat to forests [143]. Human overexploitation has been the most relevant risk to forests over millennia, and the C balance of forests is mostly affected by human activities, comprising deforestation, productive forest management, reforestation, afforestation, and others [144]. Other perturbations related to climate change and atmospheric factors also impact the carbon balance of forests. As such, the impacts of temperature rise, drought persistence, fire recurrence, or the increase in native and invasive forest pests, combined with higher nitrogen and carbon dioxide loads, influence deeper nutrient cycles to such a degree that forests have the potential to become net sources rather than sinks of CO_2 [145]. Consequently, a clear comprehension of the role of forests in C fluxes, which are strongly influenced by the bacterial and fungal activity, has been indicated as an essential requirement for making future predictions about the health of our planet [49]. If plants are primarily responsible for absorbing carbon from the atmosphere in forests, the microorganisms in forests play a major role in the carbon balance in these ecosystems.

From a global perspective, it is crucial to comprehend the diversity of fungi and bacteria within coniferous forest biomes. Fungi are dominant in decomposing litter material, whereas bacteria become increasingly important as soil depth increases [146]. Coniferous forests play a significant role in the global carbon cycle, and understanding the microbial processes in the soil is essential for predicting global carbon fluxes and their potential changes in the future.

Global change, which includes factors such as climate change, increased atmospheric carbon dioxide (CO_2) levels, altered precipitation patterns, and human activities, can have a significant impact on bacterial communities within forest ecosystems. These changes can disrupt the composition, diversity, and functioning of soil and litter bacterial communities, leading to various ecological consequences. Understanding the impact of global change on bacterial communities within forest ecosystems is critical for predicting the responses of these ecosystems to ongoing environmental changes. Scientists continue to study these interactions to better inform forest management and conservation efforts, including strategies for mitigating the negative effects of global change on forest bacterial communities and their associated ecosystem functions.

Different studies show that climate change causes significant changes in the microbial populations, although not significantly to cause a change in vegetation, so it is possible to develop a microbial indicator to assess the disturbance of accumulated pressure on the forest ecosystem. The unpredicted vegetation change can be a serious problem caused by climate change, and the soil microbial community should imitate the pressure on forest ecosystems due to climate change [147,148]. Classen et al. [149] highlighted also that the global climate change is altering species distributions, and thus interactions among organisms. J Jansson and Hofmockel [150] studied current findings about the climate change impact on the microbiome of soil and the possible ways in which soil microorganisms can be utilized to mitigate the negative consequences of climate change.

Although it is known that climate changes infer changes in soil microbial community composition, species abundance, diversity, survival and resilience, changes in enzyme production, and changes in interactions between microbes and plant roots, etc. Singhal et al. [43] found that there is still limited knowledge about the potential impacts of climate change-induced disturbances on soil microbial communities. Limitations in understanding these changes are caused by the bi-directional nature of climate change effects on microbial communities and the continuous modification of both microbial communities and their associated biogeochemical cycles. This knowledge gap is significant, as the stability of microbial communities, which refers to their ability to withstand and recover from disturbances, is likely to affect ecosystem function. Given the essential roles of forest ecosystems in climate stability, biodiversity, and economic development, microbial community structure can serve as an indicator of forest ecosystem status. Changes in microbial community structure can reveal alterations in nutrient and energy flow patterns before they cause irreversible effects on long-term soil productivity. Therefore, understanding the dynamics of soil microbial communities is vital for effective conservation and management of forest ecosystems [8]. This context imposes a continuous monitoring of both climate and microbial changes, a continuous update of forecasting models, and improvement of indicator sensitivity.

Even more than a decade following tree harvesting, the diversity and structure of soil bacterial and fungal communities remained significantly altered by the disturbances caused by harvesting [151–155]. Moreover, individual taxonomic groups exhibited different responses to varying levels of disturbances. Some taxonomic groups, such as plant symbionts like ectomycorrhizal fungi, and saprobic taxa like *Ascomycetes* and *Actinomycetes*, were highly sensitive to harvesting disturbances. As these taxa play crucial ecological roles in forest development, their fate is essential for the sustainability of forest ecosystems. While abundant bacterial populations were found throughout the study area, abundant fungal populations often exhibited patchy distributions, which is consistent with their higher sensitivity to the soil disturbances examined.

Extensive research has been conducted on the impact of climate change on soil biota abundance, diversity, and activity. However, the literature reports inconsistent effects that are highly dependent on environmental context. Elevated atmospheric CO_2 has been shown to increase microbial biomass, fungal abundance, and the abundance and body size of most faunal groups, resulting in changes in the structure of the soil food web. Bacterial diversity, on the other hand, is minimally affected by elevated CO_2. Warming, however, increases the abundance of bacteria and fungi, as well as most soil faunal groups, resulting in changes in the soil food web structure [156,157]. Drought conditions reduce the biomass and abundance of the majority of soil microbial and faunal groups, but increase the abundance of fungi compared to bacteria. However, the impacts of increased rainfall and flooding on soil biota are still poorly understood and highly context-dependent. Changes in soil organism abundance and soil food web structure caused by climate change will have impacts on soil functioning and the provision of soil-based ecosystem services [158]. As such, the relative importance of these changes in relation to the direct impacts of climate change on soil functioning is not clear, and the relationships between soil biota and soil functioning are context-dependent. Further research is therefore needed to establish the effect of climate change-induced changes in soil biota on the suppression of diseases, and also the effect of invasive microbes on soil functioning [39,159]. The effects of global change on soil food webs and their functions is a growing concern. Soil communities address several global change factors, which include land-use change, contamination, habitat fragmentation and modification, invasive species, soil sealing, and climate change [160–162]. Although significant research has been conducted on the impacts of these drivers on soil food webs, most studies have concentrated on aboveground organisms.

Climate change is causing increased temperatures and extreme events, such as droughts, heat waves, and excessive rainfall, on a global scale, and these changes are already affecting the functioning of terrestrial ecosystems.

4. Conclusions

Studying the factors that drive forest soil microbiome is a complex and dynamic field. Forest soil microbiomes are extremely diverse and complex. The multiplicity of microbial species and their interactions can make it challenging to identify the specific factors that drive their composition and functions.

This review highlights that the forest soil microbiome is influenced by various environmental, ecological, and biotic factors. The main key drivers of the soil microbiome in forest ecosystems, addressed in the various sections, are soil composition and nutrient availability, plant composition, soil microbial interactions, disturbances, succesion, and temporal dynamics. Overall, these factors interact in complex ways, shaping the diversity, function, and resilience of the soil microbiome in forest ecosystems.

Microbial communities in forest soils can vary significantly over small spatial scales due to factors such as tree distribution, topography, and microclimate. This spatial variation can further hamper efforts to generalize results to larger forest ecosystems. Microbial communities can change over time due to seasonal variations, disturbances (e.g., forest fires or logging), and long-term environmental shifts (e.g., climate change). Capturing these dynamics requires long-term and comprehensive monitoring, which can be resource-intensive.

To understand the forest soil microbiome comprehensively, combinations of field studies and advanced techniques are often used. The key approaches and techniques for measuring the changes of the forest soil microbiome due to various factors are soil sampling, metagenomic sequencing, quantitative PCR, enzyme activity assays, Biolog plates, respiration rate measurements, microbial diversity indices, and long-term monitoring. Studying these factors and their interplay, insights into how the forest soil microbiome responds to environmental changes and contributes to the overall health and functioning of forest ecosystems can be provided. While advances in high-throughput sequencing technologies have enabled more in-depth studies of microbial communities, these methods not lack limitations. They may not capture all microbial taxa, and issues such as contamination and PCR bias can affect the data quality. While DNA sequencing can provide information about microbial composition, it does not directly reveal the functional roles of these microbes. Moreover, it is fundamental to know the microbial abundance related to different abiotic and biotic factors, which must be related to the soil microbial diversity. Determining how specific microbial taxa contribute to ecosystem functions is a challenging and complex issue. Microbial communities in forest soils do not exist alone. They interact with other microorganisms and are influenced by factors like nutrient availability, soil chemistry, and plant–microbe interactions. Understanding these complex interactions is a crucial challenge. Studying the factors that drive the forest soil microbiome is fundamental for understanding ecosystem health, nutrient cycling, and resilience when faced with environmental changes. Researchers continue to develop innovative techniques, collaborate across disciplines, and employ long-term studies to address these challenges and gain deeper insights into the complex relationships within forest ecosystems. A combination of molecular, chemical, and ecological tools is necessary to comprehensively measure the changes in the forest soil microbiome. These approaches allow for the identification of shifts in microbial communities in response to various environmental factors, such as soil properties, vegetation, disturbances, and human activities. By integrating these methods, we can better understand the dynamics of forest soil microbiomes and their role in ecosystem processes.

Author Contributions: Conceptualization, P.G., C.O., V.S., V.C. and A.O.; methodology, C.O. and A.O.; data curation, P.G., C.O., V.S., V.C. and A.O.; writing—original draft preparation, P.G., C.O., V.S., V.C. and A.O.; writing—review and editing, P.G., C.O., V.S., V.C. and A.O.; supervision, P.G. All authors have read and agreed to the published version of the manuscript.

Funding: This research was partially funded by the Ministry University and Research, "Next Generation EU–Piano Nazionale Resistenza e Resilienza (PNRR)—Missione 4 Componente 2 Investimento 1.4—Notice No. 3138 16 December 2021 rectified by D.D. n. 3175 18 December 2021". Award Number: CN_00000033, Decree MUR n. 1034 17 June 2022, CUP B83C22002930006, Project title "National Biodiversity Future Center—NBFC, *Microbial diversity and functioning of freshwater and terrestrial ecosystems*".

Data Availability Statement: Data can be requested from the authors.

Conflicts of Interest: The authors declare no conflicts of interest.

References

1. Canadell, J.G.; Raupach, M.R. Managing forests for climate change mitigation. *Science* **2008**, *320*, 1456–1457. [CrossRef] [PubMed]
2. Miles, L.; Kapos, V. Reducing greenhouse gas emissions from deforestation and forest degradation: Global land-use implications. *Science* **2008**, *320*, 1454–1455. [CrossRef] [PubMed]
3. Watson, J.E.M.; Evans, T.; Venter, O.; Williams, B.; Tulloch, A.; Stewart, I.; Thompson, I.; Ray, J.C.; Murray, K.; Salazar, A.; et al. The exceptional value of intact forest ecosystems. *Nat. Ecol. Evol.* **2018**, *2*, 599–610. [CrossRef]
4. Grantham, H.S.; Duncan, A.; Evans, T.D.; Jones, K.R.; Beter, H.L.; Schuster, R.; Walston, J.; Ray, J.C.; Robinson, J.G.; Callow, M.; et al. Anthropogenic modification of forests means only 40% of remaining forests have high ecosystem integrity. *Nat. Commun.* **2020**, *11*, 5978. [CrossRef] [PubMed]
5. Baldrian, P.; López-Mondéjar, R.; Kohout, P. Forest microbiome and global change. *Nat. Rev. Microbiol.* **2023**, *21*, 487–501. [CrossRef] [PubMed]
6. Laudon, H.; Sponseller, R.A.; Lucas, R.W.; Futter, M.N.; Egnell, G.; Bishop, K.; Ågren, A.; Ring, E.; Högberg, P. Consequences of more intensive forestry for the sustainable management of forest soils and waters. *Forests* **2011**, *2*, 243–260. [CrossRef]
7. Titus, B.D.; Brown, K.; Helmisaari, H.S.; Vanguelova, E.; Stupak, I.; Evans, A.; Clarke, N.; Guidi, C.; Bruckman, V.J.; Varnagiryte-Kabasinskiene, I.; et al. Sustainable forest biomass: A review of current residue harvesting guidelines. *Energ. Sustain. Soc.* **2021**, *11*, 10. [CrossRef]
8. Cabanás, C.G.-L.; Legarda, G.; Ruano-Rosa, D.; Pizarro-Tobías, P.; Valverde-Corredor, A.; Niqui, J.L.; Triviño, J.C.; Roca, A.; Mercado-Blanco, J. Belowground microbiota and the health of tree crops. *Front. Microbiol.* **2018**, *9*, 1006. [CrossRef]
9. Baldrian, P. Forest microbiome: Diversity, complexity and dynamics. *FEMS Microbiol. Rev.* **2017**, *41*, 109–130. [CrossRef]
10. Faure, D.; Simon, J.-C.; Heulin, T. Holobiont: A conceptual framework to explore the eco-evolutionary and functional implications of host–microbiota interactions in all ecosystems. *New Phytol.* **2018**, *218*, 1321–1324. [CrossRef] [PubMed]
11. Haichar, F.Z.; Heulin, T.; Guyonnet, J.; Achouak, W. Stable isotope probing of carbon flow in plant-holobionte. *Curr. Opinion Biotechnol.* **2016**, *41*, 9–13. [CrossRef]
12. Puga-Freitas, R.; Barot, S.; Taconnat, L.; Renou, J.P.; Blouin, M. Signal molecules mediate the impact of the earthworm *Aporrectodea caliginosa* on growth, development and defence of the plant *Arabidopsis thaliana*. *PLoS ONE* **2015**, *7*, e49504. [CrossRef]
13. Gillespie, L.M.; Fromin, N.; Milcu, A.; Buatois, B.; Pontoizeau, C.; Hättenschwiler, S. Higher tree diversity increases soil microbial resistance to drought. *Commun. Biol.* **2020**, *3*, 377. [CrossRef] [PubMed]
14. Philippot, L.; Griffiths, B.S.; Langenheder, S. Microbial community resilience across ecosystems and multiple disturbances. *Microbiol. Mol. Biol. Rev.* **2021**, *85*, e00026-20. [CrossRef]
15. Dumonceaux, T. Composition and dynamics of plant- and soil-associated microbial communities in forest and agricultural ecosystems. *Microorganisms* **2023**, *11*, 1782. [CrossRef] [PubMed]
16. Osburn, E.D.; Aylward, F.O.; Barret, J.E. Historical land use has long-term effects on microbial community assembly processes in forest soils. *ISME Comm.* **2021**, *1*, 48. [CrossRef] [PubMed]

17. You, Y.; Wang, J.; Huang, X.; Tang, Z.; Liu, S.; Sun, O.J. Relating microbial community structure to functioning in forest soil organic carbon transformation and turnover. *Ecol. Evol.* **2014**, *4*, 633–647. [CrossRef] [PubMed]
18. Heo, Y.-M.; Lee, H.; Kwon, S.-L.; Yoo, Y.; Kim, D.; Han, S.-I.; Lee, A.-H.; Kim, C.; Kim, G.-H.; Kim, J.-J. Influence of tree vegetation on soil microbial communities in temperate forests and their potential as a proactive indicator of vegetation shift due to climate change. *Sustainability* **2020**, *12*, 10591. [CrossRef]
19. Gougoulias, C.; Clark, J.M.; Shaw, L.J. The role of soil microbes in the global carbon cycle: Tracking the below-ground microbial processing of plant-derived carbon for manipulating carbon dynamics in agricultural systems. *J. Sci. Food Agric.* **2014**, *94*, 2362–2371. [CrossRef]
20. Osburn, E.D.; McBride, S.G.; Aylward, F.O.; Badgley, B.D.; Strahm, B.D.; Knoepp, J.D.; Barrett, J.E. Soil bacterial and fungal communities exhibit distinct long-term responses to disturbance in temperate forests. *Front. Microbiol.* **2019**, *10*, 2872. [CrossRef] [PubMed]
21. Xin, J.; Yajun, L.; Wanjin, H.; Guobing, W.; Zhaoyu, K.; Lan, W.; Gang, G. Soil bacterial and fungal communities and the associated nutrient cycling responses to forest conversion after selective logging in a subtropical forest of China. *Forest. Ecol. Manag.* **2019**, *444*, 308–317. [CrossRef]
22. Maltz, M.R.; Treseder, K.K. Sources of inocula influence mycorrhizal colonization of plants in restoration projects: A meta-analysis. *Restor. Ecol.* **2015**, *23*, 625–634. [CrossRef]
23. Rodrigo, A.G. Modelling the evolution of holobionts: An incomplete review. *New Zealand J. Zool.* **2023**, *52*, 87–102. [CrossRef]
24. Youle, M.; Knowlton, N.; Rohwer, F.; Gordon, J.; Relman, D. Superorganisms and holobionts. *Microbe Mag.* **2013**, *8*, 152–153. [CrossRef]
25. Singh, B.K.; Liu, H.; Trivedi, P. Eco-holobiont: A new concept to identify drivers of host-associated microorganisms. *Environ. Microbiol.* **2020**, *22*, 564–567. [CrossRef]
26. Mishra, S.; Hättenschwiler, S.; Yang, X. The Plant microbiome: A missing link for the understanding of community dynamics and multifunctionality in forest ecosystems. *Appl. Soil. Ecol.* **2020**, *145*, 103345. [CrossRef]
27. Dessaux, Y.; Grandclément, C.; Faure, D. Engineering the rhizosphere. *Trends Plant Sci.* **2016**, *21*, 266–278. [CrossRef]
28. Rosenberg, E.; Zilber-Rosenberg, I. Microbes drive evolution of animals and plants: The hologenome concept. *mBio* **2016**, *7*, e01395-15. [CrossRef] [PubMed]
29. Suárez, J.; Triviño, V. What is a hologenomic adaptation? Emergent individuality and inter-identity in multispecies systems. *Front. Psychol.* **2020**, *11*, 187. [CrossRef] [PubMed]
30. Singh, Y.; Ahmad, J.; Musarrat, J.; Ehtesham, N.Z.; Hasnain, S.E. Emerging importance of holobionts in evolution and in probiotics. *Gut Pathog.* **2013**, *5*, 12. [CrossRef]
31. Schneider, T. The holobiont self: Understanding immunity in context. *Hist. Philos. Life Sci.* **2021**, *43*, 99. [CrossRef] [PubMed]
32. Mesny, F.; Hacquard, S.; Thomma, B.P. Co-evolution within the Plant Holobiont Drives Host Performance. *EMBO Rep.* **2023**, *24*, e57455. [CrossRef]
33. Vandenkoornhuyse, P.; Quaiser, A.; Duhamel, M.; Le Van, A.; Dufresne, A. The importance of the microbiome of the plant holobiont. *New Phytol.* **2015**, *206*, 1196–1206. [CrossRef] [PubMed]
34. Bordenstein, S.R.; Theis, K.R. Host biology in light of the microbiome: Ten principles of holobionts and hologenomes. *PLoS Biol.* **2015**, *13*, e1002226. [CrossRef]
35. Simon, J.-C.; Marchesi, J.R.; Mougel, C.; Selosse, M.-A. Host-microbiota interactions: From holobiont theory to analysis. *Microbiome* **2019**, *7*, 5. [CrossRef] [PubMed]
36. Peixoto, R.S.; Voolstra, C.R.; Sweet, M.; Duarte, C.M.; Carvalho, S.; Villela, H.; Lunshof, J.E.; Gram, L.; Woodhams, D.C.; Walter, J.; et al. Harnessing the microbiome to prevent global biodiversity loss. *Nat. Microbiol.* **2022**, *7*, 1726–1735. [CrossRef] [PubMed]
37. Berg, G.; Dorador, C.; Egamberdieva, D.; Kostka, J.E.; Ryu, C.-M.; Wassermann, B. Shared governance in the plant holobiont and implications for one health. *FEMS Microbiol. Ecol.* **2024**, *100*, fiae004. [CrossRef] [PubMed]
38. Huitzil, S.; Huepe, C.; Aldana, M.; Frank, A. The missing link: How the holobiont concept provides a genetic framework for rapid evolution and the inheritance of acquired characteristics. *Front. Ecol. Evol.* **2023**, *11*, 1279938. [CrossRef]
39. Lladó, S.; López-Mondéjar, R.; Baldrian, P. Drivers of microbial community structure in forest soils. *Appl. Microbiol. Biotechnol.* **2018**, *102*, 4331–4338. [CrossRef] [PubMed]
40. Lladó, S.; López-Mondéjar, R.; Baldrian, P. Forest soil bacteria: Diversity, involvement in ecosystem processes, and response to global change. *Microbiol. Mol. Biol. Rev.* **2017**, *81*, e00063-16. [CrossRef] [PubMed]
41. Fengqiao, L.; Haiyun, Z.; Christian, S.; Xiaogang, L. Microbiome sustains forest ecosystem functions across hierarchical scales. *Eco-Environ. Health* **2023**, *2*, 24–31. [CrossRef]
42. Naylor, D.; McClure, R.; Jansson, J. Trends in microbial community composition and function by soil depth. *Microorganisms* **2022**, *10*, 540. [CrossRef] [PubMed]

43. Singhal, A.; Pandey, S.; Kumari, N.; Chauhan, D.K.; Jha, P.K. Impact of climate change on soil microbes involved in biogeochemical cycling. In *Climate Change and the Microbiome*; Choudhary, D.K., Mishra, A., Varma, A., Eds.; Springer: Cham, Switzerland, 2021; pp. 63–94. [CrossRef]
44. Crowther, T.W.; Van Den Hoogen, J.; Wan, J.; Mayes, M.A.; Keiser, A.D.; Mo, L.; Averill, C.; Maynard, D.S. The global soil community and its influence on biogeochemistry. *Science* **2019**, *365*, eaav0550. [CrossRef]
45. Fatichi, S.; Manzoni, S.; Or, D.; Paschalis, A. A mechanistic model of microbially mediated soil biogeochemical processes: A reality check. *Glob. Biogeochem. Cycles* **2019**, *33*, 620–648. [CrossRef]
46. Hartmann, M.; Howes, C.; VanInsberghe, D.; Yu, H.; Bachar, D.; Christen, R.; Nilsson, R.H.; Hallam, S.J.; Mohn, W.W. Significant and persistent impact of timber harvesting on soil microbial communities in Northern coniferous forests. *ISME J.* **2012**, *6*, 2199–2218. [CrossRef]
47. López-Mondéjar, R.; Brabcová, V.; Štursová, M.; Davidová, A.; Jansa, J.; Cajthaml, T.; Baldrian, P. Decomposer food web in a deciduous forest shows high share of generalist microorganisms and importance of microbial biomass recycling. *ISME J.* **2018**, *12*, 1768–1778. [CrossRef]
48. Abdul Rahman, N.S.N.; Abdul Hamid, N.W.; Nadarajah, K. Effects of abiotic stress on soil microbiome. *Int. J. Mol. Sci.* **2021**, *22*, 9036. [CrossRef] [PubMed]
49. De Angelis, K.M.; Pold, G.; Topcuoglu, B.D.; van Diepen, L.T.A.; Varney, R.M.; Blanchard, J.L.; Melillo, J.; Frey, S.D. Long-term forest soil warming alters microbial communities in temperate forest soils. *Front. Microbiol.* **2015**, *6*, 104. [CrossRef]
50. Lauber, C.; Hamady, M.; Knight, R.; Fierer, N. Pyrosequencing-based assessment of soil ph as a predictor of soil bacterial community structure at the continental scale. *Appl. Environ. Microbiol.* **2009**, *75*, 5111–5120. [CrossRef]
51. Urbanová, M.; Šnajdr, J.; Baldrian, P. Composition of fungal and bacterial communities in forest litter and soil is largely determined by dominant trees. *Soil Biol. Biochem.* **2015**, *84*, 53–56. [CrossRef]
52. Prescott, C.E.; Grayston, S.J. Tree species influence on microbial communities in litter and soil: Current knowledge and research needs. *For. Ecol. Manag.* **2013**, *309*, 19–22. [CrossRef]
53. Chemidlin Prevost-Boure, N.; Maron, P.A.; Ranjard, L.; Nowak, V.; Dufrene, E.; Damesin, C.; Soudani, K.; Lata, J.C. Seasonal dynamics of the bacterial community in forest soils under different quantities of leaf litter. *Appl. Soil. Ecol.* **2011**, *47*, 14–23. [CrossRef]
54. Baldrian, P.; Merhautová, V.; Cajthaml, T.; Petránková, M.; Šnajdr, J. Small-scale distribution of extracellular enzymes, fungal, and bacterial biomass in Quercus petraea forest topsoil. *Biol. Fertil. Soils* **2010**, *46*, 717–726. [CrossRef]
55. Gillespie, L.M.; Hättenschwiler, S.; Milcu, A.; Wambsganss, J.; Shihan, A.; Fromin, N. Tree species mixing affects soil microbial functioning indirectly via root and litter traits and soil parameters in European forests. *Funct. Ecol.* **2021**, *35*, 2190–2204. [CrossRef]
56. Kim, M.; Kim, W.S.; Tripathi, B.M.; Adams, J. Distinct bacterial communities dominate tropical and temperate zone leaf litter. *Microb. Ecol.* **2014**, *67*, 837–848. [CrossRef]
57. López-Mondéjar, R.; Voříšková, J.; Větrovský, T.; Baldrian, P. The bacterial community inhabiting temperate deciduous forests is vertically stratified and undergoes seasonal dynamics. *Soil Biol. Biochem.* **2015**, *87*, 43–50. [CrossRef]
58. Purahong, W.; Schloter, M.; Pecyna, M.J.; Kapturska, D.; Daumlich, V.; Mital, S.; Buscot, F.; Hofrichter, M.; Gutknecht, J.L.; Kruger, D. Uncoupling of microbial community structure and function in decomposing litter across beech forest ecosystems in Central Europe. *Sci. Rep.* **2014**, *4*, 7014. [CrossRef] [PubMed]
59. Gessner, M.O.; Swan, C.M.; Dang, C.K.; McKie, B.G.; Bardgett, R.D.; Wall, D.H.; Hattenschwiler, S. Diversity meets decomposition. *Trends Ecol. Evol.* **2010**, *25*, 372–380. [CrossRef]
60. Xia, Z.; Bai, E.; Wang, Q.; Gao, D.; Zhou, J.; Jiang, P.; Wu, J. Biogeographic distribution patterns of bacteria in typical chinese forest soils. *Front. Microbiol.* **2016**, *7*, 1106. [CrossRef] [PubMed]
61. Kardol, P.; Cregger, M.; Campany, C.; Classen, A. Soil ecosystem functioning under climate change: Plant species and community effects. *Ecology* **2010**, *91*, 767–781. [CrossRef]
62. Franciska, T.d.V.; Robert, I.G.; Chapter, F. Impacts of Climate Change on Soil Microbial Communities and Their Functioning. In *Developments in Soil Science*; Horwath, W.R., Kuzyakov, Y., Eds.; Elsevier: Amsterdam, The Netherlands, 2018; pp. 111–129. [CrossRef]
63. Hussnain, M.; Rainer, F.W.; Adnan, M.; Andrianto, A.; Oleg, V.; Shipin, T.N.-D.C.; Yu-Pin, L. Soil microbiome feedback to climate change and options for mitigation. *Sci. Total Env.* **2023**, *882*, 163412. [CrossRef]
64. Yilun, H.; Hasbagan, G.; Guozheng, H.; Xuexia, W.; Zhiqiang, W.; Qingzhu, G. Seasonal patterns of soil microbial community response to warming and increased precipitation in a semiarid steppe. *Appl. Soil. Ecol.* **2023**, *182*, 104712. [CrossRef]
65. Wei, H.; Peng, C.; Yang, B.; Song, H.; Li, Q.; Jiang, L.; Wei, G.; Wang, K.; Wang, H.; Liu, S.; et al. Contrasting soil bacterial community, diversity, and function in two forests in China. *Front. Microbiol.* **2018**, *9*, 1693. [CrossRef]
66. Lili, F.; Muhammad, W.K.T.; Yangyang, Z.; Yongzhen, H.; Jundong, R.; Xinhang, C.; Liguang, C.; Chengkun, S.; Yushan, Z. Patterns of soil microorganisms and enzymatic activities of various forest types in coastal sandy land. *Glob. Ecol. Conserv.* **2021**, *28*, e01625. [CrossRef]

67. Jin, Z.-Z.; Lei, J.-Q.; Xu, X.-W.; Li, S.; Zhao, S.-F. Microbial diversities of shelter 'forest soils in the extreme arid area. *Acta Ecolc. Sinica.* **2009**, *29*, 4549–4559. [CrossRef]
68. Zhang, J.; Xin, Y.; Zhao, Y. Diversity and functional potential of soil bacterial communities in different types of farmland shelterbelts in Mid-Western Heilongjiang, China. *Forests* **2019**, *10*, 1115. [CrossRef]
69. Prada-Salcedo, L.D.; Prada-Salcedo, J.P.; Heintz-Buschart, A.; Buscot, F.; Goldmann, K. Effects of tree composition and soil depth on structure and functionality of belowground microbial communities in temperate European forests. *Front. Microbiol.* **2022**, *13*, 920618. [CrossRef] [PubMed]
70. Lovelock, C.E.; Andersen, K.; Morton, J.B. Arbuscular mycorrhizal communities in tropical forests are affected by host tree species and environment. *Oecologia* **2003**, *135*, 268–279. [CrossRef]
71. Rożek, K.; Rola, K.; Błaszkowski, J.; Leski, T.; Zubek, S. How do monocultures of fourteen forest tree species affect arbuscular mycorrhizal fungi abundance and species richness and composition in soil? *For. Ecol. Manag.* **2020**, *465*, 118091. [CrossRef]
72. Ibáñez, B.; Gómez-Aparicio, L.; Ávila, J.M.; Pérez-Ramos, I.M.; García, L.V.; Marañón, T. Impact of tree decline on spatial patterns of seedling-mycorrhiza interactions: Implications for regeneration dynamics in Mediterranean forests. *For. Ecol. Manag.* **2015**, *353*, 1–9. [CrossRef]
73. Carteron, A.; Vellend, M.; Laliberté, E. Mycorrhizal dominance reduces local tree species diversity across US forests. *Nat. Ecol. Evol.* **2022**, *6*, 370–374. [CrossRef]
74. Barceló, M.; van Bodegom, P.M.; Tedersoo, L.; Olsson, P.A.; Soudzilovskaia, N.A. Mycorrhizal Tree impacts on topsoil biogeochemical properties in tropical forests. *J. Ecol.* **2022**, *110*, 1271–1282. [CrossRef]
75. Tedersoo, L.; Bahram, M. Mycorrhizal Types differ in ecophysiology and alter plant nutrition and soil processes. *Biol. Rev.* **2022**, *94*, 1857–1880. [CrossRef] [PubMed]
76. Churchland, C.; Grayston, S.J. Specificity of Plant-Microbe Interactions in the Tree Mycorrhizosphere Biome and Consequences for Soil C Cycling. *Front. Microbiol.* **2014**, *5*, 261. [CrossRef] [PubMed]
77. Boeraeve, M.; Leroux, O.; De Lange, R.; Verbeken, A.; Jacquemyn, H. The effect of surrounding vegetation on the mycorrhizal fungal communities of the temperate tree *Crataegus monogyna* Jacq. *Front. Fungal Biol.* **2021**, *2*, 741813. [CrossRef]
78. Bainard, L.D.; Klironomos, J.N.; Gordon, A.M. The mycorrhizal status and colonization of 26 tree species growing in urban and rural environments. *Mycorrhiza* **2011**, *21*, 91–96. [CrossRef] [PubMed]
79. Singavarapu, B.; Beugnon, R.; Bruelheide, H.; Cesarz, S.; Du, J.; Eisenhauer, N.; Guo, L.-D.; Nawaz, A.; Wang, Y.; Xue, K.; et al. Tree mycorrhizal type and tree diversity shape the forest soil microbiota. *Environ. Microbiol.* **2022**, *24*, 4236–4255. [CrossRef]
80. Barra Caracciolo, A.; Grenni, P.; Garbini, G.L.; Rolando, L.; Campanale, C.; Aimola, G.; Fernandez-Lopez, M.; Fernandez-Gonzalez, A.J.; Villadas, P.J.; Ancona, V. Characterization of the belowground microbial community in a poplar-phytoremediation strategy of a multi-contaminated Soil. *Front. Microbiol.* **2020**, *11*, 2073. [CrossRef]
81. Almonacid-Muñoz, L.; Herrera, H.; Fuentes-Ramírez, A.; Vargas-Gaete, R.; Larama, G.; Jara, R.; Fernández-Urrutia, C.; da Silva Valadares, R.B. Tree cover species modify the diversity of rhizosphere-associated microorganisms in Nothofagus obliqua (Mirb.) Oerst temperate forests in South-Central Chile. *Forests* **2022**, *13*, 756. [CrossRef]
82. Yurong, Y.; Miao, D.; Yaping, C.; Jinlong, W.; Ming, T.; Yihui, B. Comparisons of soil properties, enzyme activities and microbial communities in heavy metal contaminated bulk and rhizosphere soils of *Robinia pseudoacacia* L. in the Northern Foot of Qinling Mountain. *Forests* **2017**, *8*, 430. [CrossRef]
83. Sapsford, S.J.; Paap, T.; Hardy, G.E.S.J.; Burgess, T.I. Anthropogenic disturbance impacts mycorrhizal communities and abiotic soil properties: Implications for an endemic forest disease. *Front. For. Glob. Change* **2021**, *3*, 593243. [CrossRef]
84. Gerz, M.; Bueno, C.G.; Ozinga, W.A.; Zobel, M.; Moora, M. Responses of plant community mycorrhization to anthropogenic influence depend on the habitat and mycorrhizal type. *Oikos* **2019**, *128*, 1565–1575. [CrossRef]
85. Sapsford, S.J.; Paap, T.; Hopkins, A.J.M.; Hardy, G.E.S.J.; Burgess, T.I. Habitat Fragmentation in a Mediterranean-type forest alters resident and propagule mycorrhizal fungal communities. *Pedobiologia* **2020**, *78*, 150611. [CrossRef]
86. Jo, I.; Fei, S.; Oswalt, C.M.; Domke, G.M.; Phillips, R.P. Shifts in dominant tree mycorrhizal associations in response to anthropogenic impacts. *Sci. Adv.* **2019**, *5*, eaav6358. [CrossRef] [PubMed]
87. Egli, S. Mycorrhizal mushroom diversity and productivity—An indicator of forest health? *Ann. Forest Sci.* **2011**, *68*, 81–88. [CrossRef]
88. Sapsford, S.J.; Paap, T.; Hardy, G.E.S.J.; Burgess, T.I. The 'chicken or the egg': Which comes first, forest tree decline or loss of mycorrhizae? *Plant Ecol.* **2017**, *218*, 1093–1106. [CrossRef]
89. Li, Y.; Zhang, Z.; Tan, S.; Yu, L.; Tang, C.; You, Y. Overview of vegetation factors related to the diversity of arbuscular mycorrhizal fungi and their interactions in karst areas. *Appl. Soil Ecol.* **2024**, *198*, 105387. [CrossRef]
90. Cairney, J.W.G.; Meharg, A.A. Influences of anthropogenic pollution on mycorrhizal fungal communities. *Environ. Poll.* **1999**, *106*, 169–182. [CrossRef]
91. Op De Beeck, M.; Lievens, B.; Busschaert, P.; Rineau, F.; Smits, M.; Vangronsveld, J.; Colpaert, J.V. Impact of metal pollution on fungal diversity and community structures. *Environ. Microbiol.* **2015**, *17*, 2035–2047. [CrossRef] [PubMed]

92. Ceulemans, T.; Van Geel, M.; Jacquemyn, H.; Boeraeve, M.; Plue, J.; Saar, L.; Kasari, L.; Peeters, G.; van Acker, K.; Crauwels, S.; et al. Arbuscular mycorrhizal fungi in european grasslands under nutrient pollution. *Glob. Ecol. Biogeogr.* **2019**, *28*, 1796–1805. [CrossRef]
93. Lilleskov, E.A.; Kuyper, T.W.; Bidartondo, M.I.; Hobbie, E.A. Atmospheric nitrogen deposition impacts on the structure and function of forest mycorrhizal communities: A review. *Environ. Poll.* **2019**, *246*, 148–162. [CrossRef]
94. Egerton-Warburton, L.M.; Allen, E.B. Shifts in arbuscular mycorrhizal communities along an anthropogenic nitrogen deposition gradient. *Ecol. Appl.* **2000**, *10*, 484–496. [CrossRef]
95. Pereira, S.; Leal, I.R.; Tabarelli, M.; Santos, M.G. Intense mycorrhizal root colonization in a human-modified landscape of the Caatinga dry forest. *For. Ecol. Manag.* **2020**, *462*, 117970. [CrossRef]
96. Rusterholz, H.-P.; Studer, M.; Zwahlen, V.; Baur, B. Plant-mycorrhiza association in urban forests: Effects of the degree of urbanisation and forest size on the performance of Sycamore (*Acer pseudoplatanus*) saplings. *Urban For. Urban Green.* **2020**, *56*, 126872. [CrossRef]
97. Chatterjee, A.; Vance, G.; Pendall, E.; Stahl, P.; Chatterjee, A.; Vance, G.F.; Pendall, E.; Stahl, P.D. Timber harvesting alters soil carbon mineralization and microbial community structure in coniferous forests. *Soil Biol. Biochem.* **2008**, *40*, 1901–1907. [CrossRef]
98. Smith, N.R.; Kishchuk, B.E.; Mohn, W.W. Effects of wildfire and harvest disturbances on forest soil bacterial communities. *Appl. Environ. Microbiol.* **2008**, *74*, 216–224. [CrossRef] [PubMed]
99. Dighton, J.; White, J.; Oudemans, P. *The Fungal Community: Its Organization and Role in the Ecosystem*, 3rd ed.; CRC Press: Boca Raton, FL, USA, 2005; p. 960. [CrossRef]
100. Farrar, J.; Hawes, M.; Jones, D.; Lindow, S. How roots control the flux of carbon to the rhizosphere. *Ecology* **2003**, *84*, 827–837. [CrossRef]
101. Outerbridge, R.A.; Trofymow, J.A. Forest management and maintenance of ectomycorrhizae: A case study of green tree retention in south-coastal British Columbia. *J. Ecosyst. Manag.* **2009**, *10*, 59–80. [CrossRef]
102. Paterson, E.; Gebbing, T.; Abel, C.; Sim, A.; Telfer, G. Rhizodeposition shapes rhizosphere microbial community structure in organic soil. *New Phytol.* **2007**, *173*, 600–610. [CrossRef] [PubMed]
103. Wolf, D.; Wagner, G. Carbon transformation and soil organic matter formation. In *Principles and Applications of Soil Microbiology*; Sylvia, D., Fuhrmann, J., Hartel, P., Zuberer, D., Eds.; Pearson Education Inc.: Upper Saddle River, NJ, USA, 2005; pp. 285–332.
104. Nakatsu, C. Microbial genetics. In *Principles and Applications of Soil Microbiology*; Sylvia, D., Fuhrmann, J., Hartel, P., Zuberer, D., Eds.; Pearson Education Inc.: Upper Saddle River, NJ, USA, 2005; pp. 85–100.
105. Mariani, L.; Chang, S.X.; Kabzems, R. Effects of tree harvesting, forest floor removal, and compaction on soil microbial biomass, microbial respiration, and N availability in a boreal aspen forest in British Columbia. *Soil Biol. Biochem.* **2006**, *38*, 1734–1744. [CrossRef]
106. Tan, X.; Chang, S.X.; Kabzems, R. Effects of soil compaction and forest floor removal on soil microbial properties and N transformations in a boreal forest long-term soil productivity study. *For. Ecol. Manag.* **2005**, *217*, 158–170. [CrossRef]
107. Shestak, C.J.; Busse, M.D. Compaction alters physical but not biological indices of soil health. *Soil Sci. Soc. Am. J.* **2005**, *69*, 236–246. [CrossRef]
108. Jordán, A.; Martínez-Zavala, L. Soil loss and runoff rates on unpaved forest roads in southern Spain after simulated rainfall. *For. Eco. Manag.* **2008**, *255*, 913–919. [CrossRef]
109. Wang, C.; Xue, L.; Dong, Y.; Wei, Y.; Jiao, R. Unravelling the functional diversity of the soil microbial community of Chinese fir plantations of different densities. *Forest* **2018**, *9*, 532. [CrossRef]
110. Wilhelm, R.C.; Cardenas, E.; Maas, K.R.; Leung, H.; McNeil, L.; Berch, S.; Chapman, W.; Hope, G.; Kranabetter, J.M.; Dubé, S.; et al. Biogeography and organic matter removal shape long-term effects of timber harvesting on forest soil microbial communities. *ISME J.* **2017**, *11*, 2552–2568. [CrossRef] [PubMed]
111. Dickie, I.A.; Richardson, S.J.; Wiser, S.K. Ectomycorrhizal fungal communities and soil chemistry in harvested and unharvested temperate nothofagus rainforests. *Can. J. For. Res.* **2009**, *39*, 1069–1079. [CrossRef]
112. Parladé, J.; Martínez-Peña, F.; Pera, J. Effects of forest management and climatic variables on the mycelium dynamics and sporocarp production of the ectomycorrhizal fungus *Boletus edulis*. *For. Ecol. Manag.* **2017**, *390*, 73–79. [CrossRef]
113. Wurzburger, N.; Elliott, K.J.; Miniat, C.F. Forest mycorrhizal dominance depends on historical land use and nitrogen-fixing trees. *J. Appl. Ecol.* **2023**, *60*, 1551–1561. [CrossRef]
114. Tomao, A.; Antonio Bonet, J.; Castaño, C.; de-Miguel, S. How does forest management affect fungal diversity and community composition? Current knowledge and future perspectives for the conservation of forest fungi. *For. Ecol. Manag.* **2020**, *457*, 117678. [CrossRef]
115. Marshall, V.G. Impacts of forest harvesting on biological processes in northern forest soils. *For. Ecol. Manag.* **2000**, *133*, 43–60. [CrossRef]
116. Jurgensen, M.F.; Harvey, A.E.; Graham, R.T.; Page-Dumroese, D.S.; Tonn, J.R.; Larsen, M.J.; Jain, T.B. Impacts of timber harvesting on soil organic matter, nitrogen, productivity, and health of inland northwest forests. *For. Sci.* **1997**, *43*, 234–251. [CrossRef]

117. Varenius, K.; Kårén, O.; Lindahl, B.; Dahlberg, A. Long-term effects of tree harvesting on ectomycorrhizal fungal communities in boreal Scots Pine forests. *For. Ecol. Manag.* **2016**, *380*, 41–49. [CrossRef]
118. Simard, S.W.; Roach, W.J.; Beauregard, J.; Burkart, J.; Cook, D.; Law, D.; Murphy-Steed, A.; Schacter, T.; Zickmantel, A.; Armstrong, G.; et al. Partial retention of legacy trees protect mycorrhizal inoculum potential, biodiversity, and soil resources while promoting natural regeneration of interior douglas-fir. *Front. For. Glob. Change* **2021**, *3*, 620436. [CrossRef]
119. Ellis, E.C. Anthropogenic transformation of the terrestrial biosphere. *Philos. Trans. R. Soc. A Math. Phys. Eng. Sci.* **2011**, *369*, 1010–1035. [CrossRef] [PubMed]
120. Millenium Ecosystem Assessment 2005. *Ecosystems and Human Well Being: Wetlands and Water Synthesis*; Island Press: Washington, DC, USA, 2005.
121. Watson, S.J.; Luck, G.W.; Spooner, P.G.; Watson, D.M. Land-use change: Incorporating the frequency, sequence, time span, and magnitude of changes into ecological research. *Front. Ecol. Environ.* **2014**, *12*, 241–249. [CrossRef]
122. Fierer, N. Embracing the unknown: Disentangling the complexities of the soil microbiome. *Nat. Rev. Microbiol.* **2017**, *15*, 579–590. [CrossRef]
123. James, J.; Harrison, R. The effect of harvest on forest soil carbon: A meta-analysis. *Forests* **2016**, *7*, 308. [CrossRef]
124. Kohout, P.; Charvátová, M.; Štursová, M.; Mašínová, T.; Tomšovský, M.; Baldrian, P. Clearcutting alters decomposition processes and initiates complex restructuring of fungal communities in soil and tree roots. *ISME J.* **2018**, *12*, 692–703. [CrossRef] [PubMed]
125. Shen, J.; Chen, C.R.; Lewis, T. Long term repeated fire disturbance alters soil bacterial diversity but not the abundance in an Australian wet sclerophyll forest. *Sci. Rep.* **2016**, *6*, 19639. [CrossRef]
126. Zhou, Z.; Wang, C.; Luo, Y. Effects of forest degradation on microbial communities and soil carbon cycling: A global meta-analysis. *Glob. Ecol. Biogeogr.* **2018**, *27*, 110–124. [CrossRef]
127. Osburn, E.D.; Elliottt, K.J.; Knoepp, J.D.; Miniat, C.F.; Barrett, J.E. Soil microbial response to Rhododendron understory removal in southern appalachian forests: Effects on extracellular enzymes. *Soil Biol. Biochem.* **2018**, *127*, 50–59. [CrossRef]
128. Simon, S.A.; Collins, T.K.; Kauffman, G.L.; McNab, W.H.; Ulrey, C.J. *Ecological Zones in the Southern Appalachians: First Approximation*; USDA Forest Service: Asheville, NC, USA, 2005. [CrossRef]
129. Gragson, T.L.; Bolstad, P.V. Land use legacies and the future of Southern Appalachia. *Soc. Nat. Resour.* **2006**, *19*, 175–190. [CrossRef]
130. Keiser, A.D.; Knoepp, J.D.; Bradford, M.A. Disturbance decouples biogeochemical cycles across forests of the southeastern US. *Ecosystems* **2016**, *19*, 50–61. [CrossRef]
131. Lin, W.R.; Chen, W.C.; Wang, P.H. Soil microbial community is resilient to thinning disturbance. *Trop. Ecol.* **2023**, *64*, 62–71. [CrossRef]
132. Teste, F.P.; Simard, S.W.; Durall, D.M.; Guy, R.D.; Jones, M.D.; Schoonmaker, A.L. Access to mycorrhizal networks and roots of trees: Importance for seedling survival and resource transfer. *Ecology* **2009**, *90*, 2808–2822. [CrossRef]
133. Bonfim, J.A.; Vasconcellos, R.L.F.; Stürmer, S.L.; Cardoso, E.J.B.N. Arbuscular mycorrhizal fungi in the Brazilian Atlantic forest: A gradient of environmental restoration. *Appl. Soil. Ecol.* **2013**, *71*, 7–14. [CrossRef]
134. Policelli, N.; Horton, T.R.; Hudon, A.T.; Patterson, T.R.; Bhatnagar, J.M. Back to Roots: The role of ectomycorrhizal fungi in boreal and temperate forest restoration. *Front. For. Glob. Change* **2020**, *3*, 97. [CrossRef]
135. Becerra, A.G.; Diván, A.; Renison, D. Bare soil cover and arbuscular mycorrhizal community in the first montane forest restoration in Central Argentina. *Restor. Ecol.* **2019**, *27*, 804–812. [CrossRef]
136. Neuenkamp, L.; Prober, S.M.; Price, J.N.; Zobel, M.; Standish, R.J. Benefits of mycorrhizal inoculation to ecological restoration depend on plant functional type, restoration context and time. *Fungal Ecol.* **2019**, *40*, 140–149. [CrossRef]
137. Medeiros, A.S.; Goto, B.T.; Ganade, G. Ecological restoration methods influence the structure of arbuscular mycorrhizal fungal communities in degraded drylands. *Pedobiologia* **2021**, *84*, 150690. [CrossRef]
138. Wall, C.B.; Egan, C.P.; Swift, S.I.O.; Hynson, N.A. Three decades post-reforestation has not led to the reassembly of arbuscular mycorrhizal fungal communities associated with remnant primary forests. *Mol. Ecol.* **2020**, *29*, 4234–4247. [CrossRef]
139. Aerts, R.; Honnay, O. Forest restoration, biodiversity and ecosystem functioning. *BMC Ecol.* **2011**, *11*, 29. [CrossRef] [PubMed]
140. Lance, A.C.; Carrino-Kyker, S.R.; Burke, D.J.; Burns, J.H. Individual plant-soil feedback effects influence tree growth and rhizosphere fungal communities in a temperate forest restoration experiment. *Front Ecol. Evol.* **2020**, *7*, 500. [CrossRef]
141. Keenan, R.J.; Reams, G.A.; Achard, F.; de Freitas, J.V.; Grainger, A.; Lindquist, E. Dynamics of global forest area: Results from the FAO Global Forest Resources Assessment. *For. Ecol. Manag.* **2015**, *352*, 9–20. [CrossRef]
142. Crowther, T.W.; Glick, H.B.; Covey, K.R.; Bettigole, C.; Maynard, D.S.; Thomas, S.M.; Smith, J.R.; Hintler, G.; Duguid, M.C.; Amatulli, G.; et al. Mapping tree density at a global scale. *Nature* **2015**, *525*, 201–205. [CrossRef] [PubMed]
143. Gauthier, S.; Bernier, P.; Kuuluvainen, T.; Shvidenko, A.Z.; Schepaschenko, D.G. Boreal forest health and global change. *Science* **2015**, *349*, 819–822. [CrossRef] [PubMed]
144. Alkama, R.; Cescatti, A. Biophysical climate impacts of recent changes in global forest cover. *Science* **2016**, *351*, 600–604. [CrossRef]

145. Millar, C.I.; Stephenson, N.L. Temperate forest health in an era of emerging megadisturbance. *Science* **2015**, *349*, 823–826. [CrossRef] [PubMed]
146. Bååth, E.; Anderson, T.H. Comparison of soil fungal/bacterial ratios in a pH gradient using physiological and PLFA-based techniques. *Soil Biol. Biochem.* **2003**, *35*, 955–963. [CrossRef]
147. Gardi, C.; Jeffery, S.; Saltelli, A. An estimate of potential threats levels to soil biodiversity in EU. *Glob. Change Biol.* **2013**, *19*, 1538–1548. [CrossRef] [PubMed]
148. Verheijen, L.M.; Aerts, R.; Brovkin, V.; Cavender-Bares, J.; Cornelissen, J.; Kattge, J.; Bodegom, P.M. Inclusion of ecologically based trait variation in plant functional types reduces the projected land carbon sink in an earth system model. *Glob. Change Biol.* **2015**, *21*, 3074–3086. [CrossRef]
149. Classen, A.T.; Sundqvist, M.K.; Henning, J.A.; Newman, G.S.; Moore, J.A.M.; Cregger, M.A.; Moorhead, L.C.; Patterson, C.M. Direct and indirect effects of climate change on soil microbial and soil microbial-plant interactions: What lies ahead? *Ecosphere* **2015**, *6*, 130. [CrossRef]
150. Jansson, J.K.; Hofmockel, K.S. Soil microbiomes and climate change. *Nat. Rev. Microbiol.* **2020**, *18*, 35–46. [CrossRef] [PubMed]
151. Wardle, D.A.; Bardgett, R.D.; Klironomos, J.N.; Setala, H.I.; Putten, W.H.V.D.; Wall, D.H. Ecological linkages between aboveground and belowground biota. *Science* **2004**, *304*, 1629–1633. [CrossRef]
152. Wenjing, C.; Huakun, Z.; Yang, W.; Jie, W.; Ziwen, Z.; Yuanze, L.; Leilei, Q.; Bing, Y.; Kelu, C.; Guobin, L.; et al. Loss of plant functional groups impacts soil carbon flow by changing multitrophic interactions within soil micro-food webs. *Appl. Soil. Ecol.* **2022**, *178*, 104566. [CrossRef]
153. Uroz, S.; Buée, M.; Deveau, A.; Mieszkin, S.; Martin, F. Ecology of the forest microbiome: Highlights of temperate and boreal ecosystems. *Soil Biol. Biochem.* **2016**, *103*, 471–488. [CrossRef]
154. Voříšková, J.; Brabcová, V.; Cajthaml, T.; Baldrian, P. Seasonal dynamics of fungal communities in a temperate oak forest soil. *New Phytol.* **2014**, *201*, 269–278. [CrossRef] [PubMed]
155. Allison, S.D.; Treseder, K.K. Climate change feedbacks to microbial decomposition in boreal soils. *Fungal Ecol.* **2021**, *4*, 362–374. [CrossRef]
156. Qiu, S.L.; Wang, L.M.; Huang, D.F.; Lin, X.J. Effects of fertilization regimes on tea yields, soil fertility, and soil microbial diversity. *Chile J. Agric. Res.* **2014**, *74*, 333–339. [CrossRef]
157. Naether, A.; Foesel, B.U.; Naegele, V.; Wüst, P.K.; Weinert, J.; Bonkowski, M.; Alt, F.; Oelmann, Y.; Polle, A.; Lohaus, G.; et al. Environmental factors affect Acidobacterial communities below the subgroup level in grassland and forest soils. *Appl. Environ. Microbiol.* **2012**, *78*, 7398–7406. [CrossRef] [PubMed]
158. Van der Wal, A.; Geydan, T.D.; Kuyper, T.W.; de Boer, W. A thready affair: Linking fungal diversity and community dynamics to terrestrial decomposition processes. *FEMS Microbiol. Rev.* **2013**, *37*, 477–494. [CrossRef] [PubMed]
159. Baldrian, P.; Kolařík, M.; Štursová, M.; Kopecký, J.; Valaskova, V.; Větrovský, T.; Žifčáková, L.; Šnajdr, J.; Rídl, J.; Vlček, C.; et al. Active and total microbial communities in forest soil are largely different and highly stratified during decomposition. *ISME J.* **2012**, *6*, 248–258. [CrossRef] [PubMed]
160. Graham, E.B.; Knelman, J.E.; Schindlbacher, A.; Siciliano, S.; Breulmann, M.; Yannarell, A.; Beman, J.M.; Abell, G.; Philippot, L.; Prosser, J.; et al. Microbes as engines of ecosystem function: When does community structure enhance predictions of ecosystem processes? *Front. Microbiol.* **2016**, *7*, 214. [CrossRef] [PubMed]
161. Walker, J.K.; Cohen, H.; Higgins, L.M.; Kennedy, P.G. Testing the link between community structure and function for ectomycorrhizal fungi involved in a global tripartite symbiosis. *New Phytol.* **2014**, *202*, 287–296. [CrossRef]
162. Sterkenburg, E.; Bahr, A.; Brandstrom-Durling, M.; Clemmensen, K.E.; Lindahl, B.D. Changes in fungal communities along a boreal forest soil fertility gradient. *New Phytol.* **2015**, *207*, 1145–1158. [CrossRef] [PubMed]

MDPI AG
Grosspeteranlage 5
4052 Basel
Switzerland
Tel.: +41 61 683 77 34

Forests Editorial Office
E-mail: forests@mdpi.com
www.mdpi.com/journal/forests

www.ingramcontent.com/pod-product-compliance
Lightning Source LLC
Chambersburg PA
CBHW061554210926
51850CB00009B/2100
* 9 7 8 3 7 2 5 8 5 9 7 1 9 *